Introduction to

Environmental Remote Sensing

Introduction to
Environmental Remote Sensing
(Second Edition)

E.C. Barrett
*Reader in Climatology
and Remote Sensing
Department of Geography
University of Bristol*

and

L.F. Curtis
*Exmoor National Park Officer
Honorary Senior Research Fellow
University of Bristol*

ANDERSONIAN LIBRARY
WITHDRAWN FROM LIBRARY STOCK
UNIVERSITY OF STRATHCLYDE

London   New York
CHAPMAN AND HALL

First published 1976 by
Chapman and Hall Ltd
11 New Fetter Lane
London EC4P 4EE
Second edition 1982

Published in the USA by
Chapman and Hall
733 Third Avenue
New York NY 10017

© 1976, 1982 E.C. Barrett and L.F. Curtis

Printed in Great Britain by
Fletcher & Son Ltd, Norwich
Photoset by Enset Ltd,
Midsomer Norton, Bath, Avon.

ISBN 0 412 23080 1 (hardback)
ISBN 0 412 23090 9 (paperback)

British Library Cataloguing in Publication Data

Barrett, E.C.
   Introduction to environmental remote sensing.—
   2nd ed.
   1. Remote sensing systems
   I. Title   II. Curtis, L.F.
   621.36'78   TD153

   ISBN 0–412–23080–1
   ISBN 0–412–23090–9 Pbk

Contents

Preface

In 1976, when the first edition of this volume was launched, we said that we believed it to be a timely book. Remote sensing of the environment, which had begun to experience explosive growth, has continued to grow and mature. At publication our book had few, if any, serious competitors as an integrated, specially-written general account of this young science to meet the growing needs of educators in, and students of, remote sensing. Five eventful years later, during which popular demand has prompted a reprint of our original work – and interest abroad has resulted in a Russian-language edition of it – the time has come to prepare a second edition so that new readers will be kept in touch with recent developments in this increasingly broad, technical, and practically valuable field. Indeed, we feel that there is now a greater need than ever for such a book, which aims to present a balanced and integrated introduction for the relative newcomer to the subject.

Those acquainted with our prototype volume will find that more than half of it has been substantially re-worked or re-written. New chapters have been constructed to reflect fresh areas of emphasis in remote sensing today. The whole has been brought up to date by additional or replacement material. Our aim in this edition, as in the first, has been to provide – at a size and price which time and pockets can afford – a realistic introductory survey of environmental remote sensing for students, scientists and decision makers with the need to know and understand the scope, potential, and limitations of remote sensing as an applied science for the service of mankind.

Our own experience in remote sensing embraces not only university teaching and supervision of postgraduate research, but also UN training of representatives from developing countries, land-use management and planning, operational development of resource and plant protection surveys, and the integration of remote sensing and conventional data in meteorology, hydrology, rangeland monitoring and crop preduction programmes. We have been involved in the planning, organization, implementation, and assessment of remote sensing activities at every level from the local to the international. We hope this book will prove of value to any who may be involved in some or many of these activities.

We owe a great debt of gratitude to the many friends and colleagues with whom we have discussed matters covered by this book; much of our own grasp of remote sensing has been gained from those with whom we have liaised, especially during the last decade. Where we have made direct use of the work of others due acknowledgement has been made through figure captions and references. These will also serve to point the reader to key literature which should be consulted for details of methods and results. Our special thanks are due to the University of Bristol, Somerset County Council, the staff of the National Aeronautics and Space Administration and the National Oceanic and Atmospheric Administration of the USA, the European Space Agency, the Food and Agriculture Organisation of the United Nations, and the UK Departments of Industry and Environment for support and encouragement for many of our own activities in remote sensing, and for ready access to published studies and data used in the preparation of this volume. We hope that all our associates in remote sensing will feel we have done justice to their findings, for, of course, the synthesis we have prepared is our own, and we must accept responsibility for any errors or misconceptions which may have crept in.

As before, our last and biggest votes of thanks must be reserved for our wives, Gillian and Diana, for their never-failing encouragement and unselfish help – and for our children, who have often regretted our seemingly unending preoccupation with this project. Fortunately the aphorism 'Be a remote sensor and see the world' sometimes has exciting and exotic implications for wives and families as well as us!

E.C. Barrett
Backwell
Avon

L.F. Curtis
Porlock
Somerset

The Authors

Dr Eric C. Barrett

Dr Eric Barrett is Reader in Climatology and Remote Sensing in the Department of Geography in the University of Bristol. His interest in remote sensing was aroused first by a presentation in London in 1963 by the Chief of the TIROS Project, which was then in its infancy. This led to a change of direction in his PhD programme, and a thesis on 'The contribution of meteorological satellites to dynamic climatology' in 1969. Since then, Dr Barrett's interest in the exploitation of satellite data for the solution of a wide range of practical problems has steadily grown. He has undertaken numerous consultancies in this general area, including several for the Food and Agriculture Organisation of the United Nations. He was one of only two UK Principal Investigators for NASA in the Landsat 2 Program, and is currently engaged on the US AgRISTARS Project to improve global crop prediction. He was a founder member of the UK Remote Sensing Society, founder Chairman of the Meteosat Working Group in EARSeL (the European Association of Remote Sensing Laboratories), and was elected UK National Representative on the EARSeL Council in 1980. He is a member of the Technical Advisory Committee of the UK National Remote Sensing Centre, and of the Remote Sensing Users Committee of NERC. He contributes regularly to Remote Sensing Training Courses in FAO, Rome, and is the author or editor of ten other books on remote sensing and its applications in climatology, meteorology, hydrometeorology, geography and the environmental sciences. In 1982 he was awarded the Hugh Robert Mill Medal and Prize of the Royal Meteorological Society for the development of methods for the remote sensing of rainfall and the degree of DSc by the University of Bristol for his 'sustained and distinguished contribution to geographic science'.

Dr Leonard F. Curtis

Dr Leonard Curtis, a Local Government Officer, is head of Exmoor National Park Department and an Honorary Senior Research Fellow of the University of Bristol. He was formerly Reader in Geography and Head of the Joint School of Botany and Geography in the University of Bristol. His early experience of remote sensing began with wartime air photography and was furthered by commercial experience of applied photo-interpretation on secondment to Hunting Surveys for irrigation development surveys in 1954. This interest in remote sensing continued when Dr Curtis began university teaching in 1956 and led to his appointment in 1970 as the UK delegate to the initial working group on a Post Apollo Earth Resources Programme set up by the European Space Research Organisation. He was subsequently appointed UK delegate to the Remote Sensing Working Group of the European Space Agency and to other advisory groups. He was chairman of the Natural Environment Research Council Committee for remote sensing evaluation flights in 1971, and was a member of Council of the UK Remote Sensing Society at its inception. He is a member of EARSeL's Working Group for Education and Training in Remote Sensing and the UK Remote Sensing Centre group for Education. He has acted as a consultant for the European Economic Community and for numerous other organizations in the sphere of environmental remote sensing. An author or editor of seven books concerned with remote sensing applications, soil studies and land use inventories, he is actively engaged in the use of remote sensing for environmental monitoring of protected landscapes.

Abbreviations and acronyms

Many abbreviations and acronyms are used in environmental remote sensing. This list expands such terms used more than once in this volume and not always set out in full.

AgRISTARS: Agriculture and Resources Inventory Surveys through Aerospace Remote Sensing
APT: Automatic Picture Transmission
ATS: Applications Technology Satellite
AVHRR: Advanced Very High Resolution Radiometer
CCT: Computer Compatible Tape
CEGB: Central Electricity Generating Board
CZCS: Coastal Zone Colour Scanner
DCP: Data Collection Platform
DCS: Data Collection Service
DoE: Department of the Environment (UK)
EARSeL: European Association of Remote Sensing Laboratories
EDC: EROS Data Center, Sioux Falls, SD, USA
EROS: Earth Resources Observation Service
ERTS: Earth Resources Technology Satellite
ESA: European Space Agency
ESMR: Electrically Scanning Microwave Radiometer
ESOC: European Space Operations Centre, Darmstadt, West Germany
ESRO: European Space Research Organization
ESSA: Environmental Sciences Services Administration (USA)
FAO: Food and Agriculture Organisation (UN)
GARP: Global Atmospheric Research Programme
GMS: Geostationary Meteorological Satellite
GOES: Geostationary Operational Environmental Satellite
HCMM: Heat Capacity Mapping Mission
HRIR: High Resolution Infrared Radiometer
HRPT: High Resolution Picture Transmission

IRIS: Infra-Red Interferometer Spectrometer
IRLS: Infra-Red LineScan
ITCB: Inter-Tropical Cloud Band
ITCZ: Inter-Tropical Convergence Zone
LAI: Leaf Area Index
LARS: Laboratory for Agricultural Remote Sensing (Purdue University, USA)
MRIR: Medium Resolution Infrared Radiometer
MSS: Multispectral Scanner
MSU: Microwave Sounding Unit
NASA: National Aeronautics and Space Administration (USA)
NERC: Natural Environment Research Council (UK)
NOAA: National Oceanic and Atmospheric Administration (USA)
PPI: Plan Position Indicator
RADAR: RAdio Direction And Ranging
RAE: Royal Aircraft Establishment (Farnborough, UK)
RBV: Return Beam Vidicon camera system
RPV: Remotely Piloted Vehicle
SAR: Synthetic Aperture Radar
SCMR: Surface Composition Mapping Radiometer
SCR: Selective Chopper Radiometer
SLAR: Side-Looking Airborne Radar
SMMR: Scanning Multichannel Microwave Radiometer
SMS: Synchronous Meteorological Satellite
SPOT: Satellite Probatoire de l'Observation de la Terre
SSU: Stratospheric Sounding Unit
THIR: Temperature Humidity Infrared Radiometer
TIROS: Television and Infra-Red Observation Satellite
TM: Thematic Mapper
TOVS: Tiros Operational Vertical Sounder
WMO: World Meteorological Organization
WWW: World Weather Watch

Part One: Remote Sensing Principles

Part One: Remote Sensing Principles

1 Monitoring the environment

1.1 The concept of environment

Few people today can be unaware of the existence of what is popularly called *The Environment*. Many communicators have discussed it, on radio, on television, and in print. The Environment has been drawn to the attention of the world community through gatherings like the United Nations Symposium on the Human Environment in Stockholm in 1972, and by international recognition of the first World Environment Day in 1974. At national, regional and very local levels increasing concern has been expressed about the need for improvement or protection of the better aspects of The Environment, and the urgency with which environmental exploitation and despoilation should be resisted. There is one problem, however, which is basic not only to such public discussion but also to the purpose and contents of this book: The Environment means different things to different people.

First of all we must recognize that there can be no such thing as 'environment' unless a situation is being studied from the point of view of the influence it has upon some selected objects, either animate or inanimate, considered either as individuals or a population.

Thus, at one extreme, we can speak of the environment of a single-celled organism – and at the other, of the environment of our home galaxy. For the human individual The Environment is comprised of every animate and inanimate influence which bears upon him, his life, health and livelihood. But it is not only living things which are clothed with such a concept. The Environment can mean that of a town (sometimes called its environs or hinterland), or other even more purely physical objects when they are considered in their spatial settings, and are affected by various surrounding influences and interactions. This book is concerned with the environment of *man*, either individually, or more often, as a race of men.

Sometimes man is discussed in the context of the so-called 'natural environment'. Frequently this is taken to be his *physical* environment, though, strictly speaking, it is broader than this. Other living things as well as air, rocks and water help to comprise our natural home. But, today, it is not enough to conceive our environment in even these broadened terms. Man, with his large, demanding population and great scientific and technological skills, can and does exploit consciously, and modify both consciously and unconsciously (or accidentally) the world environment in which he lives. Without doubt he is the dominant life-form on his planet – and his influence is spreading out into the further corners even of his local Solar System.

In these days which are so marked by a growing awareness of the potential good (and potential evil) we may bring upon ourselves through our interaction with our world environment it is important to appreciate also that other people are part and parcel of the smaller environments into which Earth is broken down. And we must not neglect to consider the cultural aspects of our environment, for these too are vital if a better future is to be mapped out for all the inhabitants of this increasingly overcrowded planet, which has been aptly likened to a tiny island in a great sea of space.

Not surprisingly, it is often hard to dissect out those relationships linking man and his environment which are distinctly natural, technical or cultural in type. Many chains of cause and effect are either intricately ramified, or inadequately known or understood. The great majority involve so-called *feedback effects*. These often mean in practice that one man, in trying to improve some aspect of his own environment, does so at the expense of some other aspect of it, or the environment of his neighbours. This kind of problem assumes enormous proportions in the economic sphere, through the discovery, evaluation, exploitation and use of natural resources – those constituents of the environment which are essential to the maintenance or improvement of our life-styles and standards of living.

Within the compass of this book we will be concerned with *The Environment* as if this were the environment of man as an individual or community inhabiting the Earth. In recent years man has thrust himself out into space, and has spent time on the Moon. He has propelled spacecraft to distant planets. Thus new horizons for human discovery are being opened up, and unfamiliar and extremely harsh new environments are being entered. But it is the lot of almost all of us to remain in more familiar surroundings, and, to confine our discourse to manageable proportions we must exclude discussion of extra-terrestrial activities. Of course, the principles and practices of remote sensing 'at home' on Earth largely apply also to remote sensing 'abroad' in the Solar System. Consequently this book should be of at least initial interest too in the context of remote sensing of space. But for present purposes man will be considered within his terrestrial environment, both as an influence upon it and an important part of it.

In assessing environmental matters, it has become commonplace to follow a well-established sequence of steps. These include:

(a) The *recognition* of forms, structures, and/or processes of significance.
(b) The *identification* of such phenomena in their real world situation(s).
(c) The *recording of their distributions*, often through both space and time.
(d) The *assessment of these distributions*, sometimes singly, but more often in some combinations.
(e) Attempts to *understand the nature and cause* of any specially significant relationships.

It is only when such background studies have been successfully completed that their results can be used to benefit man and his environment. Commonly their use is in one or more of the following ways:

(a) The preparation and execution of schemes of environmental or resource management.
(b) The planning and development of future projects to modify and improve the environment.
(c) The prediction of forthcoming events, especially those over which man has less direct control.

At the heart of the all-important basic or background studies are problems concerned with *observation*. Data must be obtained by appropriate means, they must be put into permanent forms, and there are great advantages if they lend themselves readily to processing for rapid analysis, comparison and interpretation. It is here that environmental remote sensing has a vital part to play.

1.2 *In situ* sensing of the environment

Man has made measurements of key aspects of his environment since an early stage in the development of his civilization and culture. Examples of measuring devices include the famous Nilometer by which water levels in the River Nile were noted, and rudimentary rain gauges used by natural philosophers in the city states of Ancient Greece. The European Renaissance of art, science and literature marked a new surge of interest in the need for, and design of, monitoring instruments. Attention began to be paid not only to environmental variables or effects which were readily visible, but also to others less directly evident in the world of nature. The invention of the thermometer by Galileo at the end of the seventeenth century is particularly noteworthy. A steady deepening of interest in environmental factors and conditions ensued in the eighteenth and nineteenth centuries.

The scientific rebirth of the Renaissance period and its aftermath was followed by a rapid acceleration in man's concern for his environment as the twentieth century unfolded. Not only could the environment be monitored by a wide range of sophisticated instruments, but advances in the related technologies of communications and recording ensured that data could be collected, collated and processed, even from quite remote locations, in 'near real-time' (i.e. very close to the time of observation). At last environmental monitoring could be organized on a wide scale so rapidly that short-term event-prediction, and dependent practices of environmental management and control, were possible. Most recently, the advance of other support technologies, especially in the related fields of electronic computing and the microchip, has begun to exercise what will doubtless grow to be a profound influence upon the design of '*in situ*' sensor networks, and through them, upon the types of questions we may hope to solve through the analysis and interpretation of such 'conventional' data.

1.3 Remote sensing of the environment

Remote sensing can be defined as *the observation of a target by a device some distance away from it*. Thus it is contrasted with *in situ* sensing, in which measuring devices are either immersed in, or at least touch, the object(s) of observation and measurement. Some authors have spoken of remote sensing systems in connection with instrument packages which are remote in the sense that they are placed in relatively inaccessible locations, or are connected to central data processing facilities by automatic data acquisition and transmission links. This arrangement does not accord with the definition above, nor with the generally accepted view of remote sensing by those who practice it. Others have spoken of remote sensing in terms of a lack of physical contact between the sensor and its target. Again, strictly speaking, this is incorrect since where no physical contact exists, no observation is possible. While no 'touch-type' contact exists between a remote sensor and its target, some physical emanation from, or effect of, the target must be found if aspects of its property and/or behaviour are to be investigated. The most important of the physical links between objects of measurement and remote sensing measuring devices involve *electromagnetic energy, acoustic waves* and *force fields* – especially those associated with gravity and magnetism, as discussed in greater detail in Chapter 2. For most surface and atmospheric remote sensing, electromagnetic energy is the supreme medium. Since man's home base is located at the interface between Earth and its atmosphere, this book is much more concerned with the exploitation of electromagnetic energy than all other energy forms together.

In Chapter 2 we will discuss ways in which we all practice some methods of remote sensing: by these we observe the world in which we live. But if the science of remote sensing is concerned as much with

recording as with observation, the remote sensing era may be said to have dawned with the invention of photography in 1826. Devices like telescopes were invented long before the nineteenth century to extend our personal observing capabilities. But it is only in the last century and a half that means have been devised whereby the environment can be observed and recorded objectively by artificial devices. The potential of photography was quickly appreciated, especially for recording scenes of special significance to the observer. New cameras and types of films were invented even extending man's view beyond the relatively narrow waveband of visible light. Then began a more systematic search through the radiation spectrum, and into the realms of acoustical, chemical, gravitational and radioactive energy to discover fresh means whereby normally insensible aspects of the natural and cultural environment might be investigated. This search was stimulated greatly by the military demands of two World Wars, especially the second. The search has gathered even more momentum since.

It was in 1960 that reference was first made by name to remote sensing as a distinctive field of study, or set of approaches to the environment of man. Since then it has passed take-off point, deriving great impetus from the opening of the satellite era, and the space race between the remote sensing superpowers, the USA and USSR. In particular, the National Aeronautics and Space Administration (NASA) has played a gigantic part as a result of American Government policy to make remote sensing data (much of it from satellites) readily available to the scientific community throughout the world. The success of NASA has led to the establishment of national space agencies in other countries as far afield as Brazil, India and Japan, and international counterparts most notably in the case of the European Space Agency (ESA), whose Member States are Belgium, Denmark, France, Federal Republic of Germany, Ireland, Italy, The Netherlands, Spain, Sweden, Switzerland and the United Kingdom. Austria is now an Associate Member of ESA, and Canada and Norway have Observer status.

Remote sensing is now accredited a policy-making Sub-Committee of the United Nations Committee for the Peaceful Uses of Outer Space.

This Sub-Committee is more concerned with Earth Observation than first appearances suggest: in UN terminology the boundary between 'Inner' and 'Outer' Space is merely 50 km above the surface of the Earth. The research-orientated UN Committee for Space Research (COSPAR) also contributes significantly to environmental remote sensing through its international conferences and working groups.

Opportunities for discussion and exchanges of research results in remote sensing have multiplied dramatically since the early 1970s. National remote sensing societies have been established by now in most major countries of the world. International links are also growing, for example through the European Association of Remote Sensing Laboratories (EARSeL), which is also serving as a catalyst in internationally-collaborative research. Spurred on by an associated upsurge of interest in the present and potential implications of remote sensing in everyday operational use, national and international remote sensing centres are multiplying too, many modelled on the Canada Center for Remote Sensing, which was one of the first in the field. Data dissemination is being organized increasingly through such centres, in which some general-purpose high-technology data processing and analysis systems (e.g. the Plessey IDP-3000 System at the new UK Remote Sensing Centre, Farnborough) are also available for use by the general user community.

Not surprisingly, in the light of such activity, remote sensing instruction has spread widely through the tertiary education sector. Here it is often correctly portrayed as the kind of integrative study which is needed in these days to bring together specialists and specialized information from many fields of environmental science, broadly defined. The result is that graduates with interests and expertise in remote sensing of the environment can find employment in such seemingly diverse areas as meteorology, pedology, hydrology, geology and geophysics, agriculture, conservation and protection, pest control, fishery development, land use planning, civil engineering and computing to mention but a few.

The needs of the growing community of remote sensing students and scientists are being met by a widening range of germane books and journals. The first area-specific journal, *Remote Sensing of Environment* was published in 1969; the *International Journal of Remote Sensing* first appeared in 1980. Newsletters have proliferated in this field which has been characterized by such rapid growth and change. Some long-established societies have come to terms with the precocious newcomer by taking him under their wings: thus, for example, the *Photogrammetric Engineering* journal has been renamed *Photogrammetric Engineering and Remote Sensing*, and *Transactions on Geoscience Electronics* has become *Geoscience and Remote Sensing*.

There is no doubt that Environmental Remote Sensing is here to stay.

1.4 Economic benefits of remote sensing

Although it is most difficult in these days of high inflation and rapidly fluctuating currency exchange rates to place meaningful figures on the costs and benefits which may accompany remote sensing of the environment some indications can be given of the kinds of proportions which are thought to be involved.

It was in meteorology that satellite remote sensing first achieved fully-operational status in 1966. In that year the significance of satellite data inputs to meteorological data-pools was acknowledged practically through the inauguration of a system of American satellites designed to yield information to any suitably-equipped and relatively modestly-priced receiver anywhere in the world – a system within which stand-by satellites were to be available for launching at short notice so that the supply of data to routine users could be reasonably assured. This system has worked well, and almost continuously ever since. Being a relatively self-contained programme, this provided the first convincing figures for cost/benefit assessments of satellite operations. Figures published in 1971 put the annual cost of American meteorological satellite *research* at around $200 million, and the annual cost of adverse weather to the American community of some $10 000 million. Of this staggering figure, it was estimated that some 20% could be eliminated given better weather information, equivalent to ten times the annual expenditure on weather satellites. The dramatically increased flow of weather data which satellites had made possible had already led to benefits of about $75 million per annum through improved hurricane forecasting alone – a figure *four times* the annual outlay on weather satellite *operations*.

More recently, evaluations of satellite meteorology have been made through a study contract of the European Space Agency, relating both to general issues and the viability of the ESA geostationary satellite system, Meteosat (see p. 76). At the outset certain difficulties were identified, of which some are common to most or all satellite remote sensing studies of Earth phenomena. The most significant of these is that, whilst the *costs* of satellite remote sensing can be evaluated rather readily, it is more difficult to estimate *returns* and *benefits*; many 'customers', e.g. members of the

agricultural or construction industries who are advantaged by improved weather forecasts, do not see themselves as users of satellite data and do not pay directly for satellite services and products as such. Fig. 1.1 illustrates the 'study logic' required for such studies.

Results revealed significant differences between Europe and Africa. For the developed countries of Western Europe, conservative estimates of expected benefits from the operational use of Meteosat-type satellites exceeded the corresponding expenditures by a global factor close to 3, rising in individual countries to as high as 7.8 in the case of Spain. Compared with the USA these figures are on the low

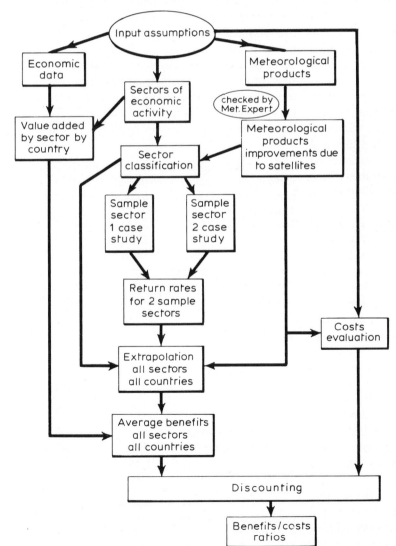

Fig. 1.1 Study Logic for an evaluation of satellite remote sensing cost/benefit ratios in meteorology. (Source: Lagarde, 1980.)

side, reflecting especially the absence of tropical cyclones from, and the lower incidence of severe convectional storms in, Western Europe. The small sizes of some of the nations involved also influence this comparison. When Africa and the Middle East are taken into account also the global benefits/cost ratio for Meteosat-type operations rose to the impressively high level of 16:1. This reflects especially the comparative sparseness of the conventional weather-observing network outside Europe and the less oblique views obtained from geostationary satellite altitudes over the very large land mass of Africa. Assessments of different sectors of economic activity revealed that, almost universally within Europe, Africa and the Middle East, Agriculture, Transport, Construction, and Utilities (Electricity, Gas, Water) were the most sensitive to the weather, in that rank order. Clear potential benefits could accrue from all of these throughout the Meteosat image area.

In the realm of Earth surface monitoring many more possibilities are found for cost-effective remote sensing from satellites – though, because of the more intricate ramifications of the issues and phenomena involved the business of benefit/cost accounting becomes even more difficult than it is in respect of meteorology. Expected annual savings in the early years of operational Earth resource satellite systems have been pitched at levels of some £60–262 million for the USA, and between £278–684 million for the World. Comparable figures, also at 1971 prices, expressed the potential benefits from a fuller use of such systems at levels of about £3300 million and £12 300 million respectively. Although guesswork plays a prominent part in the formation of such figures there is no doubt that, yet again, satellite remote sensing must be a cost-effective pursuit, indeed much more so than weather satellite operations with particular respect to oceanography, and geology and mineral resource monitoring and evaluation. Table 1.1 presents more recent figures for expected benefits in one realm of applied oceanography, namely marine transportation in and around North America, itemized by area and resource or pursuit.

At the same time it should not be overlooked that both aircraft and satellites permit terrestrial monitoring in regions, and of types, not possible or not normally effected on the ground. Land-use

mapping may be used as an example of an activity rarely completed on a national or international scale at ground level because of the magnitude and organizational problems such surveys entail. Aircraft and satellites permit detailed land-use mapping even in developing countries, many of which have never been mapped in this way on the ground. Careful cost-comparisons for aerial surveys and Landsat image analyses for land-use mapping in both developed and developing countries emphasize the great advantages of satellites over aircraft for these purposes. Table 1.2, summarizing a comparative study for three counties in the American State of Mississippi, shows that without significant loss of accuracy, computerized Landsat data analyses can be carried out for a fraction of the time and cost of conventional air-photo procedures. An attendant bonus is the repetitivity of the Landsat coverage, permitting up-dated maps to be prepared at quite short intervals. Similarly, in a study in the Sudan, an integrated aircraft–Landsat project resulted in a land system–land capability map with 83% less human effort than otherwise required in the absence

Table 1.1 Potential aggregate gross benefits of environmental surveillance (including satellites, aircraft and *in situ* sensor systems) for marine transportation (1975 $ Million). (From MacQuillan and Clough, 1978)

	1986–1990 (5-year total)	1991–2000 (10-year total)
Ocean routing:		
Pacific ports	120	370
Atlantic ports	100	320
Great Lakes ports	9	26
Ice surveillance:		
St. Lawrence	75	230
Atlantic	45	150
Arctic oil and gas	135	745
Arctic supply, minerals	95	230
West coast log towing	10	20
Totals for comprehensive surveillance systems	589	2091
Contribution of satellite systems	402	1543

Table 1.2 Cost comparison of land-use information extraction techniques for 3 counties in Mississippi, USA. (Source: Abiodun, 1978)

	Photo-interpretation, black/white 1:24 000	Photo-interpretation, colour, infrared 1:120 000	Computer implemented Landsat digital analysis
Cost of basic data per square mile	$11.97[a]	$0.36[b]	$0.06[c]
Cost of information extraction per square mile	$28.34	$8.35	$2.04[d]
Effort, man-hours	6200	3300	500
Time required	12 months	6 months	1 month
Accuracy	92–96%	92–96%	89–95%

a Commercial contract
b $8 per frame purchased at EROS Data Centre
c $160 per set of 5 tapes purchased at EROS Data Centre and prorated over 2650 square miles
d Prorated over 2650 square miles

of the Landsat component. The cost was reduced by some 96%.

Of course, such statistics presume that the costs of the remote sensing systems themselves are borne by others. Therefore, a related question is 'How much does a Landsat receiving and processing facility cost?' Tables 1.3 (a)–(c) provide some indications of station installation and operating costs. Although these may appear to be high, cost-sharing amongst nations and/or agencies is often possible. Furthermore, as Table 1.3(c) reveals, installation costs of Landsat receiving stations have shown a downward trend, running counter to inflation and movements of the value of the US Dollar. Such a trend may be expected to continue in the future, in relative, if not absolute, monetary terms.

Satellite remote sensing should become increasingly cost-effective as reception and processing facilities proliferate. There is, however, a real danger that unit costs of satellite images may rise, especially to the user who lacks his own receiving facilities, as new US policies are implemented. The time is coming when global satellite remote sensing should be organized more internationally, following the example of systems like the *World Weather Watch* (WWW). The aim should be to bring remote sensing data – subject to their general availability – to the user whoever and wherever he may be, at a cost which he can bear. Then, and perhaps only then, will the new data types be used as widely and as fully as their contents deserve.

1.5 The geographical uses of remote sensing

Since the broadest and most coordinated use of remote sensing data has been made by the geographical community (see Fig. 1.2) it is of interest lastly to review the historic growth of remote sensing practices in geography, the most comprehensive of the sciences of man's environment.

(a) Pre-1925: a period of slowly increasing recognition of the potential utility of air-photos in topographic mapping. The rise of aviation in the First World War accelerated the process to such an extent that already by the late 1920s some regions were being photographed systematically from the air.

Table 1.3 Capital, equipment and operating costs of Landsat ground stations. (From Abiodun, 1978)

(a) Initial equipment costs of Landsat ground stations (in thousands of US dollars)

	Minimum requirement station	Full ground station
Antenna and receiving equipment	400	500
MSS recording and pre-processing equipment	120	440
RBV recording and pre-processing equipment	—	550
Quick-look film processor	—	200
Timing and display equipment	10	150
Landsat 'Housekeeping' (PCM) data recording equipment	—	30
Photo lab	60	250
MSS processing equipment (applies corrections to data and converts it to CCT or film)	800	1500
RBV processing equipment (applies corrections to data and converts it to CCT or film)	—	400
Precision processing equipment	—	500
System installation and test	135	250
Manuals	50	250
Spares	75	400
Special test equipment	100	300
On-site training	150	280
TOTAL:	1900	6000

(b) Annual operating costs of Landsat ground stations (in thousands of US dollars)

	Minimum system	Full ground system
System operation and maintenance	600	1200
Expendable items (film, magnetic tapes, etc.)	100	400
Special equipment maintenance contracts	50	75
	750	1675

(c) Capital costs of existing Landsat ground stations

Country	Cost in millions of United States dollars in year of installation	Year of installation	Notes
United States of America	25	1972	C, R
Canada 1 (Prince Albert)	6	1973	C, Q
Brazil	4.2	1974	C, Q
Italy	3	1975	D, M, L
Canada 2 (East Coast)	2	1976	D, P, M, L, Q

Key to Notes:
C – Complete system including RBV, CCT and high quality photo products
R – RBV digital recording
Q – Quick-look photographic products available within one hour of acquisition
D – Digital data products emphasized
P – High volume photo production, lower quality than C
M – MSS only
L – receiving station co-located with processing element

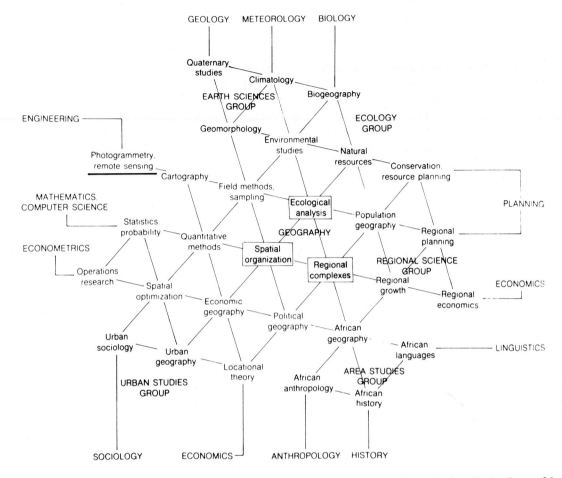

Fig. 1.2 Remote sensing as a gateway to the environmental sciences from the world of modern technology. In the cluster of Area Studies, Africa is chosen as an example only. (After Haggett, 1972.)

(b) 1925–1945: a period of widespread but superficial use of aerial photography. During this span of twenty years air photo-interpretation became a fully-fledged intelligence-gathering technique for both civilian and military purposes. Many parts of the world were subjected to scrutiny from aloft, even hostile environments, relatively inaccessible on the ground, like much of Antarctica.

(c) 1945–1955: a period of preoccupation with interpretation techniques. This was a period during which many geographers in particular became enthusiastically acquainted with this new tool of great potential, and too much emphasis was placed on methods of analysis and interpretation, and too little on the applications to which the results could be put.

(d) 1955–1960: a period of widespread application of aerial photography. Now the theory began to be put to good use not only in topographic mapping, but also in various fields of human and physical geography, in geology, forestry, agriculture, archaeology, and numerous other disciplines with interests in spatial and temporal variation in the landscape.

(e) 1960–present: a period of active platform and sensor experiment, leading to the establishment of the first fully-operational satellite remote sensing systems. The first meteorological satellite in 1960 heralded the opening of a period of intense activity, investigating the potentialities of balloons, rockets, and especially satellites for remote sensing not only by conventional photography but also a wide variety of other means.

In particular, the launching of the American Earth Resources Technology Satellite, ERTS 1*, on 23 July, 1972 heralded a new era in surface-orientated applications of satellite and remote sensing technologies.

Hopefully, the next decade will see remote sensing 'come of age' as a mature discipline with a balanced development of theory, practices, and truly operational applications. As more scientists, technologists, and even politicians have become intelligently aware of the possibilities, so the door has opened to suitably planned and integrated environmental monitoring programmes at every scale from the local to the global, based on fully-appropriate instruments and sensor packages, entailing efficient processing techniques, and yielding many potential benefits to man and his environment through expeditious use of the results.

*The ERTS programme was subsequently re-named 'Landsat'. Although much of the literature refers to ERTS 1, we have standardized references in the text by using the name Landsat throughout, for both this and further satellites in this family.

2 *Physical bases of remote sensing*

2.1 Natural remote sensing

We all use our natural senses to observe and explore the environment in which we live. Certain senses – smell, taste, and, more often than not, touch – permit us to assess environmental qualities directly, through our neurophysical responses to the gases, liquids and solids with which we have immediate contact. The others – sight and hearing – can make us aware of more distant features through the patterns of energy propagations associated with them. Feeling, manifested through the sensitivity of skin to heat, also enables us to assess some characteristics of distant phenomena. These appreciations of the behaviour of energy sources some distance from ourselves are natural forms of remote sensing.

Since man's visual powers are perhaps the most valuable he possesses for gathering information about an object or phenomenon with which he is not in direct contact, we may usefully examine them in more detail as an introduction to the principles involved in remote sensing by artificial means. The key components of the remote sensing system which makes sight possible are the eye and the brain. Visible energy, in the form of light emitted by, or reflected from, an illuminated object is detected by sensitive cells in the eye. The eye is linked by the optic nerve to a high speed, real-time (i.e. near instantaneous) data processor – the brain.

The human eye is sensitive both to the *intensity* of the energy received, and the *frequency* of the wave-like perturbations which may be taken to characterize such flows of energy. As a consequence we can differentiate both ranges of brightness and colour tone. In the brain the quickly-processed data are compressed and presented as visual images. The brain also serves as a data bank in which earlier images can be stored, albeit within frustratingly narrow time limits, and with considerable loss of accuracy and definition. However, some qualitative pattern-matching can be carried out by drawing mental comparisons of, say, the present image and selected images recalled from the past.

Our mental pictures can be modified by simple means such as optical lenses to correct vision defects, selective filters like polaroid sunglasses to reduce the glare of strong sunlight, or by chemicals such as hallucinatory drugs to modify the way we perceive the area of illumination.

Clearly, a natural remote sensing system such as the eye-brain is adapted best to the assessment of prevailing environmental patterns at the present time. In the evolutionary context this would be one of its greatest strengths. On the other hand, where conscious effort and freedom of choice are dominant this adaptation is one of its primary weaknesses. The chief requirements for artificial systems whereby man's remote sensing performance may be improved include:

(a) A broader and more selective ability to detect variations in environmental conditions. Our natural sensors have very limited performances: what is physiologically possible is restricted to the visible region of the electromagnetic spectrum.

(b) A capacity for recording more permanently the patterns which are detected. This permits a more leisurely inspection of features of special interest.

(c) A better recall system so that patterns at different points in time might be compared with greater accuracy and in greater detail.

(d) The facility to play back the past at different speeds if required. Many natural events may be examined best when studied by time-lapse methods, when reality is speeded up, or 'action replays' when recorded data are slowed down.

(e) Opportunities for automatic ('objective') analyses of observations so that the personal (psychological, educational and/or neuro-physical) peculiarities of the observer are minimized.

(f) Means of enhancing images to reveal or highlight selected phenomena.

Great strides have been made in recent years towards the fulfilment of such desiderata in remote sensing of the environment. Let us consider in more detail the range of opportunities presently being exploited and explored.

2.2 Technologically-assisted remote sensing

Over the years our limited natural capabilities for remote sensing have been much extended by the invention and steady improvement of a variety of specialized instruments. Here we shall confine our attention to the broad fields in which such instruments are designed to operate. Later in this chapter and elsewhere we shall go on to examine some of the instruments themselves, the forms and characteristics of their data, and ways in which these may be processed, analysed and interpreted.

2.2.1 Electromagnetic energy

Undoubtedly the most important medium for environmental remote sensing is electromagnetic radiation. This, the only form of energy transfer that can take place through free space as well as a medium, exhibits enormous variety in behaviour and its implications. So important is this field that some authorities have considered it to be the only one of significance for environmental remote sensing. More than one account of remote sensing contains a categorical statement such as this: 'In remote sensing, information transfer from an object to a sensor is accomplished by electromagnetic radiation'. Whilst other modes of information transfer certainly do occur (see below), we too, from Section 2.3 onwards, shall focus our attention in this book mainly on electromagnetism as a medium for remote sensing. This is justified by the preponderance of present interest and practice in this field.

2.2.2 Acoustical energy

Acoustic sounding of the Earth's oceans is well-established as a means of profiling the sea floor. In a typical acoustic echo sounder, short pulses of audible sound are emitted in a selected direction, and the returning reflected or scattered signals are collected and interpreted as indications of ocean depth and sea floor characteristics. The atmosphere is also amenable to investigation using acoustic sounding techniques, especially aspects of its wind and thermal structures (Plate 2.1). Indeed, the strength of interaction of acoustic waves and the atmosphere is far greater than for electromagnetic waves, though the operational range of acoustic waves is less, and their slower speed of propagation can be a source of error with some measurements. Consequently the applications of such techniques to the atmosphere have been predominantly in research. For the future, the simplicity and much lower cost of an acoustic echo sounder system should ensure the continued development of such systems with eventual objectives in operational meteorology.

2.2.3 Force-fields

To the Earth scientist the most familiar 'force-fields' are those of gravity and magnetism. Gravity is the force exerted by each body in the universe on every other body. It is directly proportional to the product of the masses of any pair of bodies, and inversely proportional to the square of the shortest distance between their centres of mass. We do not understand what causes gravity, but we can measure it, and use the results, for example to substantiate geophysical theories of the solid Earth and to reveal facts concerning the nature and disposition of its crust.

Magnetism is the attraction which certain natural minerals (and the Earth which contains them) have for others, especially iron. The magnetic properties of such minerals permit us to prospect for commercial deposits of them in geomagnetic surveys. Slow changes in the magnetic field of the Earth accompany the 'wanderings' of the magnetic pole; wilder variations are caused by fluctuations in the 'solar

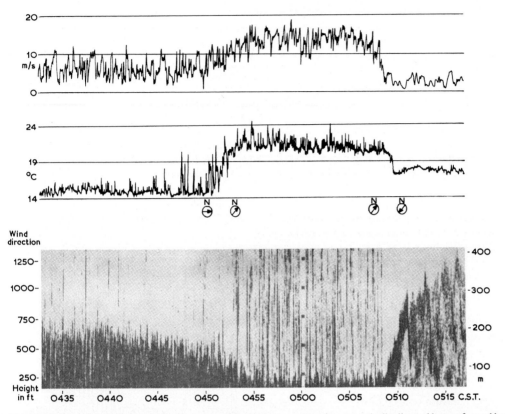

Plate 2.1 It is possible to detect the small amount of energy backscattered from a vertically directed beam of sound by inhomogeneities in the density structure of the lower troposphere. Plate 2.1 exemplifies the type of visual record obtained, as a function of height and time. This record was obtained during the passage of a cold front. The traces show wind (at top) and temperature (bottom). (From McAllister and Pollard, 1969.)

wind' of charged particles arriving from the Sun. These are responsible for 'magnetic storms' and the associated interference to radio communications, and the aurorae. They can be monitored by magnetometers to reveal and record changes in the behaviour of the Sun.

2.2.4 *Active remote sensing*

A distinction is commonly drawn between 'active' and 'passive' remote sensing. In the latter (illustrated earlier by the eye-brain system) the sensor detects the energy or force which emanates from the target itself, or from a 'third party' source. Active remote sensing involves the detection of a signal which is artificially produced. Since such signals are generated under controlled conditions much can be learned from the ways in which they are affected by objects in the environment, and/or by the media through which they are transmitted. Radar (Radio Direction and Ranging) is a common example of an active system exploiting electromagnetic radiation. Sonar (Sound Navigation and Ranging) is an active system utilizing acoustical energy.

2.3 Electromagnetic energy

2.3.1 *The nature of radiation*

Energy is the ability to do work. It can exist in a variety of forms including chemical, electrical, heat, and mechanical energy. In the course of work being done, energy must be transferred from one body or one place to another. Such transfers are effected by:

(a) Conduction. This involves atomic or molecular collisions.

Table 2.1 Key terms associated with electromagnetic radiation

Processes and phenomena	Entities	Processes	Properties[1]	Behavioural characteristics[2]	Performances[3]
Suffixes	-er or -or	-ion	-ivity	-ance	
Terms	Emitter/Radiator	Emission	Emissivity	Emittance	Exitance
	Absorber	Absorption	Absorptivity	Absorptance	Absorptive power
	Reflector	Reflection	Reflectivity	Reflectance	
	Transmitter	Transmission	Transmissivity	Transmittance	
		Extinction			

1. Usually related to theoretical (model) maxima on 0–1 scales. Each observed value can be viewed as a natural limitation on the associated performance.
2. Expressing actual source, medium, or target performance as a ratio of the possible maximum. Sometimes expressed as a % (e.g. albedo (the percentage reflectance of natural objects)).
3. Expressed as radiant fluxes per unit area.

(b) Convection. This is a corpuscular mode of transfer in which bodies of energetic material are themselves physically moved.

(c) Radiation. This is the only form in which electromagnetic energy may be transmitted either through a medium or a vacuum.

It is this third type of transfer with which we are primarily concerned in remote sensing studies. Table 2.1 introduces key concepts and related terminology.

In keeping with many other areas in the environmental sciences, remote sensing employs *models* (simplified or idealized representations of reality) to describe complex phenomena, situations or interactions. In the case of electromagnetic radiation two models are necessary to describe and elucidate its most important characteristics:

(a) The wave model. This typifies radiation through regular oscillatory variations in the electric and magnetic fields surrounding a charged particle. Wave-like perturbations emanate from the source at the speed of light (3×10^{10}cm s^{-1}). They are generated by the oscillation of the particle itself. The two associated force-fields are mutually orthogonal, and both are perpendicular to the direction of advancement (see Fig. 2.1).

(b) The particle model. This emphasizes aspects of the behaviour of radiation which suggest that it

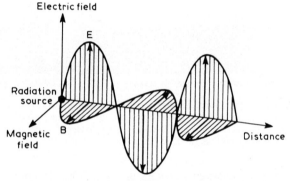

Fig. 2.1 Electric (E) and magnetic (B) vectors of an electromagnetic wave, viewed at a given instant. The intensity of radiation varies with the square of the peak amplitude of the electric field, and is proportional to the number of photons in the field. The type of radiation (see Fig. 2.2) is governed by wavelength, measured in absolute units.

is comprised of many discrete units. These are called 'quanta' or 'photons'. These carry from the source some particle-like properties such as energy and momentum, but differ from all other particles in having zero mass at rest. It has been hypothesized consequently that the photon is a kind of 'basic particle'.

Whatever the true nature of electromagnetic radiation, we know that it results whenever an electrical charge is generated. In terms of the wave model, the wavelength of the resulting ray of energy is determined by the length of time that the charged particle is accelerated; the frequency of the radiation waves depends upon the number of accelerations per second to which the particle is subjected. The relationship between wavelength (λ), frequency (f), and the speed of light (a universal constant, c) is:

$$\lambda f = c, \text{ or } \lambda = c/f \qquad (2.1)$$

This tells us that wave frequency is inversely proportional to wavelength, and directly proportional to its speed of wave advancement.

We may relate this expression of wave theory to the particle theory through the statement:

$$E = hf \qquad (2.2)$$

where E is the energy of a quantum, h is a constant (named after Planck, who proposed the quantum theory in 1900), and f the frequency of the radiation waves. If we multiply ($\lambda = c/f$) by h/h (which does not alter its value), and substitute E for hf, we are left with a new expression, namely:

$$E = hc/\lambda \qquad (2.3)$$

This tells us that the energy of a photon varies directly with wave frequency and inversely with radiation wavelengths. This is borne out by experiment, for it can be demonstrated that the longer the wavelength of radiation the lower the energy involved, and conversely, the higher the frequency of radiation, the greater the energy involved. These relationships are fundamental to the appreciation of the behaviour of electromagnetic radiation.

Probably the most familiar form of electromagnetic radiation is *visible light*. Radiation detected by the human eye ranges through the well-known sequence of component colours from red through orange, yellow, green, blue and indigo to violet. This range, the so-called 'visible spectrum', should not be confused with the much broader 'electromagnetic spectrum', of which it forms but a tiny part. The eye-brain system is capable of detecting only a minute section of the total spectrum of radiation, all of which travels at the speed of light. Some knowledge of the breakdown of the electromagnetic spectrum is essential for an adequate appreciation of many remote sensing studies.

Largely for convenience of reference, the electromagnetic spectrum is subdivided into a number of sections. The boundaries between them are expressed in several different ways (see Fig. 2.2):

(a) Wavelength. The spectrum stretches from extremely short waves (cosmic rays) to very long

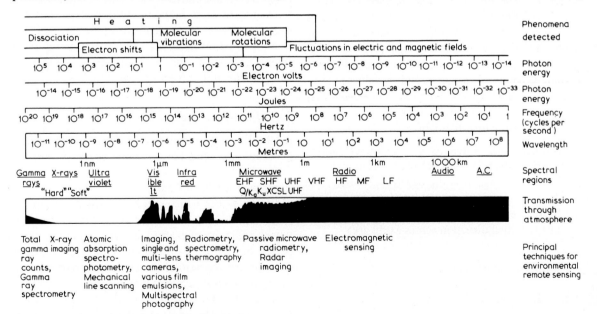

Fig. 2.2 The electromagnetic spectrum. The scales give the energy of the photons corresponding to radiation of different frequencies and wavelengths. The product of any wavelength and frequency is the speed of light. Phenomena detected at different wavelengths are shown, and the principal techniques for environmental remote sensing. The significance of the results for different environmental applications constitute a greater part of the latter chapters of this book: the student might usefully tabulate them for himself.

('electromagnetic') waves. The waves range from the microscopic to hundreds of kilometres in length.

(b) Frequency. We have seen that wavelength and frequency are always inversely related, since all waves advance at a common speed, the speed of light. Longwave radiation is a low frequency propagation, whereas short wave radiation is characterized by much higher frequencies. The units of measurement are Hertz (Hz), or cycles per second.

(c) Photon energy. Equation 2.3 stated that the energy of a photon is directly proportional to wave frequency. Fig. 2.2 shows that long wave radiation has low photon energy, whereas short-wave radiation has high photon energy. This is expressed in joules or watt-seconds.

The whole spectrum ranges from almost infinitely short cosmic rays to the long waves of radio and beyond. It includes a number of familiar everyday manifestations such as gamma rays, X-rays, ultra-violet light, infrared rays and radar waves. Fig. 2.2 indicates some of their more common uses in remote sensing. It should be understood that the whole spectrum is a continuum which is subdivided in a rather inconsistent way. The boundaries shown between different regions of the electromagnetic spectrum are inexact. There is considerable overlap between some neighbouring regions, no fully-accepted terminology applied to them, and different authorities prefer different boundary values.

Lastly we should note that radiation *amounts* may be quantified in several different ways. These involve the following distinctions:

(a) Radiant flux, which is a measure of radiant energy passing a reference plane per unit time.
(b) Irradiance, which is the density of radiant flux incident upon unit area of a surface.
(c) Radiance, which is the intensity of radiant flux in a specified direction across a unit area.

2.4 Radiation at source

2.4.1 The generation of radiation

An electromagnetic source, whether natural or artificial, may emit:

(a) A broad continuum of wavelengths of radiation,

(b) Radiation within a narrow (single spectral) band, or
(c) Radiation of a single wavelength.

The intensity of such radiation varies with the square of the peak amplitude of the electric field, and is proportional to the number of photons in the field. Variations in the intensity, and perhaps also the wavelength or frequency, of radiation from a source may occur through time. The electrically-charged particles involved in generating the emitted radiation are basic units of matter, namely atoms, electrons and ions. All are in a state of constant motion under normal circumstances, when the temperatures of the objects they constitute are above absolute zero. Since the molecules of these objects are built of electrically-charged particles they possess natural resonance, either vibrational or rotational, so that they act as small oscillators which accelerate the electrical charges. The emitted photons of energy have frequencies which cor-respond to the resonances of the molecules. The distribution of molecules among energy states is a function of temperature, so that at higher temperatures there are proportionately more molecules in the higher energy states. This is to say that all particles do not increase their energy equally as the temperature of a radiation source rises, but the overall effect is an increase in molecular activity, accompanied by an increase in emitted energy and a related shift in the dominant wavelength of radiation towards the higher frequencies.

2.4.2 Emission of radiation

All bodies with temperatures above absolute zero generate and send out, or emit, energy in radiant form. Each radiation source, or radiator (whether natural or artificial), emits a characteristic array of radiation waves which can be assessed in terms of their wavelengths and intensities. For many real-world radiators the array is very complex, com-prised of a number of different contributions from the various constituents. Hence a characteristic curve or *spectral signature* may be obtained for each type of natural remote sensing target by plotting the intensities of the emitted radiation against appro-priate wavelengths in the electromagnetic spectrum. Chapter 3 provides illustrations of a number of spectral signatures. For the present we must concern ourselves more with the principles which cause such signatures to differ.

A useful concept, widely used by physicists in radiation studies, is that of the *blackbody*, a model (perfect) absorber and radiator of electromagnetic radiation. A blackbody is conceived to be an object or substance which absorbs all the radiation incident upon it, and emits the maximum amount of radiation at all temperatures. Although there is no known substance in the natural world with such a performance, the blackbody concept is invaluable for the formulation of laws by comparison with which the behaviour of actual radiators may be assessed. These laws include the following:

(a) Stefan's (or Stefan–Boltzmann's) law. This states that the total emissive power of a black-body is proportional to the fourth power of its absolute temperature (*T*). This can be expressed as:

$$W = \sigma T^4 \qquad (2.4)$$

where W is radiant exitance in watts cm^{-2}, and σ is the 'Stefan-Boltzman Constant'. This relationship applies to all wavelengths shorter than the microwave. In the microwave region radiant exitance varies as a direct function of T^0. The significance of Stefan–Boltzmann's law is that hot radiators emit more energy per unit area than cooler ones.

(b) Kirchhof's law. Since no real body is a perfect emitter, its exitance is less than that of a blackbody. Clearly it is often useful to know how the real exitance (*W*) of a radiator compares with that which would be anticipated from a corresponding perfect radiator (W_b). This may be established by evaluating the ratio W/W_b, which gives the emissivity (ϵ) of the real body. Thus, for the general case, we may say that:

$$W = \epsilon W_b \qquad (2.5)$$

The emissivity of a real blackbody would be 1, whilst the emissivity of a body absorbing none of the radiation upon it (not unexpectedly known as a 'white body') would be 0. Between these two limiting values, the 'greyness' of real radiators can be assessed, frequently to two decimal places. If we plot emission curves for real radiators against curves for corresponding blackbodies, we usually find that the idealized curves are much smoother and simpler than the observed patterns. In effect the emissivity index is a measure of radiating efficiency across the spectrum as a whole.

(c) Wien's (displacement) law. This states that the wavelength of peak radiant exitance (λ_{max}) of a blackbody is inversely proportional to its absolute temperture (*T*):

$$\lambda_{max} = C_3/T \qquad (2.6)$$

C_3 is a constant equal to $2897\mu m$ K. This equation tells us that, as the temperature of a blackbody increases, so the dominant wavelength of emitted radiation shifts towards the short wavelength end of the spectrum. This can be exemplified by reference to the Sun and Earth as radiating bodies. For the Sun, with a mean surface temperature of about 6000K, $\lambda_{max} = 0.5$ μm. For the Earth, with a surface temperature of about 300K on a warm day, $\lambda_{max} = 9.0$ μm. Empirical observations have confirmed that the wavelength of maximum radiation from the Sun falls within the visible waveband of the electromagnetic spectrum whilst that from the Earth falls within the infrared. We experience the former dominantly as light, and the latter as heat.

(d) Planck's law. This more complex statement describes accurately the spectral relationships between the temperature and radiative properties of a blackbody. In one form, Planck's law may be expressed by

$$W_\lambda = C_1 \lambda^{-5}/(e^{C_2/\lambda^T} - 1) \qquad (2.7)$$

where W_λ is the energy emitted in unit time for our unit area within a unit range of wavelengths centred on λ, C_1 and C_2 are universal constants, and e the base of natural logarithms (2.718). A useful feature of Planck's law is that it enables us to assess the proportions of the total radiant exitance which fall between selected wavelengths. This can be useful in remote sensor design, and also in the interpretation of remote sensing observations.

In consequence of these four laws we see in Fig. 2.3 that emission curves are characteristically negatively skewed, with their peak intensities falling in the lower quartiles of their spectral bands. The wavelengths of maximum emission are progressively shorter for the hotter radiators, and the total

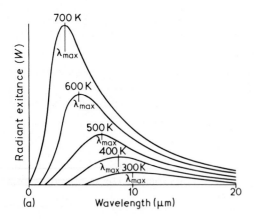

(a)

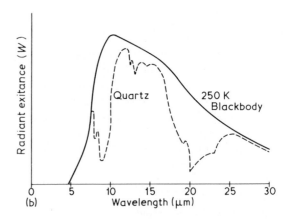

(b)

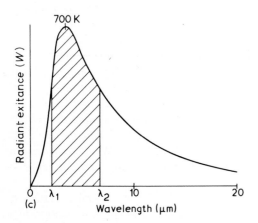

(c)

Fig. 2.3 (a) Selected blackbody radiation curves for various temperatures. Note that the areas under the curves diminish as temperature decreases. Note, too, that λ_{max}, the wavelength of peak radiant exitance, shifts towards the longer wavelengths with decreasing blackbody temperature. (b) The total radiant exitance of a real body (e.g. quartz) is less than that of a blackbody at the same temperature. The general relationship is expressed by Kirchhof's Law. (c) One form of Planck's Law of radiation permits the assessment of that portion of the total radiant exitance falling between selected wavelengths, e.g. λ_1 and λ_2.

emissions from the hotter radiators are greater than those from the cooler radiating bodies.

2.5 Radiation in propagation

2.5.1 Scattering

It is unfortunate (for remote sensing) that considerable complications arise in practice where the Earth's atmosphere intervenes between the sensor and its target. Although the speed of electromagnetic radiation is unaffected by the atmosphere, this medium may affect several of the other characteristics of this form of energy propagation. These include:

(a) The direction of radiation.
(b) The intensity of radiation.
(c) The wavelength and frequency of the radiation received by a target, at the base of the atmosphere.

(d) The spectral distribution of this radiant energy.

Frequently both the direction and intensity of radiation are altered by particles of matter borne in the atmosphere. These redirect, in a rather unpredictable way, the radiation en route through a turbid medium. An understanding of radiation scatter is necessary for the selection of sensors or filters where certain effects are required, and where image degradation due to atmospheric impurities may be avoided, or at least reduced, by sensing at the most appropriate waveband(s).

The *attenuation* which results from scattering by particles suspended in the atmosphere is related to the wavelength of radiation, the concentration and diameters of the particles, the optical density of the atmosphere (discussed under Refraction on p. 21) and its absorptivity. The common types of scatter are:

(a) Rayleigh scatter. This mostly involves

molecules and other tiny particles with diameters much less than the radiation wavelength in question. It is characterized by an inverse fourth power dependence on wavelength. Hence, for example, ultraviolet radiation (about one-quarter the wavelength of red light) is scattered sixteen times as much. This helps to explain the dominance of orange and red at sunset when the sun is low in the sky: the shorter wavebands of visible light are cut out by a combination of atmospheric absorption and powerful scattering.

(b) Mie scatter. This occurs when the atmosphere contains essentially spherical particles whose diameters approximate to the wavelengths of radiation in question. Water vapour and particles of dust are the main agents which scatter visible light.

(c) Nonselective scatter. Here particles with diameters several times the radiation wavelengths are involved. Water droplets, for example, with diameters ranging commonly from 5–100 μm scatter all wavelengths of visible light (0.4–0.7 μm) with equal efficiency. As a consequence clouds and fog appear whitish, for a mixture of all colours in approximately equal quantities produces white light.

2.5.2 Absorption

This is the retention of radiant energy by a substance or a body. In the real world it involves the transformation of some of the incident radiation into heat, and the subsequent re-emission of that energy at a longer wavelength.

Considering the ideal case, it will be recalled that a blackbody is a perfect radiator; it is also a fully efficient absorber of radiant energy which is incident upon it. Just as the *emissivity* of a real body can be defined as an inherent characteristic of its material, expressing the ease with which it gives up energy by radiation, so the *absorptivity* (α) of a real body is an expression of its ability to absorb radiant energy. Logically, for a blackbody, $\alpha_b = \epsilon_b$, and both ϵ and α equal unity. For 'grey' bodies, ϵ and α are also equal (but have values of less than 1) if they are opaque, in which case both absorption and emission are confined to the (physically simple) surface layer. For

all other cases $\alpha_b \neq \epsilon_b$, and modelling the processes involved can be very complicated and difficult.

In the atmosphere (which can be decidedly murky, but never opaque) radiation absorption takes place not so much at its surface, but in transit. Three gases: water vapour, carbon dioxide, and ozone, are particularly efficient absorbers of radiation from the Sun. Consequently incoming solar radiation (often abbreviated to 'insolation'), which is by far the most important natural source of radiation for passive remote sensing, is attenuated significantly by its passage through the atmosphere. Often the combined effects of absorption and scattering by particulate matter are expressed in terms of an *extinction coefficient*. This enables us to evaluate the energy which reaches the surface of the planet Earth as a ratio of the radiation incident upon the outer limits of the atmosphere. This permits us to evaluate the *transmittance* of the atmosphere. For any such medium the transmission of radiant energy through it is inversely related to the product of the thickness of the layer and its extinction coefficient. In practice, therefore, the transmittance decreases as the combined effects of absorption and scattering accumulate.

2.5.3 Refraction

When electromagnetic radiation passes from one medium to another, bending or 'refraction' occurs in response to the contrasting densities of the media (Fig. 2.4). A measure of this is given by the Index of Refraction (n), where:

$$n = c/c_n \tag{2.8}$$

The index is the ratio of the speed of light in a vacuum (c) to its speed in the substance (c_n). In a non-turbulent atmosphere (which can be conceived as a series of layers of gases each of a different density) the refraction of radiation is predictable. For, as Snell's law states, there is a constant relationship for a given frequency of light between the index of refraction and the sine of the angle between the ray and the interface between each pair of adjacent layers. Problems do arise, however, where a turbulent atmosphere is involved. Turbulent motions are essentially random, and their effects upon radiation are consequently unpredictable.

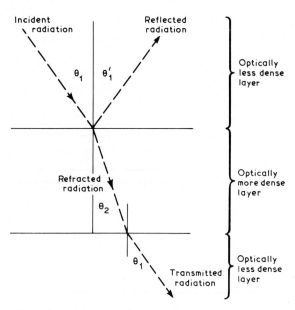

Fig. 2.4 The relationships between incident radiation, reflected radiation, refracted radiation and transmitted radiation, and a stratification of optically different layers.

2.6 Radiation at its target

2.6.1 Reflection

So far we have discussed the effect of particles on radiation as if its resulting redirection is another variable we cannot predict. In the real world this is not always so. The critical factors are the smoothness and orientation of the object lying in the path of an electromagnetic ray. If the object has smaller surface irregularities than the wavelength of the impinging energy, it will act as a mirror and the angle at which the energy is directed away from the object will equal the angle of its incidence upon it. This process of reflection is sometimes called *backscattering*. The angles of incidence and reflection, and a perpendicular to the reflecting surface from which those angles are measured, all lie in the same plane. Conversely scattering may be described in terms of reflection, when it is labelled *diffuse reflection* in contrast to *specular reflection* which is the simpler ideal case. Although we might expect some reflection by particles in the atmosphere, reflection is much more important for remote sensing where continuous (e.g. land or sea) surfaces are involved.

A useful expression of the reflectivity of different terrestrial surfaces is the reflection coefficient, or *albedo* (Table 2.2). The albedo of a surface is the percentage of the insolation incident upon it which is reflected back towards space.

We referred earlier to a white body as one which absorbs none of the radiation which impinges on it. We are now able to say that, whereas a blackbody is conceived as a perfect absorber of radiation, a white body is a perfect reflector. Since none of the incident radiation is absorbed, the surface temperature of a white body remains unchanged.

Because visible spectrum wavelengths are so short, most surfaces reflect light diffusely regardless of the angles at which this radiation strikes them. Some of the reflected energy returns in the direction of its source. Longer wavelengths, e.g. microwaves, create more specular reflection off the same surfaces, the reflected energy being directed generally away from its source. Some of the consequences of such differences in performance may be illustrated by reference to the remote sensing of a building at night. Using visible light – a torch or a searchlight – the facing wall of the building will be discernible whatever its angle to the light beam. Using radar to locate the same feature much depends upon the angle between the radar beam and the wall it strikes. Since most of the reflection in this case is specular rather than diffuse, much stronger returns are obtained from a head-on, rather than an oblique, orientation.

Reflected radiation is of considerable importance to remote sensing since so many observing systems are based upon it (Plate 2.2). The natural remote sensing system of the eye and the brain observes and perceives many aspects of the natural environment through reflection of the solar radiation by which it is illuminated. The early, now well-developed, artificial technique of photography records phenomena through the light which is reflected from them, whether in the studio or out of doors. Active remote sensing systems such as radar often measure the reflection of specially-propagated energy so that the sizes, positions and/or densities of natural or man-made reflectors can be assessed thereby.

2.6.2 Absorption

Not all the radiation incident upon the surface of a

Table 2.2 Values of albedoes for various surfaces and types of surfaces. (Source: Lockwood, 1974)

Type of surface	Surface	Albedo % of incident shortwave radiation
Soils	Fine sand	37
	Dry, black soil	14
	Moist ploughed field	14
	Moist black soil	8
Water surfaces	Dense, clean and dry snow	86–95
	Woody farm, snow-covered	33–40
	Sea ice	36
	Ice sheet with water covering	26
Vegetation	Desert shrubland	20–29
	Winter wheat	16–23
	Oaks	18
	Deciduous forest	17
	Pine forest	14
	Prairie	12–13
	Swamp	10–14
	Heather	10
Geographic locations	Yuma, Arizona	20
	Winnipeg (July)	13–16
	Washington, DC (September)	12–13
	Great Salt Lake, Utah	3

target is necessarily scattered or reflected by it. Some of the incident energy may enter the target to be propagated through it as a refracted wave front. As we noted earlier, with a gaseous medium like the atmosphere, energy in transit may be attenuated or depleted, at least partially, by the process of absorption. The ability of a substance to absorb radiation (i.e. its absorptance) depends upon its composition and its thickness.

Absorptance also varies with wavelength of radiation. A target may behave very differently if exposed to radiation of different wavelengths. For example, an object or a medium may be highly absorptive in the visible range yet transparent in the infrared (like some 'semi-conductors'), or it might be transparent in the visible, yet opaque in the infrared (like glass).

Clearly some knowledge of the absorptance of a target is often invaluable in choosing the best remote sensing means to solve a specific problem, and in interpreting correctly the results of a remote sensing survey.

The net effect of absorption of radiation by most substances – including the atmosphere and terrestrial surfaces – is that the greater part of the absorbed energy is converted into heat. Thereby the temperatures of the substances are raised. This heat energy may enable the original target of radiation to become a secondary radiation source itself, emitting some of the energy absorbed. Since the peak intensity of solar radiation is in the waveband of visible light, and the Earth/atmosphere system is by no means a blackbody, it is not surprising that its radiation temperature – as we saw earlier – is much less that of the Sun; the radiation peak of the Earth is in the infrared (see Fig. 3.1). Fortunately for environmental remote sensing, infrared energy (a form of 'thermal' radiation) is emitted from the surface of the Earth by both day and night, and is affected little by atmospheric particles such as haze and smoke. If we take care to view such remote sensing targets through atmospheric window wavebands in the infrared portion of the spectrum (where absorption by certain gaseous constituents of the atmosphere is negligible) and under cloud-free conditions, we are able to add much to our knowledge of their physical and chemical properties, and their diurnal cycles, which we could *not* have learnt by conventional photography. As we shall see in later chapters, thermal mapping by infrared and microwave techniques is becoming increasingly significant in remote sensing applications today.

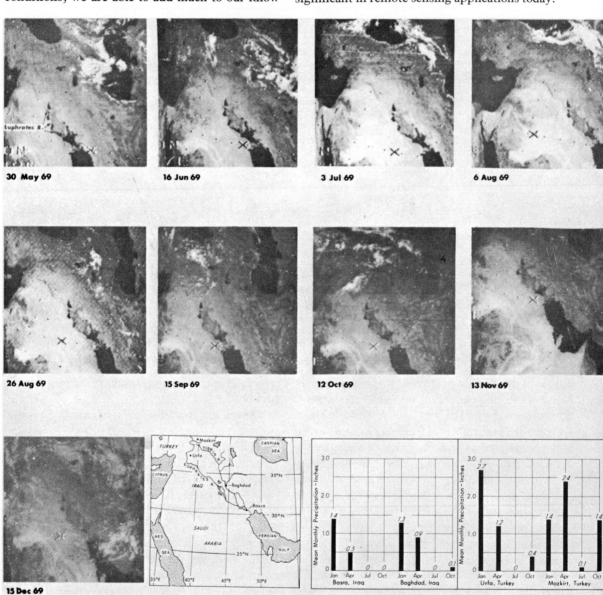

Plate 2.2 Nimbus 3 daytime HRIR pictures showing the effects of the summer minimum of precipitation on the levels and patterns of the 0.7–1.3 μm band reflected solar radiation over the Middle East. In the image for 30 May the Euphrates valley is still dark and wide with vegetation and soil moisture remaining from winter precipitation. As the summer progressed and precipitation decreased the river areas assumed an appearance like the surrounding steppes and deserts. By 6 August, the Euphrates in Iraq has become almost indistinct. The graphs show rainfall for Basra and Baghdad, along the river in Iraq, and for Urfa and Mazkirt in the highlands of Turkey. (Courtesy, NASA.)

2.6.3 Transmission

The transmittance of a target (or a medium like the atmosphere) is defined as the ratio of radiation at distance x within it to the incident radiation. Since we have already mentioned transmission in relation to other effects, we may, in this section, conveniently conclude our review of physical principles by revising some of the interplay between reflection, absorption and transmission at the target. Clearly the sum of the three must equal the incident radiation, much of the detail depending upon the nature of the target itself, whether it be transparent or opaque. We must not forget, however, that a target which is relatively transparent at one wavelength may be relatively opaque at another, so that the relations between reflection, absorption and transmission generally vary across the electromagnetic spectrum. Lastly, it should be remembered that the angle of incidence of radiation may, if low, cause the proportion of reflected energy to exceed the combined proportion of absorbed and transmitted energy, whereas, if high, that angle may allow more of the energy to enter the target providing its detailed surface characteristics permit (see Fig. 2.4).

2.7 General conclusions

Although it may have seemed at the beginning of this chapter that the Earth scientist wishing to employ remote sensing techniques to improve his perception and appreciation of the human environment need do no more than choose the most appropriate instrument and the best platform for his purposes, it must be clear by now that many factors may influence that choice. In later chapters we shall review some of the more common sensors, sensor packages, and remote sensing platforms. However, it is clear that if we are to understand this rapidly expanding field of scientific activity, rather than just know of it, we must try to relate experimental and operational practice to the principles we have examined thus far. Since this might be difficult without further explanation and illustration, Chapter 3 is concerned with the observed radiation characteristics of some natural phenomena. This will provide a bridge between the rather abstract theory in Chapter 2 and the very technical array of practical possibilities discussed later in this book.

3   *Radiation characteristics of natural phenomena*

3.1 Radiation from the Sun

3.1.1 The spectrum of solar radiation

Solar radiation reaches the outer limit of the Earth's atmosphere undepleted except for the effect of distance. Since the total radiation from a spherical source such as the Sun passes through successively larger spheres as it radiates outward, the amount passing through a unit area is inversely proportional to the square of the distance between the area and the radiation source. Although the planet Earth, 93 million miles from the Sun, intercepts less than one fifty millionth of the Sun's total energy output, this intercepted energy is vital not only to most remote sensing techniques (either directly or indirectly) but also to terrestrial life itself.

Observations of the radiation output from the Sun are made difficult by the attenuation effects of the Earth's atmosphere, at or near whose base most measurements have traditionally been made. Clearly the best vantage point for evaluating the spectrum of solar radiation is outside the atmosphere. Although some data from satellite-borne sensors are now available, probably the most complete and reliable evaluation of it is still that obtained many years ago by staff members of the Smithsonian Institute from the elevated observatory at Mount Wilson. Rocket measurements of ultraviolet radiation have been added to extend the spectrum towards the short wavelength end, where solar radiation is almost completely blocked by absorption in the upper atmosphere. Fig. 3.1 summarizes the results, including a curve extrapolated to the outer limits of the atmosphere, and corrected for mean solar distance.

3.1.2 The atmospheric absorption spectrum

We remarked in Chapter 2 that solar radiation is significantly attenuated by the envelope of gases which surrounds the Earth. Consequently the curve of solar radiation incident on the Earth's surface stands below that for the top of the atmosphere, as shown in Fig. 3.1. Some of the attenuated radiation is quickly lost to space by scattering processes, whilst the remainder of it is absorbed within the atmosphere itself. Almost all the solar radiation which penetrates the atmosphere and reaches the surface of the Earth is of wavelengths shorter than 4.0 μm, whereas that emitted by the Earth (see Fig. 3.1) is principally in the broad waveband from 4.0–40 μm. There is practically no overlap between these two types of radiation, which are sometimes referred to as *shortwave* and *longwave* radiation respectively. The atmosphere has the ability to absorb not only direct solar radiation, but longwave (terrestrial) radiation also. In considering the atmospheric absorption spectrum it is more convenient to review the impact this has upon solar and terrestrial radiation together, rather than separately.

The patterns of absorptivity of different constituents of the atmosphere across part of the electromagnetic spectrum are portrayed in Fig. 3.2. The most striking feature of this set of curves as a whole is the great variability observed from one wavelength to another. In many cases sharp peaks of absorptivity at some wavelengths are separated by equally abrupt troughs where the ability of the gases to absorb radiation is very low. The most important absorbers of radiation within the mixture of gases we call the atmosphere are:

(a) Oxygen and ozone. Radiation of wavelengths less than about 0.3 μm is not observed at the ground. Almost all of that reaching the top of the atmosphere is absorbed high in the atmosphere: the rest is backscattered to space. Energy of $\lambda < 0.1$ μm is highly absorbed by 0 and 0_2 (also by N_2) in the ionosphere; energy of 0.1–0.3 μm is absorbed efficiently by O_3 in the ozonosphere. Further, but less complete, ozone absorption occurs in the 0.32–0.36 μm region, and at a minor level around 0.6 μm (in the visible), and 4.75 μm, 9.6 μm and 14.1 μm (in the infrared).

(b) Carbon dioxide. This is of chief significance in the lower stratosphere. It is more evenly distributed than the other absorbing gases and it is over-shadowed at high altitude by oxygen and ozone, and in the troposphere by water vapour. It has weak absorption bands at about 4 μm and

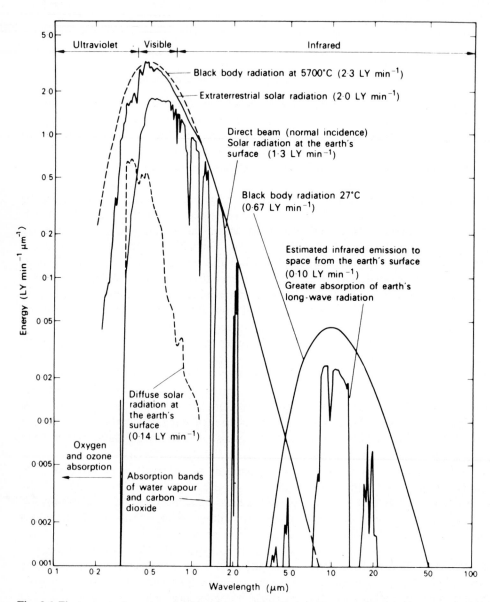

Fig. 3.1 Electromagnetic spectra of solar and terrestrial radiation. The curve for extra-terrestrial solar radiation represents that radiation which is incident on the top of the Earth's atmosphere for the mean distance between the Earth and the Sun. (Source: Sellers, 1965.)

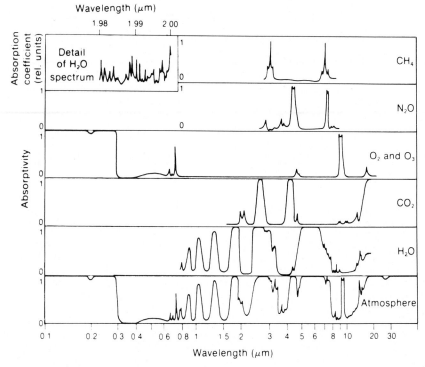

Fig. 3.2 Spectra of absorptivity by constituents of the atmosphere, and the atmosphere as a whole. (Source: Fleagle and Businger, 1963.)

10 μm, and a very strong absorption band around 15 μm which is well known and widely exploited for atmospheric sounding.

(c) Water vapour. Among the atmospheric gases this absorbs the largest amount of solar energy. Several weak absorption lines occur below 0.7 μm, while important broad bands of varying intensity have been identified between 0.7–8.0 μm. The strongest water vapour absorption is around 6 μm, where approaching 100% long-wave radiation may be absorbed if the atmosphere is sufficiently moist.

In Fig. 3.2, below the curves for individual constituents of the atmosphere, we see the total atmospheric absorption spectrum. At the short end of this spectrum the atmosphere effectively absorbs almost all the incidence radiation. However, it is largely transparent in the visible band from 0.3–0.7 μm. Thereafter in the direction of increasing wavelengths we see a sequence of more or less sharply-defined absorption bands alternating with relatively transparent regions. These transparent regions are the most useful 'windows' in the absorption spectrum.

Broadly speaking, practitioners of environmental remote sensing whose major interests are in Earth surface features avoid those wavebands in which atmospheric absorption is strongly marked, especially if they plan to use high-level observation platforms. The wavebands of maximum absorption may, however, be chosen deliberately for some kinds of meteorological and climatological studies. For example, sounding of the atmosphere in depth by multiband radiometers has exploited the 15 μm CO_2-absorption waveband. In studies of this kind the aim is to identify and measure the radiation emitted from a number of levels in the atmosphere, so that vertical profiles of the structure of the atmosphere may be obtained (see Chapter 10).

The basic principle on which satellite depth sounding rests is that the molecules of a gas emit electromagnetic radiation at the frequencies at which they absorb energy. Absorption by a gas such as CO_2 is due to many different vibrational modes of the gas molecules. These modes occur in a series of

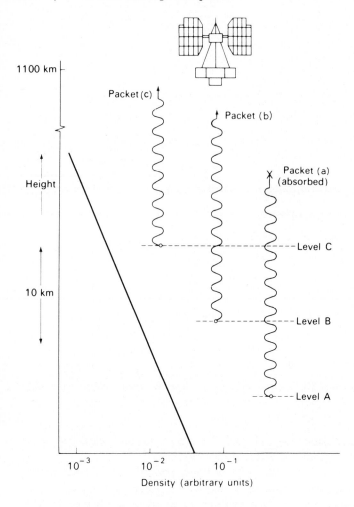

Fig. 3.3 The behaviour of packets of radiation emitted by carbon dioxide from different levels of the atmosphere. (Source: Barnett and Walshaw, 1974.)

more or less well-defined bands across the spectrum. So-called 'spectroscopic' sounding by satellite-borne multichannel radiometers and spectrometers involves the measurement of radiant emissions from the underlying atmosphere across a series of frequencies in each of which the atmosphere absorbs in a known way. Given an observed value for radiation from each of a number of different air layers a vertical profile of atmospheric temperature can be retrieved. Fig. 3.3 illustrates this approach through reference to the upward emission of energy of a single frequency. This has been chosen to give a 50% transmittance from level B to the satellite, i.e. radiation packet (b) reaches the satellite half attenuated. Packet (a) is attenuated more on account of its passage through a greater depth of the atmosphere. Packet (c) arrives at the satellite virtually un-attenuated, but emission from this level is small: the

emission of radiation by CO_2 is proportional to the concentration of the gas, which decreases exponentially with height. Hence the radiation measured by the satellite-borne sensor is a weighted mean of that emitted from various layers, with the greatest contribution from around level B. If the emitted radiation is measured simultaneously at several frequencies, suitable weighting functions can be calculated to specify the atmospheric pressure levels from which most of the measured radiations originate (Fig. 3.4).

Radiometric soundings of the atmosphere were pioneered by experimental sensors on Research and Development (R and D) Nimbus satellites. Their operational successors are flying now on National Oceanic and Atmospheric Administration (NOAA) satellites. Further details of these systems are set out in Chapter 10. Broad-band 'spectroscopic' sensors

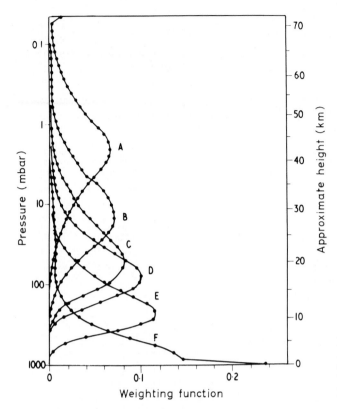

Fig. 3.4 The weighting functions of the Selective Chopper Radiometer (SCR) on Nimbus 4. The height scale is approximate, since the heights of the pressure surfaces vary by up to a few km according to the temperature profile. (Source: Barnett and Walshaw, 1974.)

such as the Infrared Interferometer Spectrometer (IRIS) of Nimbus have been used to retrieve thermal emission spectra for various environments on the planet Earth, in addition to vertical profiles of temperature, humidity, and ozone concentration, as illustrated by Fig. 3.5. IRIS measured planetary radiation in a broad waveband from 6.25–22.50 μm using a modified Michelson interferometer. Radiation from the target is split into two approximately equal beams to give an interferogram which is eventually transmitted to the ground. Here a Fourier transform is performed to produce thermal emission spectra of regions on the Earth (Fig. 3.5(a)). From the water vapour, ozone and carbon dioxide absorption bands in the spectrum, vertical profiles of the concentrations of these gases in the overlying atmosphere can be derived (Fig. 3.5(b)).

Infrared Interferometer Spectrometers have been employed not only to investigate the Earth's atmosphere, but also the atmospheres of other planets, for example Mars. Still further spectrometers have been tested or proposed, and, as with many other remote sensing systems designed initially for investigating

our terrestrial home, there must be a bright future for at least some of them for very long range remote sensing to observe other units in the solar system. We already have a firm basis for studying in considerable detail the dynamics of atmospheres other than our own.

Fig. 3.5 also suggests a general principle of fundamental importance in terrestrial remote sensing, namely that the best spectral regions for observing the Earth on the one hand and its atmosphere on the other do not coincide. We must proceed to explore in greater detail why this is so.

3.1.3 Atmospheric transmission

In view of the almost perfect inverse relationship between the atmospheric absorption spectrum and the spectrum of atmospheric transmission the best spectral regions for observing the surface of Earth are the 'atmospheric windows' in which absorption is low and transmission is high: these are the relative peaks in Fig. 3.5(a) where observed thermal emission spectra most closely approach related blackbody spectra. Conversely, the best spectral regions for observing the atmosphere must be the 'window frames' or pillars of high absorption and low transmission. So the profiles in Fig. 3.5(b) were retrieved from the CO_2, O_3 and H_2O absorption bands respectively whose positions are indicated in the Sahara Desert emission spectrum.

Broadening the discussion, a typical atmospheric transmission curve for the atmosphere for radiation wavelengths from 0–14 μm appears as portrayed in Fig. 3.6. Its upside-down resemblance to the atmosphere's absorption spectrum in Fig. 3.2 is very striking. However, when we come to consider remote sensing from aircraft instead of satellites, it is

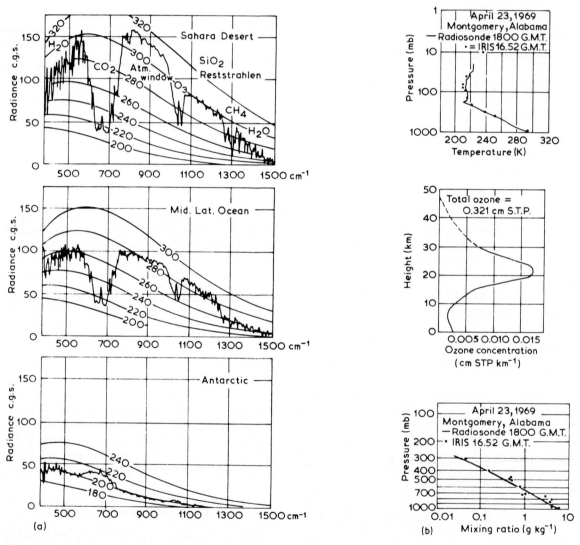

Fig. 3.5 Data received from the Nimbus Infrared Interferometer Spectrometer (IRIS) experiment have been processed to give: (a) Thermal emission spectra of the Earth from 5.25–22 μm. Here representative spectra on one day for three regions of the world are compared with blackbody curves. The departures are due to both atmospheric absorption, and changes of emissivity of the source with wavelength. Within these wave numbers the primary absorption constituents of the atmosphere are water vapour (H_2O), carbon dioxide (CO_2), ozone (O_3), and methane (CH_4). In the broad atmospheric window surface temperatures in the three graphs (from top to bottom) are 312 K, 285 K and 200 K respectively. (b) Temperature, humidity and ozone profiles. Excellent correspondence is found between these profiles and radiosonde data and measurements from ground-based spectrometers. In sharp contradistinction to data from the scatter of surface weather stations IRIS returns continuous strips of data practically encircling the globe. (Source: Aracon, 1971.)

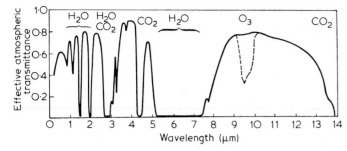

Fig. 3.6 The spectral transmittance of 2000 yards of sea-level atmosphere at low humidity and low haziness. The principal regions of absorption by certain gaseous constituents of the atmosphere are labelled.

important to remember that the effective transmittance of any column of the atmosphere is dependent not only on absorption but on scattering too. The situation we have chosen to illustrate is one in which humidity is low, and there is little particulate matter in the column. Under hazy or foggy conditions attenuation by scattering may have a larger effect upon the spectral transmittance of the atmosphere at sea-level than attenuation by absorption. Consequently if one wishes to view the surface of the Earth from aloft, especially at the short end of the spectrum (e.g. visible air photographic missions), care must be taken to choose those weather conditions under which attenuation will be as low as possible, for then transmittance will be at its peak. Often the angle of view will be critical for the decision whether or not to undertake any specific photographic mission. Under given attenuation conditions transmission is best for vertical paths, deteriorating through slant paths to the horizontal, where the view at the surface of the Earth is entirely within the layer characterized by the highest concentrations of atmospheric water and suspended particles. For this and other reasons vertical air photography is often strongly preferred to oblique air photography.

3.2 Radiation from the Earth

3.2.1 Atmospheric window wavebands

We have seen how much of the planning of airborne or spaceborne missions in remote sensing of the planet Earth as distinct from its atmospheric envelope involves the identification and exploitation of the windows between bands of peak attenuation. Since it is often sufficient for the purposes of introductory studies in climatology and meteorology to refer to atmospheric window radiation in the singular it is worth stressing one fact which has been implicit in earlier discussion and illustration in this chapter, namely that the atmosphere contains not one, but several, important windows for radiant energy transmission. Indeed, in spectral regions above radiation wavelengths of approximately 1 m, little absorption takes place excepting under extreme weather conditions. Otherwise radio and television reception would be greatly impaired.

Even in the infrared ('heat') region of the electromagnetic spectrum there are not one, but several, atmospheric windows. (See Figs 3.2 and 3.6). Of these the most valuable for use in remote sensing are those between about 3.0–4.5 μm and 8.5–14 μm. The chief advantage of the first is that it is the more sharply defined. The chief advantage of the second is that it lies wholly within the thermal emission spectrum of the Earth, and embraces the wavelength of peak infrared emission from that source (approximately 10 μm). Unfortunately there is a marked atmospheric absorption band around 9.6 μm but this is usually disregarded in all but satellite and high-altitude rocket missions. As Fig. 3.6 reveals, this absorption peak is caused by ozone in the stratosphere. This is why its influence on the spectral transmittance of the atmosphere is shown in Fig. 3.6 (which illustrates a sea-level case) by a broken line.

Both these windows in the infrared have been used for remote sensing studies of the surface of the Earth (and its cloud cover) from satellite altitudes. For example, weather satellites like the American Nimbus and Noaa, and some members of the Russian Cosmos and Meteor families have been equipped with infrared sensors designed in part to explore window waveband radiation. We will illustrate in greater detail the relationships between choice of waveband and object of enquiry through reference to the Advanced Very High Resolution

Radiometer (AVHRR) system on the current family of (Tiros-N) operational weather satellites (see also Chapters 10 and 11). Table 10.2 summarizes the wavebands which are involved, and the objects of their enquiry.

Differences in the transparency of the atmosphere in two commonly-exploited multichannel radiometer wavebands are illustrated by Plate 3.1. The Temperature Humidity Infrared Radiometer (THIR) on Nimbus 5 was designed for temperature evaluation of radiating surfaces of the Earth in the 10–12 μm region of the broadest infrared window, and for assessment of atmospheric moisture in the water vapour absorption waveband centred on 6.7 μm. A noteworthy complication in the first case is that, in the two principal infrared windows, clouds often obscure the surface of the Earth. Water vapour is transparent to radiation but aggregations of water droplets reflect and scatter incident radiant energy. These attenuating effects are so strong that even quite shallow clouds reduce the effective trans-

mittance to zero. Therefore, where clouds are present in an atmospheric column, their upper surfaces act as the effective radiating surfaces so far as sensors designed to exploit the atmospheric window wavebands are concerned. An important benefit of this is that, in cloudy areas, the equivalent blackbody temperatures derived from the recorded radiation levels can be used to map the heights of the cloud tops: temperatures usually decline upwards through the troposphere. By physical or statistical techniques cloud top temperatures can be translated into heights of the cloud tops above the ground.

The Surface Composition Mapping Radiometer (SCMR) of Nimbus 5 was interesting in that it measured target radiation in spectral regions between 8.4–9.5 μm and 10.2–11.4 μm. Both of these fall within the broader infrared window, but both avoid the ozone absorption band around 9.6 μm, (see Fig. 3.2). Simultaneous data from the two channels yielded different equivalent temperatures for the same targets, for as Chapter 2 suggested,

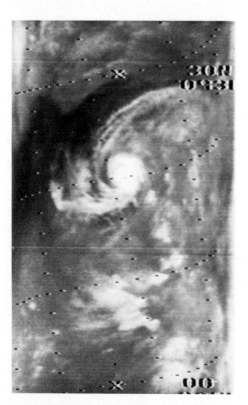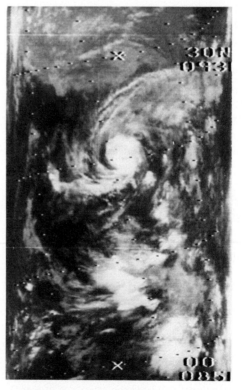

Plate 3.1 Significant differences are evident in these simultaneous views of the Bay of Bengal obtained from the THIR (11.5 μm and 6.7 μm channels) on Nimbus 5. The 11.5 μm image (right) indicates surface and cloud top temperatures; the 6.7 μm image (left) reveals concentrations in the moisture content of the upper troposphere and stratosphere. (Courtesy, NASA.)

radiant exitance varies with the wavelength of radiation. Intercomparisons of measurements from the two channels of SCMR gave general clues as to the types and variations of mineral surfaces as viewed from space.

Other window wavebands well worth exploiting in remote sensing include the microwave region of the electromagnetic spectrum. The opaque regions there are due to water vapour and atmospheric oxygen. The intervening windows have clear advantages over those in the infrared: they can be used even when weather conditions are quite severe. The relatively long wavelength rays in the microwave region (ranging from about 1 mm to 1 m) are able to penetrate thick clouds and even rain storms. Microwave sensing, whether active or passive, therefore has to some degree the 'all-weather' capability which other systems lack. At these longer wavelengths scattering is relatively insignificant, and for normal atmospheric conditions absorption can be read for attenuation with little resultant error.

Before leaving, for the present, the question of atmospheric windows it remains to be stressed that the first to be exploited was in the visible region of the spectrum, and this is still of great importance today, although it is true to say that conventional photography, both colour and black and white, is hindered more by the common attenuating agencies of atmospheric water in droplet form and particulate matter than absorption by gases in the atmosphere. We shall discuss conventional photography in more detail later.

3.2.2 Recording reflected and/or emitted radiation

Not forgetting the significance of the altitude of remote sensing platforms, nor the special opportunities for remote sensing within atmospheric window wavebands, we can say that phenomena on the surface of the Earth can be investigated by several passive remote sensing means:

(a) Broad waveband sensing. Non-specific sensors can be used to integrate the energy from many wavelengths into a composite image. An example is the common ('panchromatic') camera-film combination. This records radiation across the visible portion of the spectrum within the upper and lower wavelength limits of the film emulsion.

(b) Narrow waveband sensing. Here radiation from the target is recorded only in a single selected waveband of the electromagnetic spectrum. We have seen how most objects reflect or emit energy over a broad range of individual wavelengths. There is, however, normally a peak wavelength at which the maximum amount of energy is being reflected or emitted. Under such circumstances objects can best be differentiated from their backgrounds by measurements made at their radiation peaks.

(c) Bispectral sensing. Sometimes the location and identification of environmental phenomena is made easier by simultaneously recording radiation from the target in two non-adjacent wavebands. The data are then used comparatively. The Nimbus THIR and SCMR sensors performed such a type of operation.

(d) Multispectral sensing. A series of sensors arranged to operate at several very narrow bandwidths (often equally spaced across a selected waveband of the radiation spectrum) permit the compilation of three or more simultaneous images of the target area (Plate 3.2). Much current work in remote sensing is now concerned with the compilation of *spectral signatures* derived from multispectral data, and their interpretation by comparison with 'fingerprint banks' of unique signatures known to be associated with particular objects or environmental phenomena. Unfortunately too little is known of the spectral responses of many targets in the natural environment. Many factors affect them, including temporal variations such as time of day and season of the year. Some multispectral scanners have employed as many as 24 separate channels, but the analysis of the signatures based on so many target responses is almost impossible without the aid of some mechanical or electrical back-up system. For manual analyses, sensors with as few as four carefully selected channels may suffice to provide maximum contrast for ready comparison.

Since we have already considered introductory examples of remote sensing by the first three of these four means we may fruitfully turn our attention finally to more examples of the fourth. As the

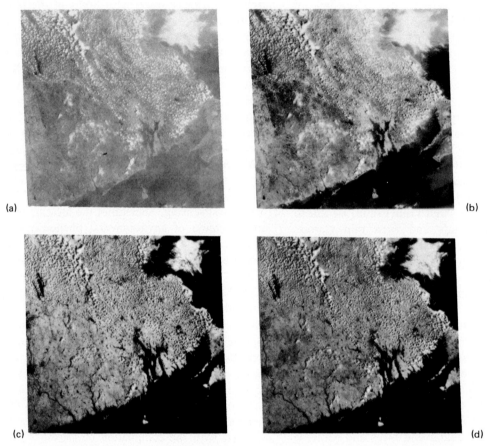

Plate 3.2 A set of Landsat 1 multispectral scanner film transparencies for New England, 28 July 1972, (a) Band 4 (0.5–0.6 μm); (b) Band 5 (0.6–0.7μm); (c) Band 6 (0.7–0.8 μm); (d) Band 7 (0.8–1.1 μm). (Courtesy, NASA.)

analysis of multispectral signatures from the environment is hedged about by many difficulties and uncertainties we should consider first the nature of the responses from some individual constituents of that environment.

3.3 Spectral signatures of the Earth's surface

3.3.1 Signatures of selected features

Rocks, the most fundamental constituents of the solid surface of the planet Earth, can be distinguished from each other under ideal conditions by their spectral signatures in the thermal emission region of the spectrum (see Fig. 3.7). We noted earlier that the emission spectrum from the Sun deviates in some details from the spectrum for a blackbody with the same mean surface temperature.

Similarly many elements of the surface of the Earth have emissivities which vary both in temperature and frequency, and act more like grey bodies than blackbodies. In Fig. 2.3(b), illustrating the distribution of energy emitted by quartz (SiO_2), there are wide differences between the two curves, especially around 9 μm and 20 μm. At these wavelengths incident energy is absorbed sharply by quartz, producing lower rates of emission here. Natural substances often behave more like perfect absorbers and radiators at some frequencies than others, revealing the so-called 'reststrahlen' or residual ray effect, when the actual emission curve is compared with the ideal. In the remote sensing of geological constituents of the environment basic rocks can be distinguished generally from acidic rocks by their signatures in the infrared, since the exact wavelength of the absorption peak varies from the one to the other (Fig. 3.7). Fortunately such radiation differences can be observed even from satellite altitudes, since they fall within the broad atmospheric window waveband from about 8–13 μm. It has become possible to differentiate rough rock specimens in the laboratory by sole use of the spectrum of infrared energies emitted by the samples. The same method has been found equally applicable in the field and from the air. A simplified version of the multispectral method has been flown in aircraft using a spectral sensor with filters in the 8.0–9.5 μm and 10.0–12.0 μm wavebands (cf. the SCMR on Nimbus 5). Further refinements may be expected in the future.

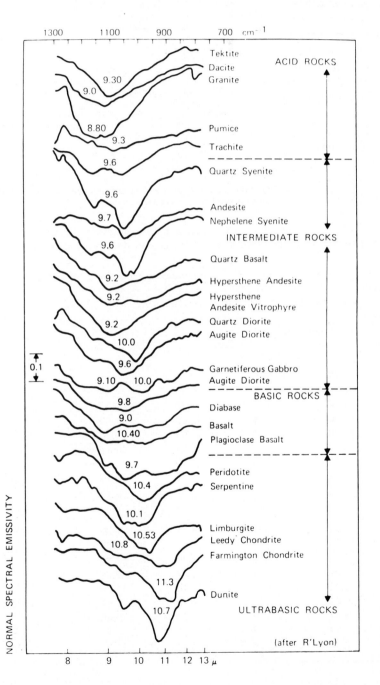

Fig. 3.7 Spectral signatures of rocks. Many types of rocks are differentiated from one another by the spectra of their radiation emissions in the thermal infrared. (Source: Laing, 1971.)

Similar differences between spectral signatures of different rocks have been described for the ultraviolet, visible and photographic infrared regions of the electromagnetic spectrum. One key source of variability in the observed spectra from particular types of rocks is the water content of the samples. Another is the carbon dioxide content. Fig. 3.8 illustrates reflectance spectra of red sandstone surfaces under different moisture conditions. It is notable that the percentage reflection from the wet surfaces is much less than that from the dry surfaces, especially around 1.4 and 1.9 μm which are strong water absorption bands. Clearly detailed 'ground truth' data must be obtained if such remote sensing data are to be correctly interpreted.

Broadening our discussion from rocks and weathered rocks to a wide variety of other environmental constituents, we may deduce from Table 3.1 that many types of soils, crops, and other active surfaces like ice and snow should also be amenable

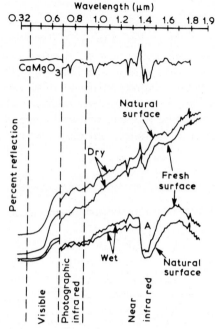

Fig. 3.8 Reflectance spectra of wet and dry, fresh and weathered (salt-encrusted) surfaces of a red sandstone.

Table 3.1 Aircraft and laboratory studies reveal differences in the percentage reflectances of different types of surfaces and crops. The four bands indicated here were those covered by the Multispectral Sensor (MSS) on Landsat 1

Band	1 (0.5–0.6 μm)	2 (0.6–0.7 μm)	3 (0.7–0.8 μm)	3 (0.8–1.1 μm)
Rock and soil materials and covers	Reflectance (%)			
Sand	5.19	4.32	3.46	6.71
Loam 1% H_2O	6.70	6.79	6.10	14.01
Loam 20% H_2O	4.21	4.02	3.38	7.57
Ice	18.30	16.10	12.20	11.00
Snow	19.10	15.00	10.90	9.20
Cultivated land	3.27	2.39	1.58	(not given)
Clay	14.34	14.40	11.99	(not given)
Gneiss	7.02	6.54	5.37	10.70
Loose soil	7.40	6.91	5.68	(not given)
Vegetation				
Wheat (low fertilizers)	3.44	2.27	3.56	8.95
Wheat (high fertilizers)	3.69	2.58	3.67	9.29
Water	3.75	2.24	1.20	1.89
Barley (healthy)	3.96	4.07	4.47	9.29
Barley (mildewed)	4.42	4.07	5.16	11.60
Oats	4.02	2.25	3.50	9.64
Oats	3.21	2.20	3.27	9.46
Soybean (high H_2O)	3.29	2.78	4.11	8.67
Soybean (low H_2O)	3.35	2.60	3.92	11.01

to multispectral identification from airborne or spaceborne sensor systems. These data, compiled during an Earth Resources Technology Satellite (ERTS) design study, represent aircraft and laboratory experience. Actual results from Landsat satellites will be examined later.

3.3.2 Signatures of complex environments

A general feature of Landsat images (e.g. Plates 3.2(a)–(d)) is that they represent the land surface in great detail. Although the spectral reflectance properties of some surface constituents can be readily identified by eye, and reported in qualitative terms (e.g. the greater propensity for water to show differences in reflection from Band 4 to Bands 5 and 7 than is found over land), more detailed and precise statements are required for many practical purposes. A wide range of automatic and semi-automatic procedures have been developed to meet the resulting need for more objective image analysis, comparison, and interpretation. Whilst later chapters will be concerned with the more valuable of these methods, it is worth noting here that many involve some element of spectral signature recognition. Table 3.2 indicates that many influences have a bearing on the spectral signatures which may be derived from multispectral aircraft or satellite data. Often it is not possible to evaluate all the influences acting on a given set of data. It is rarely (if ever) possible to assess the effects of such influences on each other. Not surprisingly, therefore, we must confess that the real world is so complex and variable that satellite mapping of environmental patterns is always subject to some doubt and error. Such mapping is often more accurate when multispectral data are available than when uni- or bi-spectral data must be used, but even multispectral feature recognition and delimitation is usually inexact.

We may illustrate the complexity of the real world, and multispectral investigations of it, by reference to a programme carried out during March 1971 over Bear Lake on the border between Utah and Idaho. Plate 3.3(a) (see colour section) shows the flight path of the instrument platform, on that occasion the NASA Convair 990 Airborne Observatory. In all, eight sensors were employed simultaneously to view the frozen lake and its surrounding snowfields and bare rock. The sensors are

Table 3.2 Sources of variation in multispectral signatures of vegetation. (Source: Polcyn *et al.*, 1969)

Illumination conditions
Illumination geometry (sun angle, cloud distribution)
Spectral distribution of radiation

Site environmental conditions
Meteorologic
Micrometeorologic
Hydrologic
Edaphic
Geomorphologic

Reflective and emissive properties
Spatial properties (geometrical form, density of plants, and pattern of distribution)
Spectral properties (e.g., reflectance or colour)
Thermal properties (emittance and temperature)

Plant conditions
Maturity
Variety
Physiological condition
    Turgidity
    Nutrient levels
    Disease
    Heat-exchange processes

Atmospheric conditions
Water vapour, aerosols, etc. (absorption, scattering, emission)

Viewing conditions
Observation geometry (scan angle, heading relative to Sun)
Time of observation
Altitude

Multichannel sensor parameters
Electronic noise, drift, gain change
Accuracy and precision of measurements on calibration references and standards
Differences in spectral responses of systems

---

summarized in Table 3.3. One viewed in the infrared at 10 μm through the broad atmospheric window. The remainder were all microwave sensors, of which the 1.55 cm sensor was different from the rest in that it was a scanning device, whereas the others all viewed at fixed angles relative to the nadir angle of the aircraft.

Table 3.3 Characteristics of the radiometers employed in a programme of multispectral sensing over Bear Lake, Utah/Idaho, March, 1971. (Source: Schmugge *et al.*, 1973)

Frequency (GHz)	Wavelength (cm)	Pointing relative to nadir (deg.)	3dB beam width (deg.)	RMS temp. sens. (K)
1.42	21	0	15	5
2.69	11	0	27	0.5
4.99	6.0	0	5	15
10.69	2.8	0	7	1.5
19.35H	1.55	Scanner	2.8	1.5
37V	0.81	45	5	3.5
37H	0.81	45	5	3.5
Infrared	1.0×10^{-3}	14	< 1	< 1

Stripchart results from all the sensors are shown in Fig. 3.9. The plot for the 1.55 cm scanner is the average of the five central beam positions across the flight path of the aircraft. Considerable differences are evident from one curve to another. Some of these concern their general forms. For example, the curves of the data from channels 2 and 3 are approximately the inverses of those from channels 5 and 6. Other differences relate to detail embroidered on the basic shapes.

One of the conclusions of this study was that at the longer wavelengths (especially 21 cm) the observed variability in brightness temperatures, especially over the lake, was related to thickness variations in the ice. Results indicate that the transparency of ice and snow is a function of wavelength, and that ice and snow depth may be assessed by these longer microwave emissions.

Plate 3.3(b) (see colour section) is the 1.55 cm microwave image of the pass over Bear Lake at an altitude of 3400 m. The area it covers corresponds with part of Plate 3.3(a). The outline of the frozen, variably snow-covered lake can be clearly seen. The low brightness temperatures of the frozen surface contrast sharply with the higher temperatures along the steep slopes of the eastern edge of the lake. Although a considerable amount of cloud was present over the target area when the flight was made, useful results were still obtained since microwaves, as we noted earlier, penetrate the clouds.

In conclusion it should be stressed that multispectral scanning, almost a science in its own right, is still at a relatively early stage in its development. Doubtless, data like those from the Bear Lake survey contain much more of interest and significance than we can recognize at present. The idea of investigating and evaluating the radiation characteristics of specific phenomena and environments simultaneously through a large number of wavebands is highly attractive. For the most part, however, operational work today concerns itself with the simpler – but by no means simple – tasks of analysing integrated radiation or radiation patterns obtained through one, two, three or at most four, wavebands at one time. We know quite a lot about the basic theory of radiation reflection and emission. We need to learn much more about the highly complex reflection and emission patterns which are associated with real phenomena and situations.

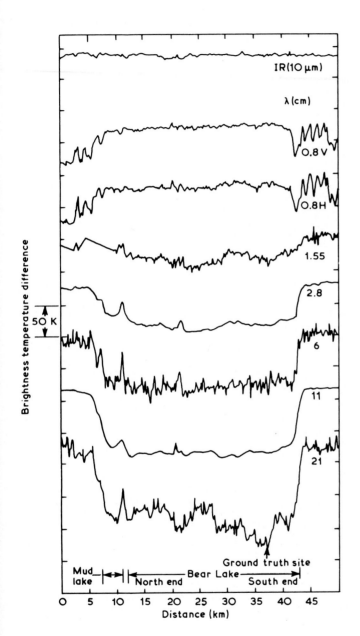

Fig. 3.9 Multispectral data obtained over Bear Lake, Utah/Idaho, 3 March 1971. H and V refer to the horizontal and vertical channels of the 0.8 cm radiometer, which viewed the surface at an angle of 45°. The remaining radiometers were nadir viewing. Each graph is related to the average temperature value at the frequency in question. (Source: Schmugge *et al.*, 1973.)

4 Sensors for environmental monitoring

4.1 Introduction

In the discussion of radiation theory and radiation characteristics of natural phenomena some examples of sensors have already been introduced. In this chapter the sensors available for remote sensing studies will be examined in more detail in order to explain the advantages and disadvantages attached to each sensing system. It will be convenient to discuss the sensors in relation to the wavebands in which they can be applied. Two broad sensing categories can, however, be identified at the outset namely *photographic* and *non-photographic* (scanner dependent) systems (Fig. 4.1). Photographic systems operate in the visible and near infrared parts of the spectrum (0.36–0.9 μm) whereas non-photographic sensors can range from X-ray to radio wavelengths (see Fig. 2.2).

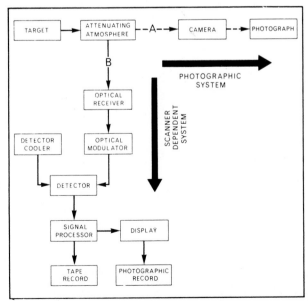

Fig. 4.1 Photographic and non-photographic (scanner dependent) systems. (Source: Curtis, 1973.)

4.2 Sensors in the visible wavelengths

4.2.1 Photographic cameras

The photographic camera is a well-known remote sensing system in which the focused image is usually recorded by a photographic emulsion on a flexible film base. There are three basic types of camera: framing, panoramic and strip cameras. The framing camera usually provides a square image with an angle of view up to 70°. The panoramic camera gives a very wide field of view which can extend to the horizons on either side, with consequent distortion of the image. There are various types of panoramic camera systems and the commonest type uses a reciprocating, or continuously rotating, narrow field lens. The strip camera consists of a stationary lens and slit together with a moving film of width equal to, or greater than, the slit length. The film is moved across the slit at a speed which ensures no blurring of the image at the height flown (i.e. it is image-motion-compensated). The exposure is controlled by the slit width. Such cameras have been used mostly in low level sorties by high speed aircraft in military reconnaissance flights.

In the framing and panoramic cameras the exposure interval can be chosen so that a sequence of overlapping images can be obtained (Fig. 4.2(a) and (b)). This overlap is essential if stereoscopic viewing is desired. In order to avoid gaps in the stereoscopic cover it is necessary to ensure that the camera is oriented in the direction of track of the aircraft or space platform (Fig. 4.2(c)). The scale of the photograph obtained by a framing camera is dependent on the focal length of the lens and the height of the camera above ground. Mapping and reconnaissance cameras commonly use 6 inch lenses but 12, 3½, 3 and 1½ inch lenses are used in special camera systems. Where the 6 inch (150 mm) lens is

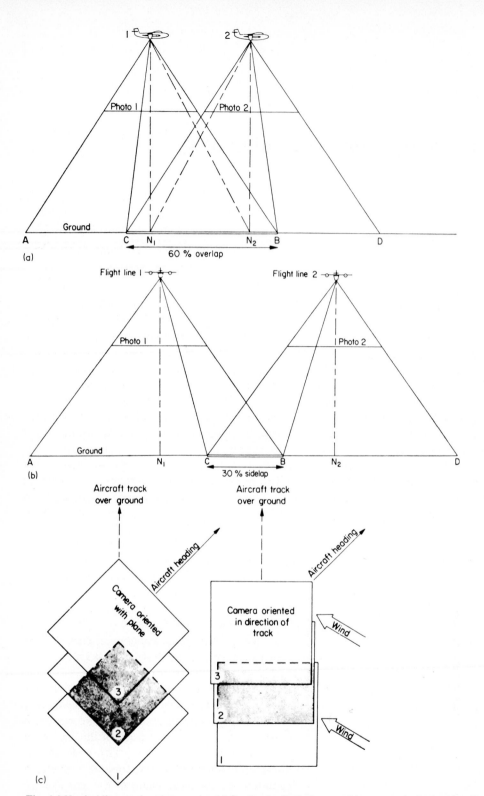

Fig. 4.2 Vertical line overlap photography. (a) Overlap (endlap) photographic coverage is obtained by photography at time intervals which provide coverage AB at position 1 and CD at 2. Overlap BC is 60% and the nadir points N1 and N2 are within the overlap area. (b) Sidelap (lateral) coverage is obtained by flight lines overlapping by 30%. (c) The effect of crabbing. Overlap areas are shaded to show reduced overlap when camera is oriented with plane and the correction when camera is rotated to the direction of the aircraft track. (Source: Curtis, 1973.)

used the calculation of scale of the photograph is a simple operation. For example, if the camera used was a Wild RC8 Universal Aviogon with a focal length of 6 inches and the flying height was 10 000 feet above ground the photographic scale can be calculated as follows:

$$\text{Photographic scale} = \frac{\text{Focal length}}{\text{Height of the camera above ground}}$$

$$= \frac{6}{10\,000 \times 12}$$

$$= \frac{1}{20\,000}$$

The principal factors which limit resolution of the camera system are:

(a) Lens resolution.
(b) Film resolution.
(c) Film flatness and focal plane location.
(d) Accuracy of image-motion-compensation (IMC).

(e) Control of roll, pitch, yaw, vibration.
(f) Optical quality of any filter or window placed in front of the lens.

The definition of photographic image quality has conventionally been quoted in terms of resolving power. This is usually measured by imaging a standard target pattern (Fig. 4.3) and determining the spatial frequency (in lines mm^{-1}) at which the image is no longer distinguishable. Resolving power can also be determined by examining the contrast and distortion of the image of a sinusoidal grating object. This measurement, as a function of the spatial frequency of the grating gives what is termed the modulation transfer function (MTF) which is commonly used to express the resolution characteristics of different films. Examples of the sensitivity and resolving power of Kodak aerial films are given in Table 4.1.

A summary of the characteristics of the different types of film available is given in Table 4.2 from which it will be apparent that the versatility of the camera sensing system is increased by the availability of different films with different spectral ranges, sensitivities and resolving powers.

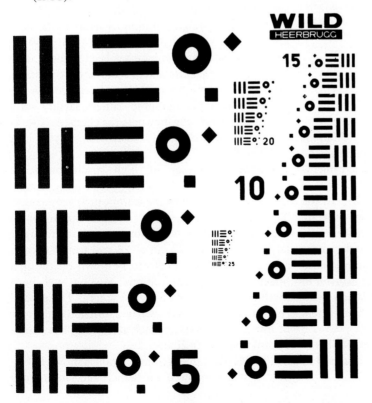

Fig. 4.3 Three line resolution test to determine resolving power. (Source: Smith, 1968.)

Table 4.1 Sensitivity and resolution for Kodak aerial films. (Source: EMI, Department of Trade and Industry, 1973)

Type	Sensitivity (aerial film speed)*	Resolving power (lines mm^{-1}) at test object contrast:	
		1000:1	1.6:1
High definition Aerial 3414	8	630	250
Panatomic X 3400 Plus X	64	160	63
Aerographic 2402 Double X	200	100	50
Aerographic 2405 Tri X	320	80	40
Aerographic 2403 Infrared	640	80	20
Aerographic 2424 Aerochrome	200	80	32
Infrared 2443 Ektachrome	40	63	32
Aerographic 2448	32	80	40

*Aerial film speed for monochrome negative material is defined as $3/2E$ where E is the exposure (in metre-candela-seconds) at the point on the characteristic curve where the density is 0.3 above base plus fog density, under strictly defined conditions given in ANSI Standard PH2.34–1969. A doubling of aerial film speed number denotes a doubling of sensitivity.

Where images are required at different wavelengths multispectral photographs may be taken either by mounting several identical cameras with different film emulsions or using a special multispectral camera (Plate 4.1(a) and (b)). In the latter several identical lenses record their separate images on a common film. The four lens system is the most common arrangement, but the Itek camera has nine lenses. An example of multispectral photography by the $I^2 S$ camera system is given in Plate 4.2. This photograph was obtained with Infrared Aerograph 2424 film using four filters which provided images at 0.4–0.5 μm (1); 0.5–0.6 μm (2); 0.6–0.7 μm (3); and 0.7–0.9 μm (4). These separate images can be recombined and analysed as discussed in Chapter 7.

The principal advantages of the photographic camera are its large information storage capacity, high ground resolution, relatively high sensitivity, and high reliability.

The main shortcomings of photographic surveys from spacecraft are that exposures can only be made in daylight, cloud obscures ground detail, and the photographic film cannot be re-used. If scanning systems had not been used on Landsat the film weight required to cover one year of filming operations to obtain 60 m ground resolution, was estimated to be 300 kg. On board processing would have nearly doubled this weight, and improving the resolution to 20 m would have increased the film weight to 3000 kg. Such weight factors must be considered alongside the difficulty in transferring the information on the film to Earth. One method of transfer involves processing the film on the satellite and electronically scanning it for picture transmission to ground stations by broadband telemetry. The other method is by ejecting the film from the satellite for recovery in the air or on the ground. Capsules of 39 kg (air retrieval) and 136 kg (ground retrieval) have been used in this way.

There is constant research aimed at improving the sensitivity and resolution of films, whilst reducing the thickness (weight) of films. Sensitivity has been approximately doubled each decade since 1850. In the last decade the film base has been reduced from 0.32 mm to 0.063 mm and in some satellite missions even thinner base materials have been used. As an example of a photographic camera facility on a recent spacecraft one may note the characteristics of the S190 Multispectral Facility on Skylab (see Table 4.3).

Further examples of high resolution space photography will be provided by the Metric Camera under development by the European Space Agency. This camera is designed to test the mapping capabilities of high-resolution space photography on large film format (23 cm × 23 cm) from Spacelab (Table 4.4). The camera to be flown on the first Spacelab Land Applications Mission will be a modified Zeiss Aerial Survey Camera. The ground resolution of the images will be a function of the type of film used (colour or black and white) but will be approximately 10–20 m.

The Metric Camera configuration is shown in Fig. 4.4. It will be stowed for launch and landing and

Table 4.2 Summary of film characteristics. (Source: Curtis, 1973)

Film	Disadvantages	Advantages
Panchromatic	(1) Limited tonal range	(1) Sharp definition (2) Good contrast (3) Good exposure latitude (4) Inexpensive
Infrared	(1) Limited tonal range (2) Very high contrast (3) Slight resolution fall-off (without correction for focal distance (4) Difficult to determine correct exposure (5) Loss of shadow detail	(1) Vegetation evident (2) Tracing of water courses facilitated (3) Inexpensive
Colour	(1) Expensive (2) Special processing facilities necessary (3) Diffusion of image under high magnification and slightly less definition than panchromatic	(1) Excellent all-round interpretation properties due to good contrast and great tonal range (2) Good exposure latitude when used as a negative film (3) Good-quality black and white prints can be produced from negatives
False colour	(1) Expensive (2) Special processing facilities necessary (3) Diffusion of image under high magnification (4) Critical exposure (5) Low ASA rating (6) Less tonal range than colour (7) Duplicates expensive and difficult from positive film	(1) Sharp resolution (2) Superior rendering of vegetation and moisture

mounted at Spacelab's optical window during flight. It can only be operated when the Shuttle payload bay is pointed towards the Earth, thus careful optimization of the observation along the orbital track will be required.

Examination of past Earth photography shows that there has been progressive improvement in the resolution of the photographic systems employed (Table 4.5). There is continuing interest in high resolution photography for mapping programmes.

4.2.2 Vidicon cameras

The development of the vidicon camera has provided satellites with a method for obtaining very high resolution pictures (0.35–1.1 μm) without the need for film replacement or converting a film image into a video signal. The principle of operation of a vidicon is shown in Fig. 4.5. The optical image is focused on the photo-conductive target and a charge pattern is built up on the target surface which is a replica of the optical image. This positive pattern is retained until the electron beam again scans the surface and restores it to equilibrium by depositing electrons. An analogue television signal is thereby created which is then amplified. In the case of the Return Beam Vidicon (RBV) camera the signal is derived from the unused portion of the electron beam. This returns along the same path as the forward beam and is amplified in an electron multiplier. The RBV tube provides greater sensitivity at low light levels and higher resolution than the ordinary vidicon tube.

Unlike the photographic film the slow scan vidicon photoconductive target may retain the previous image in vestigial form. Thus a 'priming' cycle is needed to allow complete discharge of the image and restoration of the target for the next exposure. In the Landsat 1 RBV camera this needed

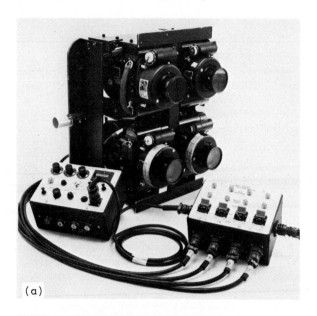

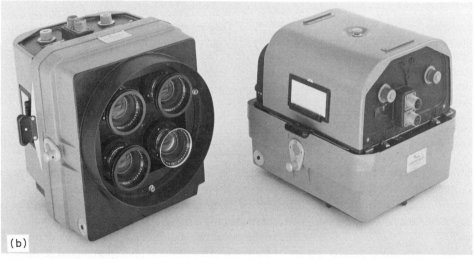

Plate 4.1 (a) Fairey multispectral camera system consisting of four Vinten cameras with lens focal length of 102 mm (4 in) loaded with 70 mm film. (Courtesy Clyde Surveys Ltd., Maidenhead, Berks.) (b) International Imaging Systems (I² S) multispectral aerial camera. Four Schneider Xenotar 150 mm or 100 mm lens. Film capacity 242 mm in width by 76.2 m long. (Courtesy John Hadland (PI) Ltd., Bovington, Herts.)

about 7 s and the complete cycle time could not be less than 11 s.

The RBV cameras which were installed in Landsat 1 failed to function satisfactorily and early problems in RBV systems included focusing and stability of the optical system. The focus of a tube can fail due to lack of sufficient stability in the supply circuits over a period of a year in unmanned satellites. Another danger is that the optical system may become dislodged in the launch period.

The vidicon camera can provide multispectral data either with a series of band pass filters on the optical system or with a number of vidicon sensors operating simultaneously to cover each band. This latter approach was used in Landsat 1 where three vidicon cameras covered different spectral ranges

Plate 4.2 Multispectral photograph of part of Thetford, Norfolk. Apart from the differentiation of deciduous and coniferous trees Band 4 (infrared) also provides considerable additional detail concerning roof structures. Road details are better portrayed in Band 3 (red) at top left. A combination of Band 3 and 4 offer advantages in studies of urban landscapes. (Courtesy NERC.)

between 0.475–0.830 μm. The data transmission rate from this camera assembly was about 35 megabits s^{-1}. A ground resolution of 80 m was achieved with this system with a field of view of 185 × 185 km.

Landsat 2 carried a similar RBV system to Landsat 1, but Landsat 3 provided much improved resolution of c. 30 m. It also provided for one spectral band (Stereo) in the range 0.5–0.75 μm with two cameras aligned to view adjacent ground

scenes with a 14 km overlap giving 183 × 98 km per scene pair.

Vidicon cameras have proved extremely valuable for producing real time and near real time imaging of the Earth's surface, especially in weather satellites. Experience with satellite borne television cameras carried in Tiros, Nimbus 1 and 2, the Ranger Moon shots, Apollo 8–16 and Mariners 1–9 has led to great improvements in Vidicon systems. In the Mariner 6 and 7 systems good quality images were obtained of

Table 4.3 S190 Multispectral photographic facility (Skylab A). (Source: EMI, Department of Trade and Industry, 1973)

Six channels, six lenses (f/2.8, 150 mm), 70 mm film, 400 frames per cassette, 2.5 or 4 mil base, 57 mm square frame format, aperture stops f/2.8 to f/16 in ½ stop increments ($\pm 1.5\%$). Shutter speeds 2.5, 5, 10 ms ($\pm 2.5\%$), 4 ms synchronization. IMC: 10 to 30 mrad s$^{-1} \div \pm 5\%$). Boresighting -60 arc s.
Sequence rate: 1 frame/2 s to 1 frame/20 s.
Dimensions, $570 \times 610 \times 460$ mm; weight, 134 kg; power, 28 V d.c., 22 A peak.

Wavelength (μm)	Film type	Expected dynamic system resolution (line pairs mm^{-1})	Expected dynamic ground resolution (m)
0.5–.06	Pan-X B and W	53	54
0.6–0.7	Pan-X B and W	63	54
0.7–0.8	i.r. B and W	26	107
0.8–0.9	i.r. B and W	26	107
0.5–0.88	i.r. false-colour	31	91
0.4–0.7	High resolution colour	53	55

System cost: approx. $400 000.
i.r. = Infrared; B and W = Black and White

Table 4.4 Operational characteristics for the Large Format Camera on Spacelab. (Source: European Space Agency)

Parameter	Specification
Lens: Focal length	30.5 cm
Aperture	f/6.0
Spectral range	400–900 nm
Image format	22.8×45.7 cm
Field of view	
Along track: degrees	73.7
altitude ratio	$1.5 \times$H
Across track: degrees	41.1
altitude ratio	$0.75 \times$H
Forward overlap modes	80, 70 or 60%

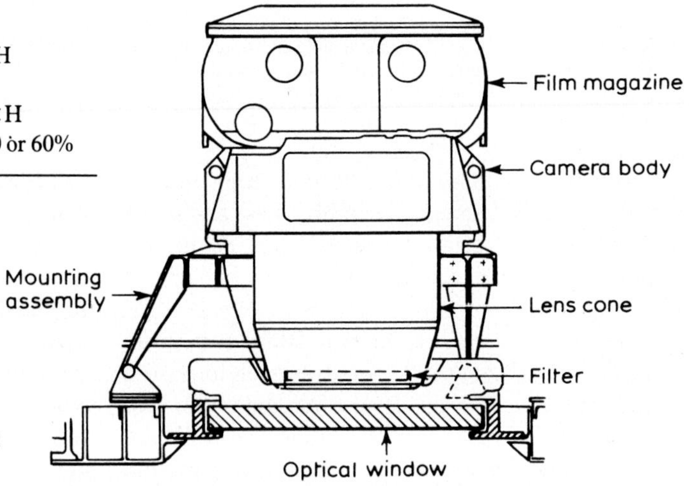

Fig. 4.4 Metric camera for the first Spacelab payload (FSLP). (Source: European Space Agency.)

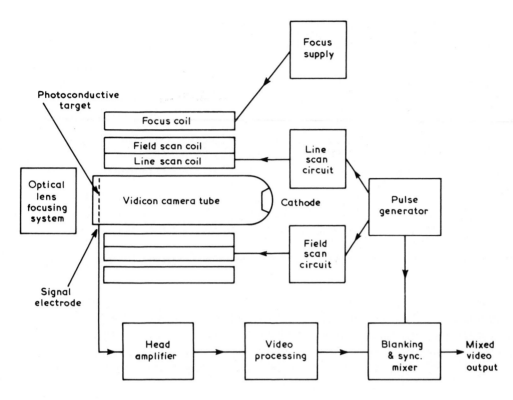

Fig. 4.5 Schematic diagram of a Vidicon Camera. The incoming radiation is focused on to the photoconductive target on which a replica of the optical images is formed. (Source: EMI, 1973.)

the surface of Mars. One camera (A) gave a ground resolution of 1 km and coverage of 900×700 km whereas a second camera (B) gave resolution of about 100 m in an area of 70×90 km.

4.3 Sensors outside the visible wavelengths

4.3.1 Infrared sensors

Passive sensors in the infrared can be broadly classified as 'image forming' and 'non-image forming'. The *image forming* sensors include Infrared Vidicon, Infrared Linescan and Thermal Imager devices. The *non-imaging* sensors include *radiometers* and *spectrometers*. All the sensors use the same basic components and the non-imaging systems of radiometers may be considered first.

An infrared radiometer is a radiation measuring instrument with substantially equal response to a relatively wide band of wavelengths in the infrared region. It measures the difference between the source radiation incident on the radiometer detector

and a radiant energy reference level. The source reflection results from reflected and scattered solar radiation and self emission from the Earth's surface and atmosphere.

A spectrometer permits selection and isolation of a desired wavelength (or band of wavelengths) in the infrared spectrum. This can be achieved by dispersing the incoming radiation by means of prisms, gratings, dichroic mirrors or filters. The dispersed radiation can then be measured in various wavebands using a number of detectors (see further discussion in Section 4.4).

Detectors for use in scanning systems can be classed as thermal or photon detectors, but Earth resource scanners invariably use photon detectors because of their greater sensitivity. A selection of photon detectors can be used to cover all the high atmospheric transmission wavebands from 0.4–14 μm. (Table 4.6). Detector cooling is necessary to achieve satisfactory results for all detectors except for silicon (0.4–1.1 μm) and germanium (1.1–1.75 μm). Since long term stability of the detector cannot

Table 4.5 Summary of past USA and USSR manned missions comprising an important Earth photography element. (Source: ESA)

Mission	Year	Camera	H	C	Format	Resolution	Ground resolution
Gemini 4–7	1965	Hasselblad C. Zeiss-Optik	200 km	80 mm	5.7×5.7 cm	20 Lp mm^{-1}	125 m
Gemini 10–12	1966	Maurer	200 km	80 mm	5.7×5.7 cm	20 Lp mm^{-1}	125 m
Apollo 7	1968	R220 Maurer	225–420 km	80 mm	5.7×5.7 cm	35 Lp mm^{-1}	70 m
Apollo 9	1969	Hasselblad C. Zeiss-Optik	192–496 km	80 mm	5.7×5.7 cm	35 Lp mm^{-1}	70 m
Skylab (S190A)	1973	ITEK	435 km	152 mm	5.7×5.7 cm	29 Lp mm^{-1}	99 m
Skylab (S190B)	1973	ETC Acton	435 km	460 mm	11.5×11.5 cm	25 Lp mm^{-1}	38 m
Sojuz 22–30	since 1976	MKF-6 Jena	250 km	125 mm	5.5×8.1 cm	80 Lp mm^{-1}	25 m

be assumed it is also necessary to provide some built-in calibration system. Calibrating signals are normally obtained using tungsten lamp, diffused solar radiation, or controlled blackbody temperature sources.

In Table 4.6 the relative cooling ability obtainable by thermoelectric, radiative, closed cycle engines and solid cryogens are shown. Also the range of chemical elements which can be used for detectors in the range 1.5–12 μm are shown, together with the

Table 4.6 Optimum detector choice and waveband detectivities. (Source: EMI., Department of Trade and Industry, 1973)

Cooling	Thermoelectric → Radiative → Closed cycle engines and solid cryogens →			
Temperature (K)	300	195	77	35
Waveband (μm)	*Detectivities* (W^{-1} cm Hz$^{1/2}$)			
1.5–1.8	(1) InAs 4.5×10^{10} (2) HgCdTe (3) InSb 1.5×10^{8}	InAs 7.5×10^{10} HgCdTe	InAs 1.5×10^{11}	
2.0–2.5	InAs 6×10^{9} HgCdTe InSb 2×10^{8}	InAs 2×10^{11} HgCdTe 1.5×10^{10} InSb 3.5×10^{9}	InAs 3×10^{11} InSb 3.5×10^{10} HgCdTe 2×10^{10}	
3.5–4.1	HgCdTe InSb 3×10^{8}	HgCdTe 2.5×10^{10} InSb 6×10^{9}	InSb 6.5×10^{10} (PV) InSb 4.5×10^{10} (PC) HgCdTe 4×10^{10}	
8–10	(TGS 6×10^{8}) (Thermistor 4×10^{8}		HgCdTe 3×10^{10} PbSnTe 7×10^{9}	GeHg 2.5×10^{10}
10–12	(Thermal detectors)		HgCdTe 2×10^{10} PbSnTe 9×10^{9}	GeHg 3×10^{10}

approximate cooling required for particular sensors, and for given detectivities.

Since the maximum achievable radiance resolution can be limited by the degree of available detector cooling, considerable attention has been given to cooling systems. Cooling down to 100–77K is required for adequate resolution within the 8–13 μm waveband. Detectors used in aircraft usually use liquid coolant (liquid nitrogen) but for satellite multispectral scanners solid cryogen, closed cycle refrigerators or passive 'radiation to space' methods are used. With the solid cryogen a pre-cooled solidified gas is stored in a high performance dewar system and vented to space. Closed cycle refrigerators (e.g. Stirling) provide higher cooling capacity/weight/size ratios than solid cryogens but are demanding in terms of power requirements. Passive radiation cooling is an attractive proposition because of its lack of power requirements and long life. However, the cooling capacity of this system may be too low for some large detector arrays.

Examples of radiometers such as the THIR and SCMR systems used on the Nimbus 5 satellite have been given in Chapter 3. It should be noted that simple infrared hand held radiometers are used also from aircraft and helicopter platforms quite extensively.

Turning now to the *imaging* sensors, the simplest

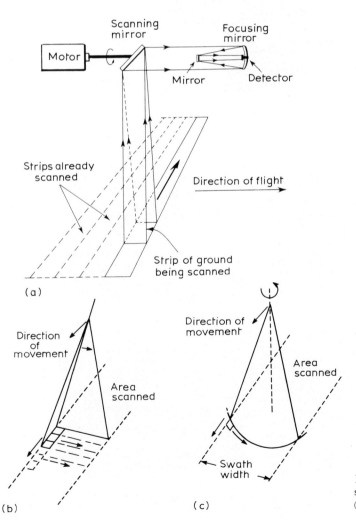

Fig. 4.6 Various forms of scanning (a) the linescan system (b) array scanning – side and forward motion (c) conical scanning.

Plate 4.3 Infrared Linescan Type 212 complete with cooling pack and control unit. (Courtesy Hawker Siddeley Dynamics Ltd., Hatfield, UK.)

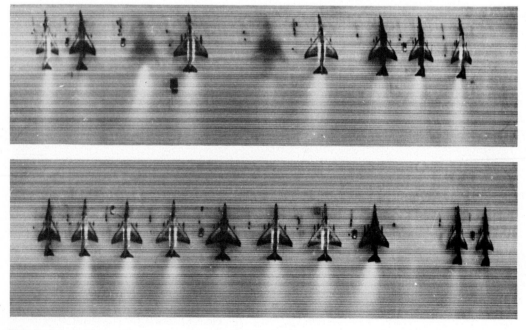

Plate 4.4 Infrared linescan image of aircraft on an airfield. The heat from engines is shown in light tones. Note also the thermal shadows of recently departed aircraft at the top of the picture. (Courtesy Hawker Siddeley Dynamics Ltd., Hatfield, UK.)

form of imager is one which carries a single detector and scans a series of lines or narrow strips across the scene so that an image can be progressively built up. The infrared linescanner (IRLS) is a simple form of imager (see Fig. 4.6(a)). Some systems allow for an array of sensors to be swept across the area (see Fig. 4.6(b)) and this can be advantageous in that it makes for easier data presentation, but conical scanning (see Fig. 4.6(c)) can also be employed. It will be apparent that scan rate must be related to the altitude and velocity of the sensor when assessing the performance of the system.

Infrared line scanning systems are used extensively from aircraft and helicopter platforms. For example the Infrared Linescan Type 212 (Plate 4.3) manufactured by Hawker Siddeley Dynamics has a temperature resolution of 0.25°C at 0°C

background temperature, thus surface temperatures can be detected to an accuracy of less than 1°C. When detector signals are processed to form an image the resultant thermal maps can provide great detail concerning temperature variations. Two examples are shown in Plates 4.4 and 4.5 where the heat from aircraft engines and the warmer areas in the shelter of hedgerows can be easily recognized. Infrared radiometers are also used by engineers of the Electricity Board to detect 'hot spots' in power lines of the electricity grid (see also p. 295).

Typically, such airborne infrared line scanning systems generate data which indicate *relative* thermal energy differences between objects scanned. An experienced observer, therefore, can look at such data and say that Point A has a greater

Plate 4.5 Infrared linescan imagery of land near Mark Yeo, Somerset. Note shelter effect of hedges showing in light tones of area with higher temperature. Grazing animals are easily seen by reason of their high body temperature. (Courtesy Royal Signals and Radar Establishment, Malvern. Crown Copyright.)

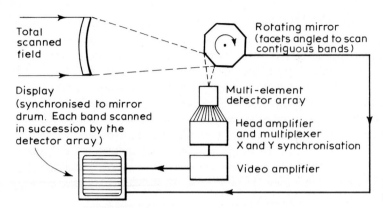

Fig. 4.7 Thermal Imager system – block schematic. (Source: Laird, 1977.)

apparent temperature than Point B, however, the difference in temperature can only be estimated in such cases.

To achieve *absolute* temperature relationships from the scanner data several additional constraints must be met. First, the video electronics must have DC (Direct Current) response to follow faithfully each thermal level variation from the detector. Secondly, radiation received by the detector must come predominantly from the area of interest being sensed. Finally, the complete electronic chain from detector through to final recording must be 'drift stabilized' so that the system responsivity, V/W (volts output/watts input) is constant.

The Daedalus DS–1230 is a quantitative scanner fulfilling the absolute temperature requirements set out above and Plate 17.4 shows an example of such imagery. This scanner provides high spatial resolution using a detector size of 1.7 milli-radians combined with temperature resolution of 0.2°C.

Thermal imagers produce images very similar to those obtained by linescanners, but with sensors in this case providing scanning in both dimensions. A high quality display can be obtained by using an array of detector elements matched to a suitable scanning system as shown in Fig. 4.7. This is the so-called *banded scan* system in which the master scan is generated by a rotating drum with progressively angled facets. This type of sensor is essential for geostationary satellites where there is no relative vehicle motion. In any case, as scanning height increases the thermal imager may become preferable to the simple line scan in that it can be more efficiently matched to the scan angle to be covered, and its scanning rate can be controlled to provide the desired compromise between sensitivity, resolution and scanning frame time. This type of sensor lends itself readily to the use of a range of detector elements, thereby increasing sensitivity for particular conditions of optics and scan rate. The HCMM and CZCS radiometers (see Chapter 9) are examples of space systems using infrared.

4.3.2 Microwave radiometers

The measurement of radiation in the range 1–100 mm forms the basis of remote sensing by microwave radiometry. Although the microwave radiation intensities are much lower then those in the infrared, resulting in poorer temperature resolution, the longer wavelengths have the advantage that they allow sensing through cloud cover.

The microwave radiometer is made up of a directional aerial, a receiver (for selection and amplification) and a detector. A block diagram of a typical microwave system is shown in Fig. 4.13(b) and the applications of some radiometers are given in Table 4.7. The ground resolution achieved by microwave radiometers depends on the aerial size and the orbiting altitude (Fig. 4.8). Since it is important that aerial size should be limited in order to avoid undue weight or distortion of the aerial array the lowest altitude orbit should be used for maximum resolution.

The principal characteristics of passive microwave radiometer (PMR) systems are, therefore:

(a) Low spatial resolution capability of the order of 1 km.
(b) Wide range in the observed surface emissivity.
(c) All weather capability.

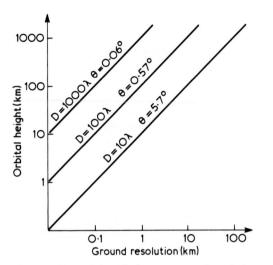

Fig. 4.8 The relationship between ground resolution, orbital height and effective aerial diameter, D for microwave radiometers. The aerial beamwidth is given by θ. (Source: EMI, 1973.)

The emissivity of an observed body changes with observation angle, polarization, frequency and surface roughness. An example of an object with low emissivity is a smooth water surface for which $\epsilon = 0.4$ when observed at normal incidence. Rough soil on the other hand has an emissivity close to unity. It is convenient when considering the measurements made by microwave radiometers to introduce the term *brightness temperature, T_B*. This is the temperature of a blackbody with the same emitted radiation as the physical object sensed by the radiometer (cf. p.19).

Thus $T_B = \epsilon T$
where ϵ = emissivity and
T = physical temperature (K).

Both natural and man made surfaces present a large range of brightness temperatures. Whereas for a blackbody the brightness temperature is equal to the physical temperature and ϵ is unity for all incidence angles, natural objects vary in brightness temperature and in their response to incidence angles. Natural objects (grey bodies) show variations in emissivity depending on incidence angle, wavelength and the degree of surface roughness. Where surfaces are very rough the brightness temperature is independent of angle of incidence, but if roughness is limited T_B will become lower as the angle of incidence increases (Fig. 4.9).

An important question in relation to microwave radiometry concerns the depth to which ground properties contribute to the observed brightness temperature (i.e. the penetration depth). The penetration depth depends on the wavelength used and the dielectric properties of the material. Typical values of the penetration depth are 20 wavelengths for asphalt and sand and only 0.5–0.1 wavelengths for water. Thus wet materials provide little penetration whereas dry substances e.g. desert sands may show substantial (approximately 1 m) penetration. The wavelength dependence of the penetration depth makes profiling of the surface layer possible by using multifrequency radiometry.

Another important parameter characterizing the

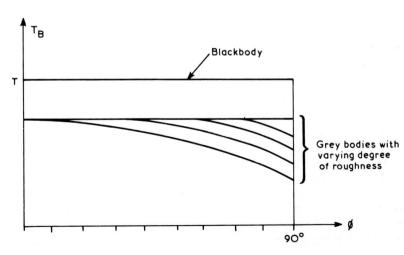

Fig. 4.9 Variations in brightness temperatures of black and grey bodies which are producing diffuse scattering.

radiation properties of an object is the *polarization* i.e. the distribution of the electrical field in the plane normal to the propagation direction. As a rule the radiometer is only sensitive to the field in one direction. Blackbody radiation is completely unpolarized, but the emission of many natural features shows pronounced polarization effects which can be useful for identifying the nature of the feature.

The measured radiation of a surface feature depends on the following characteristics of the microwave system:

(a) Polarization direction of the radiometer.
(b) Observation angle related to the surface plane of the object.

(c) Frequency (wavelengths) of bands used.

The radiation received by the microwave system is also influenced by characteristics of the material which is being sensed. The important characteristics are:

(a) Electrical and thermal properties of the material.
(b) Surface roughness and size of the object.
(c) Temperature and its distribution in the body.

At present the chief limitation of passive microwave sensing from space is the low resolution of the system at orbital altitudes. However, the potential applications are considerable, as for example in the fields of hydrology and oceanography (Table 4.7).

Table 4.7 Microwave radiometry and its application to Earth resources surveys. User requirements in hydrology and oceanography (Source: Ohlsson, ESRO 1972).

Application	Spatial resolution (km)	Thermal resolution (K)	Frequency of coverage
1. Temperature monitoring			
Ocean wide, synoptic	150	1	Bimonthly
Regional	20	1	Weekly
Coastal and great lakes	5	1★	On demand
Temperature gradient, ocean	150	†	
2. Soil Moisture			
Water content	0.2	5★	Daily to monthly
Gradient	0.2	†	Daily to monthly
Horizontal moisture	0.2	5★	On demand
3. Sea state	150	5★	4 per day to daily
4. Oil film detection	0.1	10★	On demand
5. Apportionment of water surfaces	1	10★	Daily to weekly
6. Ice/water boundaries	0.2	20★	Weekly
7. Snow			
Water equivalent of snow cover	0.2	†	Weekly
Snow/land limits	0.2	5★	Weekly
Snow/ice limits	0.2	1★	Weekly
Snow temperature	0.1	1	On demand
Snow moisture	0.1	1★	Daily

★Required accuracy in measured brightness temperature, judged from data in the literature relating primary quantity to brightness temperature.
†No satisfactory technique available; research needed.

The Nimbus 7 programmes afford examples of space applications. The Nimbus 7 radiometer consisted of a Scanning Multichannel Microwave Radiometer (SMMR) operating on four wavelengths of 0.8, 1.4, 1.7, and 2.8 cm. The sensor was dual polarized and so effectively provided eight channels. The temperature sensitivity ranged from 1.5 K at 0.8 cm to 0.9 K at 2.8 cm wavelength.

Lastly, one may note that the dividing line between imaging and non-imaging sensors is not clear cut and all sensors use the same components. Factors such as target emission and shape, background emission, environment and the positional accuracy required affect their design. The heat sensors in the shorter infrared wavelengths (1.5–12.5 μm), together with those in the microwave (1–100 mm) just described, offer great flexibility for the measurement of thermal radiation at the Earth's surface.

4.3.3. Microwave radar

The microwave systems discussed so far have been *passive* systems, that is those receiving Earth emissions. It is possible, however, to devise an *active* sensing system in which waves are propagated near the sensor and are bounced off the Earth surface to be recorded on their return. This is the essence of the Radar (Radio Direction and Ranging) system. It was first operated in the VHF radio band (30–300 MHz) but later improvements have led to the use of microwave radar (300 MHz-100 GHz).

Images of landscape derived from airborne side-looking radar resemble air photographs with low angle sun illumination in that shadow effects are produced (Plate 4.6(a)). This enhances the impression of morphological relief in the imagery but also leads to some obscure areas where the radar shadows occur. The geometry of radar pictures is entirely different from that of conventional air photographs. This is because radar is essentially a distance ranging device which consequently produces lateral distortion of elevated objects as shown in Fig. 4.10(a).

The basic elements of a *Real Aperture* Side-Looking Radar system (SLAR) are shown in Fig. 4.10(b). A long aerial mounted on the airborne platform scans the terrain by means of a radar beam (pulse). The radar echoes are recorded by a receiver and processor so that a cathode ray tube (CRT) glows in response to the strength of each echo. The resultant CRT display is then recorded on film. These radars can be subdivided into six variants:

1. Monofrequency. Monopolarized, transmitting a radar carrier polarized in one plane, usually horizontal.
2. Multipolarized. This normally transmits a horizontally polarized electromagnetic wave but can receive horizontally and vertically polarized waves separately.
3. Circular polarized. This transmits a circularly polarized wave and can receive right hand or left hand circularly polarized ground reflections.
4. Multifrequency. This transmits two or more modulated radar carriers at widely differing wavelengths to show up differences in penetration and scattering from various surfaces.
5. Panchromatic. This transmits a broad-band carrier in order to minimize diffraction effects (speckling).
6. Polypanchromatic (a multifrequency panchromatic radar).

In most Real Aperture (SLAR) systems the resolution across the film record is up to ten times better than that achieved in the along track direction. The latter can only be improved by increasing the size of the antenna used. There is a limit, however, to the size of the antenna which can be mounted on aircraft and spacecraft.

In fact, further improvements in Real Aperture resolution became increasingly hampered by limitations in:

(a) Millimetric radar technology, where it becomes difficult to acquire components for engineered radars at wavelengths less than 8 mm; and
(b) Mechanical engineering tolerances on antenna manufacture. Even at 8 mm the demanded tolerances are approximately 0.1 mm over a 5 m aperture. Furthermore, these must not only be achieved in manufacture but maintained in flight in the face of temperature and vibration effects.

So it was that attention was drawn to the technique of *aperture synthesis* whereby the movement of the aircraft (or spacecraft) along its track can be used to form a long synthetic aerial

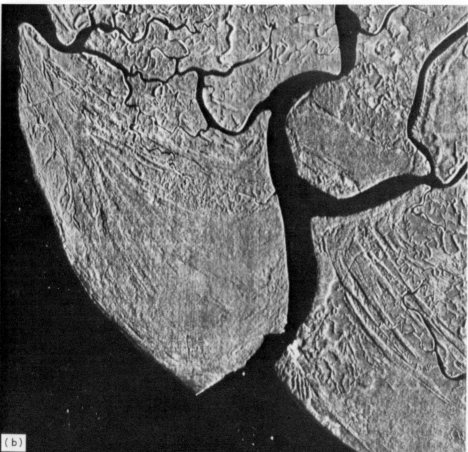

Plate 4.6 (a) Radar imagery of the Malvern Hills. (Courtesy Royal Signals and Radar Establishment, Malvern. Crown Copyright.)
(b) Radar imagery of coastal area of Nigeria. Note different textures of woodland vegetation and the evidence of an emergent coastline shown by the lines running parallel to the coast. (Courtesy Hunting Technical Services Ltd.)

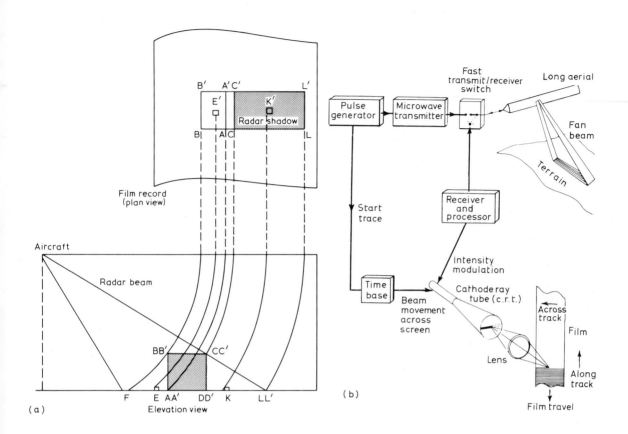

Fig. 4.10 (a) Image distortions due to the directional and ranging characteristics of radar. (b) Basic elements of a Real Aperture Radar System. (Side Looking radar). (Source: Grant, 1974.)

(Fig. 4.11). This system depends on a particular method of processing the Doppler information associated with the radar returns, and is termed the Synthetic-Aperture system. The synthetic-aperture system is most easily understood by analogy with more familiar optical concepts. If one imagines that a plane wave of monochromatic light falls on a narrow slit the wave on the far side of the slit will diverge to form the familiar Fresnel zone pattern for a slit (Fig. 4.12(a)). The Fresnel zones on the screen are alternately dark and light in tone and are caused

by wave interference patterns. The light wave amplitude that causes this observed variation can be shown to be both positive and negative. This represents the variation in the *phase* of the wave fronts at these points relative to the *phase* of the wave arriving at the zero point, i.e. the centre of the pattern on the screen.

It would be possible to place a film at the screen B and record the complete intensity image. Alternatively the same photograph could be obtained by traversing a narrow slit across the film

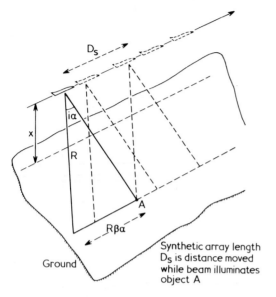

Fig. 4.11 Schematic diagram to show elements of a Synthetic Aperture Radar system. (Source: Grant, 1974.)

(Fig. 4.12(b)) to expose it a piece at a time. If the phase of the waves could be added back it would then be possible to reconstruct an image of the slit A by shining a monochromatic (laser) light through the film record.

Now if one compares this example with the synthetic-aperture system radar the slit A can be regarded as a small part of the ground terrain. The slit C (Fig. 4.12(b)) is the recording system (CRT) for one item of information obtained by the radar receiver. The screen B (Fig. 4.12(b)) is the complete film record obtained from the CRT display for the pulse striking the ground element represented by slit A.

If the image of the element (A) is to be reconstructed it is necessary to record both the phase and the amplitude of the signals received. This is achieved by carrying a very stable reference oscillator in the aircraft or spacecraft and comparing the phase of the return at each position of the sampling slit C with the phase of the stable oscillator. A radar with this type of stable oscillator is known as 'coherent' radar and it is an essential part of the synthetic radar system.

The Seasat mission launched in 1978 carried a Synthetic Aperture Radar (SAR) with a frequency of 1.37 GHz providing a ground resolution of 25 m and a swath width of 100 km (see Chapter 12).

The *advantages* of side-looking radar compared with conventional aerial photography may be summarized as follows:

(a) It has all weather and day/night capability.
(b) It is capable of measuring surface roughness and dielectric properties.
(c) It provides active illumination, with ability to select wavelength.

The *disadvantages* of the radar system can be:

(a) Its poor resolution, although at satellite altitudes the potential resolution of a synthetic-aperture system is superior to that of an optical system.

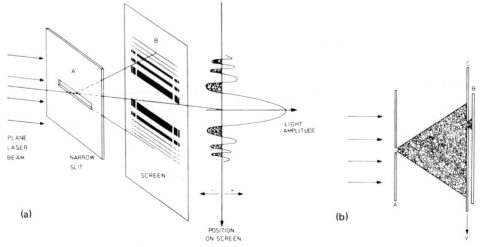

Fig. 4.12 (a) Fresnel zones appearing on screen and diagrammatic appearance of the phases of the waves. (b) Slit (C) moved across the face of the film (B). (Source: Grant, 1974.)

(b) Its small scale.

(c) Its image distortion.

(d) Shadowing in areas of pronounced relief.

A particular type of radar system designed to obtain a measure of backscattered energy for various incident angles of the radar beam is termed the *Scatterometer*. The data obtained for each resolution element is presented in a graph of reflected energy versus incident angle. The variation of radar returns can be used by the scientist to determine roughness, texture, and orientation of the terrain. Different surface materials may also be identified.

Scattering signatures have been investigated experimentally for many materials and theory has been developed for statistical roughness parameters and electromagnetic parameters of various materials. Operational scatterometers are used to gather data along a strip underneath the spacecraft or aircraft. Resolution capability is of the order of 1 m. On Skylab a narrow round beam was used which discretely shifted from the maximum angle to the nadir of the spacecraft as it moved over a given area.

More recently the European Space Agency has shown interest in a Two Frequency Scatterometer (2FS). In this approach use has been made of the principle of the 'beat frequency oscillator'. The single frequency transmission is replaced by a pair of frequencies, the difference frequency having the required value to give suitable resonance. The advantage of this arrangement is that both transmitted frequencies can be in the microwave region where good directional antenna beams can be formed with quite modest sized antennas. The scanning of the difference frequency can be done either by keeping one frequency fixed, as a sort of reference, and varying the other or by varying both keeping the mean frequency constant.

4.3.4 Lidar (or laser radar)

The lidar is an active system similar to a microwave radar, but operating in that part of the spectrum comprising u.v. to near i.r. regions. It consists of a laser which emits radiation in pulse or continuous mode through a collimating system. A second optical system collects the radiation returned and focuses it on to a detector.

The lidar system is only effective in clear sky conditions due to the atmospheric absorption at the short wavelengths used. It is, however, a sensor of higher resolution than microwave radar. Three types of lidar are at present available: an altimeter type which can plot a terrain profile, a scanning type which can be used as a mapping instrument, and a third type employing spectroscopic techniques which may be used for mapping air pollutants.

4.4 Absorption spectrometers

In conclusion it may be noted that spectroscopic techniques can be applied to various types of sensors and where this is done the instrument is commonly referred to as a *spectrometer*. Spectrometers are essentially instruments used to determine the wavelength distribution of radiation. This can be done by dispersing the radiation spatially according to wavelength. The prism spectrometer relies on the dependence of the index of refraction of various prism materials on the wavelength of radiation entering the prism. The grating spectrometer achieves dispersion by diffraction and interference. One of the most popular grating spectrometer systems used today was devised by Ebert some sixty years ago. Scanning of the spectrum is achieved by applying a sawtooth type of oscillatory motion to the grating.

A rotating, circularly variable filter may replace the prism or grating and thereby gain an increase in the rate at which incoming radiation can be scanned. When a circular filter of this kind is used with a rotating beam modulator (chopper) and an internal blackbody reference source, calibration of the spectrometer measurements can be achieved (Fig. 4.13(a)). Commercial circularly variable spectrometers are available with the capability of interchanging detectors. These instruments allow coverage of waveband frequencies in the range from 0.35–23 μm and can accommodate both cooled and uncooled detectors. Infrared spectrometers have been used to obtain atmospheric profiles – generally values of temperature and water vapour as a function of height. An infrared sensor which has demonstrated the feasibility of remote profiling of the atmosphere is the Satellite Infrared Spectrometer (SIRS) which uses an Ebert type grating spectrometer (see p. 180). The more advanced Tiros Operational Vertical Sounder covers seventeen infrared bands from 3.7 μm to

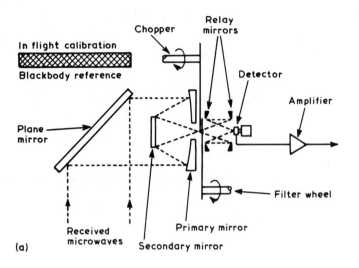

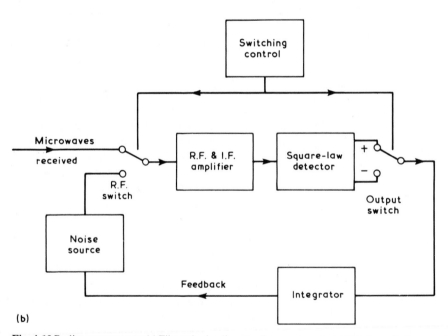

Fig. 4.13 Radiometer systems: (a) Filter wheel radiometer (i.e. spectrometer); (b) Schematic diagram of microwave radiometer system.

29.4 μm. It is designed to measure temperature (1K accuracy), relative humidity (10% accuracy) and total ozone content (± 0.01 cm).

The use of absorption spectrometry in remote gas sensing depends on the unique spectral signature of each gas (Fig. 4.14). At present, spaceborne applications of remote gas sensing have been mainly limited to meteorological studies and to Martian atmospheric studies. However, the principles and instruments described could also be applied to the detection of pollutants and emitted gases and vapours from Earth sources. Up to the present no space instruments have been flown specifically for the detection of Earth pollutants or Earth resources. However, modifications of existing meteorological instruments and further technological development should enable some of the applications listed in Table 4.8 to be investigated.

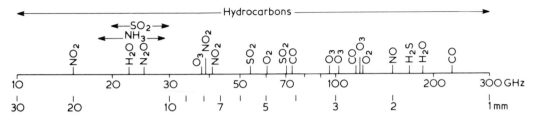

Fig. 4.14 Principal absorption frequencies of relevant gases in the microwave region. (Source: EMI, 1973.)

Table 4.8 Applications of remote gas sensing (Source: EMI, Department of Trade and Industry, 1973).

Discipline	Applications	Ground resolution desirable
Pollution	Determination of total vertical amount and vertical distribution of atmospheric gases (both man-made and natural). Can then determine	
	(a) the global pollution background,	
	(b) regional pollution dispersion and circulating patterns,	
	(c) spatial and temporal variations of pollutants over urban, rural and ocean areas,	
	(d) sink mechanisms,	
	(e) atmospheric chemical mechanisms,	
	(f) pollution flows across international boundaries.	L
	Mapping of small area pollutant sources, e.g. paper/sand/pulp mills, iron and steel plants, petroleum refineries, smelters and chemical plants, housing estates (where fuel is burnt for domestic purposes), busy road junctions (CO).	H
	Monitoring residues from large scale spraying of pesticides.	M
	Mapping of small area pollutant sources, e.g. paper/sand/pulp mills,	M
Detection of Earth resources	Detection of metals from vapours emitted (e.g. arsenic, cadmium, zinc, sodium, mercury).	H
	Detection of micro gas seeps (especially NH_3) associated with natural gas fields.	H
	Detection of trace gases emitted from oxidizing ore deposits.	H
	Detection of iodine (iodine occurs in high concentrations in water in petroleum bearing strata, and is also associated with marine life which concentrates in primary fish food areas in oceans).	M
	Detection of chlorine (chlorine is indicative of bioproductivity in the oceans).	M
Other disciplines	Detection of volcanic activity by detection of SO_2.	H
	Determination of the composition of volcanic gases (enables predictions to be made of the composition of the molten mantle).	H
	Monitoring of volcanic emissions of SO_2 and H_2S to give warnings of eruptions.	H

Key. L—Low ground resolutions   (10 km)
     M—Medium ground resolution (2 km)
     H—High ground resolution   (0.1–0.5 km).

5 Sensor platforms, sensor packages and satellite data distribution

5.1 Introduction

Remote sensing techniques can be applied from different types of observation platform and each platform, mobile or stationary, has its own characteristics. Generally speaking, three types of platforms are of interest for remote sensing: ground, airborne and spaceborne observation platforms (see Fig. 5.1).

Each of these platforms has particular advantages and disadvantages. For example, the satellite platform offers great synoptic viewing potential, stability of the platform and orbital movements of predictable character. Yet the space platform is difficult to reach and instrumental failure may be hard to correct. Airborne platforms offer easier maintenance of sensors, lower operating cost and

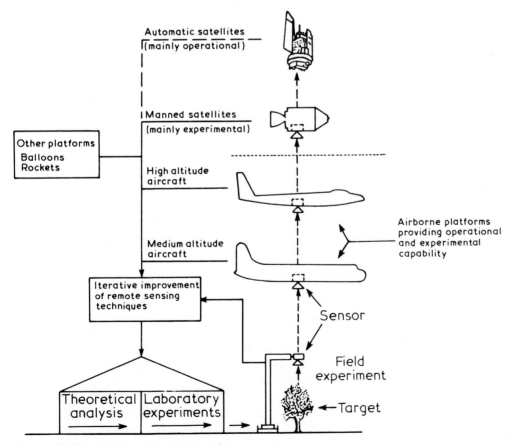

Fig. 5.1 Multi-platform remote sensing operations.

good accessibility for manned flights. Yet airborne platforms may suffer from instability and flight lines may be difficult to maintain. The following sections discuss each type of platform with these issues in mind.

5.2 Ground observation platforms

A scientific field and laboratory programme is necessary to provide a good understanding of the basic physics of the object/sensor interaction. Measurements of physical characteristics such as spectral reflectance and emissivity of different natural phenomena are necessary in order to design and develop sensors. Normally fundamental studies are of two kinds:

1. Laboratory studies where soils, plants, building materials, water bodies etc., are subjected to an external source of radiation and detectors of various kinds are used to measure reflection and emission spectra.
2. Field investigations in which the spectral characteristics of surface phenomena or crops are investigated in different atmospheric conditions.

In field investigations the most popular platform to date has been the 'Cherry-Picker' platform (Fig. 5.2). These platforms can be extended to approximately 15 m. They have been used by various American and European research organizations to carry spectral reflectance meters, photographic systems and scanners in the infrared or radar wavelengths.

Frequently these platforms are linked to automatic recording apparatus in field vans (Fig. 5.3). The limitation of this type of platform is that it is not a cross-country vehicle and is limited to roads.

Portable masts are also available in various forms and can be used to support cameras and sensors for testing. For example, a Land Rover fitted with an extending mast was used successfully in conjunction with airborne studies (see Plate 5.1(a)). However, such masts can be very unstable in windy conditions.

Towers are also available which can be dismantled and moved from one place to another (Plate 5.1(b)). These offer greater rigidity than masts, but are less mobile and require more time to erect. Where heavy instruments such as radar sensors are undergoing

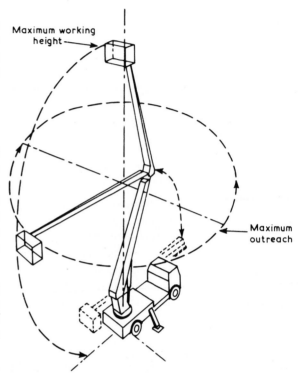

Fig. 5.2 Mobile hydraulic platform of the 'Cherry-Picker' type. (Source: EMI, 1973.)

tests it may be economic to place towers on a wheeled platform so that they can be moved on rails.

5.3 Balloon platforms

Balloon platforms can be considered under two headings: tethered and free flying. Tethered balloons were used for remote sensing observations (air photography) as long ago as the American Civil War (1862) and very recently for nature conservancy studies on beaches in Britain. Modern studies are, however, often carried out by free-flying balloons which can offer reasonably stable platforms up to very considerable altitudes. For example the balloons developed by the Société Européenne de Propulsion can carry a photographic nacelle (Plates 5.2(a) and (b)) to altitudes of 30 km. This nacelle consists of a rigid circular base plate supporting the whole equipment to which is nested a tight casing, externally protected by an insulating and shock-proof coating. It is roll-stabilized and the inner temperature is kept at 20°C and at ground atmospheric pressure. Standard equipment includes two cameras, multispectral photometer, power supply

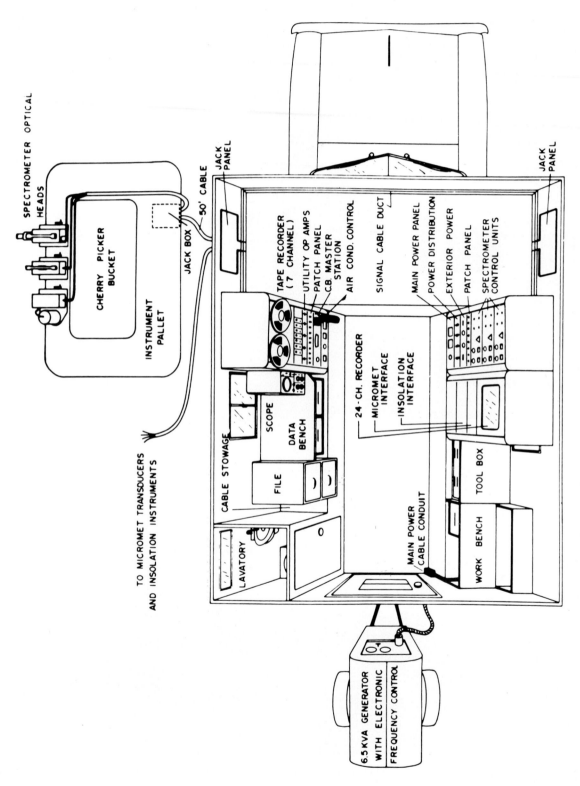

Fig. 5.3 Cherry-Picker instrumentation arrangement. (Source: LARS, 1968.)

Plate 5.1 (a) An extending mast fitted to a Land Rover. The mast can be extended to 10 m height and is equipped with camera and photo-diode sensors for studies of crop reflectance. (Courtesy University of Bristol.) (b) Tower for supporting sensors. (Courtesy Hawker Siddeley Dynamics Ltd., Hatfield.)

(a)

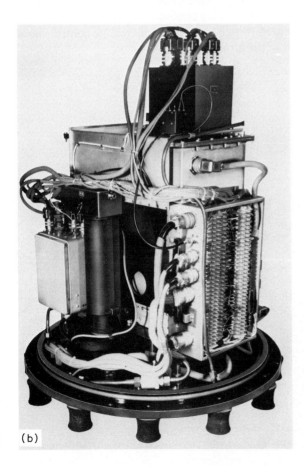

(b)

Plate 5.2 (a) Balloon nacelle before flight. (b) Sensor equipment inside balloon nacelle. Photometer (black) and cameras in base plate. (Courtesy Société Européenne de Propulsion.)

units and remote control apparatus. Its return to Earth is achieved by parachute, after remotely controlled tearing of the carrying balloon.

5.4 Aircraft and remotely piloted vehicles

Remote sensing from aircraft has been carried out for more than fifty years and, more recently, helicopters and autogyros have been widely used. Aircraft have been used mainly for obtaining stereoscopic black and white photography for subsequent interpretation and photogrammetric mapping. Many resource development and planning programmes employ aerial photography from altitudes of approximately 1500–3000 m. These missions at low or medium altitudes are appropriate for surveys of local or limited regional interest. Although they are widely used for periodic surveys it must be recognized that flights at specified times at short intervals or for defined repetition rates over long periods are rarely achieved. Nevertheless, aircraft platforms offer an economical method of testing sensors under development. Thus photographic cameras and various scanners in the infrared, radar

and microwave wavelengths have been flown over ground truth sites in many countries, especially in the NASA and ESA programmes.

In order to obtain a greater areal coverage and to test atmospheric effects on sensors special high altitude aircraft have been used as platforms. These have often been developed from military reconnaissance aircraft such as the American U2 plane (see Fig. 5.1). The altitudes attained by such aircraft are normally about 15 km from which ground coverage of 100–400 km^2 per frame can be obtained.

A new type of platform has emerged as a result of military requirements for target systems. This is usually termed the 'remotely piloted vehicle' (RPV), or more commonly the 'drone'. A drone is a radio-controlled miniature aircraft. Since it is controlled from the ground its relatively short control range of 5 km is dictated by the visual range obtainable. However aerial photography of local, inaccessible, areas is highly practicable. For example, surveys of oil pipelines, fence lines, power transmission lines or forest areas are possible using an underwing robot camera which is triggered by remote control signal from the ground.

Plate 5.3 Remotely Piloted Vehicle (drone) which is launched from a trailer (Courtesy AERO Electronics (AEL) Ltd, Horley, Surrey.)

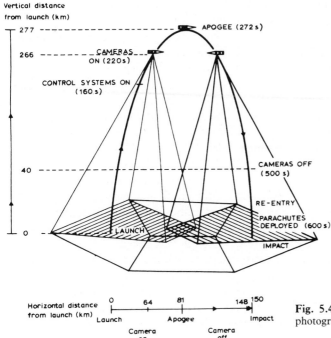

Fig. 5.4 Schematic diagram of Skylark rocket trajectory and photographic coverage. (Source: Savigear *et al.*, 1974.)

An example of such a system is the SNIPE manufactured by Aero Electronics Ltd. (Plate 5.3). This drone has a guaranteed flight time of 45 min, and a maximum speed of 177 km h^{-1}. It is launched by a catapult system, and is retrievable by parachute. Such systems can obtain high-definition photographs from altitudes up to 900 m. Typically the 35 mm camera has automatic wind system, f 2.8 lens and variable, high-speed, shutter.

5.5 High-altitude sounding rockets

Experiments in the use of sensors at space altitudes can be carried out by using high altitude sounding rockets. Synoptic imagery has been obtained from rockets for areas of some 500 000 km^2 (40 000–90 000 km^2 per frame). The apogee of existing European rockets is approximately 300 km but this may be extended to 400 km with 200 kg payload (cf. Space Shuttle, 185 km orbit, 18 200 kg payload).

The Skylark Earth Resource Rocket (Fig. 5.4) is an example of such a platform. The rocket is fired from a mobile launcher to altitudes between 90–400 km. During the flight the rocket motor and payload separate and its sensors are held in a stable attitude by an automatic control system. By stepping the payload round six times the sensors can scan a 360° field of view. The payload and the spent motor are returned to ground slowly by parachute enabling speedy recovery of the photographic records. The sensor payload consists of two cameras, normally of the Hasselblad 500 EL/70M 70 mm type. This rocket system has been used in surveys over Australia and Argentina.

The main limitation upon the use of such rocket systems is that of ensuring that the descending rocket assembly does not cause damage. Therefore, its use tends to be limited, to sparsely populated areas. Also, the complex geometric properties of the data may restrict its general use.

5.6 Satellite platforms

In this section attention will be focussed mainly on low-level satellite systems applied to land and ocean monitoring. In succeeding chapters key characteristics of the meteorological satellite programmes, in which both low and high-level satellites are employed, will be outlined. It is clear that much environmental monitoring can be achieved best by a combination of satellite missions embracing a variety of sensors. Some aspects of the future development of sensor platforms will be discussed in Chapter 19.

As well as providing a synoptic view of regional relationships the satellite platform can be put into orbit in such a fashion that it will provide repeated coverage of the whole of the Earth's surface. The interval between satellite observations at a particular location can, to some extent, be selected by choice of an appropriate orbit. Furthermore, such orbits can be so defined that observations can be made under comparable or identical conditions of illumination over comparatively long periods. Therefore, some description and explanation of orbital configurations is appropriate before considering details of the platforms and sensor packages themselves.

The orbital relationships of satellites are affected by many factors but one may note first that there is a basic relationship between the *heights* of satellite orbits and their lives before re-entry to the Earth's atmosphere. Satellites in low orbits are affected much more by upper atmospheric drag and resulting increases in the effects of gravitational pull (see Table 5.1).

Table 5.1 Relationships between average altitudes of satellites in near-circular orbits, and their expected lives before re-entry

Height above Earth surface	Life before re-entry
250 km	12 days
500 km	10 years
600 km	50 years
1000 km	1000 years
10 000 km	Indefinite

Thus there are lower limits below which Earth observation satellites will not normally be operated. Since the design lives of their payloads are usually within the span from 1–5 years, a minimum operational altitude of about 450 km is indicated. Actual orbital altitudes of unmanned remote sensing satellites are characteristically well above that level; whilst most operate between 800–1500 km some are at approximately 36 000 km altitude. The first of these groups are mostly polar or near-polar orbiting satellites occupying so-called 'sun-synchronous' orbits (crossing the equator at the same sun time each day); the second group are mostly 'geostation-

ary' satellites which appear to 'hover' over pre-selected points on the equator.

Although all satellite orbits are elliptical, for most Earth observation purposes it is desirable to achieve as nearly a circular orbit as possible. Noaa 6, which was launched on June 27th 1979, met this requirement well. The apogee (top) of the Noaa 6 orbit was 826 km, and the perigee (bottom) was 810 km. A Russian satellite launched on the same day was Cosmos 1109. Its orbit was highly-elliptical with an apogee of 40 058 km and a perigee of 613 km. Such orbits may be preferred for satellites whose viewing missions differ from one region of the world to another. For example, in design studies for Meos (a proposed ESA Multispectral Earth Observation Satellite) account was taken of the advantage to Europe of a particular imaging payload which would return relatively low-resolution, synoptic-scale, data from a high Southern Hemisphere apogee and much higher-resolution data of the small countries of Western Europe from a low Northern Hemisphere perigee. We will return to the Meos design studies a little later, after considering simpler orbital configurations in greater detail.

5.6.1 *Low-level satellites*

Low-level (800–1500 km) Earth observation satellites encircle the Earth whilst the Earth itself revolves on its polar axis within their orbits. These satellites fall into three broad groups on the basis of detailed orbital differences. These are (see Fig. 5.5(a)):

(a) Equatorial-orbiting satellites, whose orbits are wholly within the plane of the equator (Fig. 5.5(a)).

(b) Polar-orbiting satellites, whose orbits are in the plane of the Earth's polar axis; each successive orbit crosses the Equator at a different sun time (Fig. 5.5(b)) True polar orbits are preferred for missions whose aim is to view longitudinal zones under the full range of illumination conditions.

(c) Oblique-orbiting (near polar orbit) satellites (by far the largest group) whose orbital planes cross the plane of the equator at an angle other than 90°. Because the range of possibilities here is much greater than in (a) and (b), further discussion of oblique orbits is required (Figs. 5.5(c) and (d)).

Oblique-orbiting satellites may be launched eastwards into direct, or *prograde*, orbits (so called because the movement of such satellites is in the same direction as the rotation of the Earth), or westwards into *retrograde* orbits. The inclination of any orbit is specified in terms of the angle between its ascending track and the equator, i.e. at the ascending node. Because the Earth is not a perfect sphere it exercises a gyroscopic influence on satellites in oblique orbits such that those in prograde orbits (Fig. 5.5(c)) *regress* whilst retrograde orbits *advance* or *precess* (Fig. 5.5(d)) with respect to the planes of their initial orbits.

Fig. 5.5(e) shows that a perpendicular from the Sun to the Earth also advances around the globe in the direction of Earth rotation on its polar axis, because of the behaviour of the Earth itself as a satellite travelling around the Sun. It is clear that a

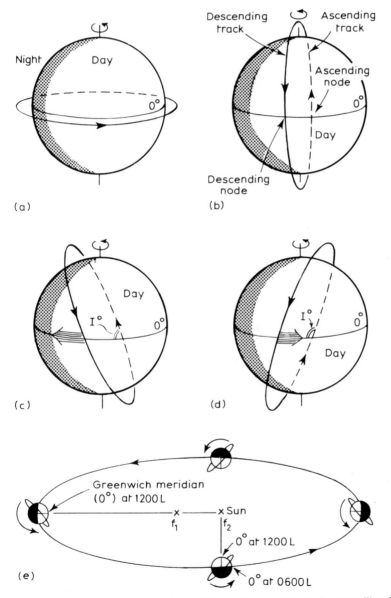

Fig. 5.5 Orbital characteristics of polar-orbiting satellites, and the Earth as a satellite of the Sun. (a) Equatorial orbit. (b) True polar orbit. (c) Prograde oblique orbit (launched Eastwards, regresses Westwards). (d) Retrograde oblique orbit (launched Westwards, precesses Eastwards). (e) Perpendicular from Sun to Earth, precesses Eastwards throughout the year.

special type of Earth satellite orbit can be achieved if the rate of precession of an orbit is geared to the orbiting of the Earth around the Sun: this is the *sun-synchronous* configuration (a retrograde oblique orbit) which ensures that satellites view the same point on the surface at the *same local time* at predetermined intervals (twice each day with near polar-orbiting weather satellites, once every 18 days with Landsat). Thus the altitude of the sun in the sky is approximately the same *within similar seasons* at every crossing of the same parallel. However, it should be noted that differences in the elevation of the sun affect each local area through the annual cycle of the seasons as a result of the apparent motion of the Sun relative to the planet Earth.

It is in this context that the observation of the surface from a satellite must be seen to be a function not only of the satellite altitude and orbital configuration, but also of the 'instantaneous field of view' (IFOV) of its sensor system. One very desirable sun-synchronous orbit is a retrograde orbit inclined at 99.1° to the equator, at an altitude of 1100 km. This orbit is completed once every 100 minutes. From this information the number of orbits which will be completed in 24 h can be calculated (approx. 14.5), and also the breadth of the IFOV required to

ensure complete global imaging cover even in equatorial latitudes. In the case of Noaa satellites once-daily coverage is achieved in the visible by a scanning radiometer with a viewing angle of 112°; twice daily coverage is achieved in the infrared by the same sensor system because heat radiation is emitted from both the daytime (descending track) and nighttime (ascending track) hemispheres. In the case of Landsat, narrower swaths of imagery are obtained from scanning radiometers like the Multispectral Scanner (MSS) of Landsat 3, for this had a viewing angle of only 11.6° and the satellite occupied a lower orbit, at about 900 km. In this way more detailed target information is acquired, but at the expense of breadth and frequency of coverage. Such trade-offs are commonplace, for communication links between satellites and ground stations constrain the volumes of data which can be transferred from the one to the other whilst they are intervisible.

5.6.2 High-level satellites

Whilst the sun-synchronous orbit is the most popular low orbit, *geosynchronous* orbits are virtually the only useful high orbits. In these the satellites precess around the Earth at rates related to the rotation of

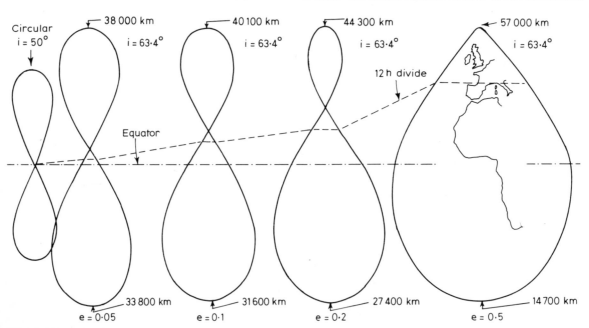

Fig. 5.6 Ground tracks of inclined geosynchronous orbits to exemplify possible variations of eccentricity, inclination and orientation. Associated (possibly advantageous) variations would result in imaging resolutions but also in image geometries (generally disadvantageous). (Source: Velten, 1976.)

the Earth on its polar axis. The most useful single orbit is at approximately 36 000 km, at which altitude a prograde satellite orbiting in the plane of the equator 'hovers': i.e. it is apparently 'fixed' above a given point on the surface. This special type of geosynchronous orbit is known as *geostationary*. It is exploited by many communications satellites, and some Earth observation satellites, especially within the programme of the World Weather Watch (WWW) of the World Meteorological Organisation (WMO) (see Chapter 10). These have included ATS, SMS, Goes, Meteosat and GMS satellites. The key advantage of the geostationary orbit is that it permits satellites to view the visible disc (limited by the curvature of the Earth) much more frequently than any low orbit satellite can do. Other geosynchronous possibilities exist but have yet to be exploited for Earth observation missions. Fig. 5.6 exemplifies some of these, drawn from the Meos design studies mentioned earlier.

5.7 The Landsat system

The Landsat Satellite System was originally called ERTS (Earth Resources Technology Satellite) but the name was changed to Landsat in 1974. Three Landsat satellites have been launched. Landsat 1 launched 22 July 1972 retired on 6 January 1978; Landsat 2 launched 22 January 1975 retired 22 January 1980 due to insufficient attitude control but was restored as a prime Earth resource satellite on 21 June 1980 based on magnetic stabilization; Landsat 3 launched 3 March 1978 developed problems in the MSS sensor in August 1978 due to a delay in the line-start pulse.

The current and probable Landsat ground stations are shown in Fig. 5.7. It will be seen that the principal American and European receiving stations in 1980 were Fairbanks, Goldstone and Greenbelt in the USA; Prince Albert and Shoe Cove in Canada; Cuiaba in Brazil and Fucino and Kiruna in Europe.

The Landsat Data Collecting System consists of a combination of ground stations, operations control centre and NASA Data Processing Facility (NDPF) as shown in Fig. 5.8. The scale of the data processing operation can be assessed by considering the data archived at the EROS Data Center, Sioux Falls, USA. By September 1, 1977, it was estimated that the Center had almost 262 000 images (four bands

for each) acquired by Landsat 1 and 2. At that time approximately 5 000 new images were being acquired each month.

The overall Landsat system was designed to make automatic observations using a payload consisting of a return beam vidicon (RBV) camera system and multispectral scanner (MSS) (Fig. 5.9(a)). The RBV system on Landsat 1 and 2 operated by shuttering three independent cameras simultaneously (Fig. 5.9(b)), each sensing a different spectral band in the range 0.48–0.83 μm. On Landsat 3 the RBV system was changed to two RBV panchromatic cameras operating in the range 0.51–0.75 μm (see Fig. 5.9(c)). These cameras produce two side-by-side images, each covering a ground scene of approximately 99×99 km. By using a focal length of 25 cm the ground resolution obtained is approximately 30 m. Digitization of the analogue signal leads to an incoming data stream of 45 megabits s^{-1}.

The MSS system is a line scanning device using an oscillating mirror to scan at right angles to the spacecraft flight direction (Fig. 5.9(d)). Optical energy is sensed simultaneously by an array of detectors in four spectral bands from 0.5–1.1 μm. The area of coverage is approximately equal to that of the RBV images.

The Landsat satellite platform operates in a near circular, sun-synchronous, near-polar orbit at an altitude of 915 km. It circles the Earth every 103 minutes, completing 14 orbits per day and views the entire Earth every 18 days. The orbit has been selected so that the satellite ground trace repeats its Earth coverage at the same local time every 18 day period to within 37 km of its first orbit. Each day the paths shift 160 km westward (see Fig. 5.10). The amount of overlap between successive passes varies from 14% sidelap at the equator to 70% at polar latitudes. For example, at 40°N the sidelap of successive images is 62 km (see Fig. 5.10(b)). Images are acquired between 9.30 and 10.00 a.m. local sun time, except at high latitudes.

The Landsat programme is planned to include Landsat D which will be equipped with the Thematic Mapper. The Thematic Mapper (TM) consists of a six band, Earth-looking scanning radiometer with a 30 m ground element resolution. It will cover a 185 km swath from 705 km altitude. The spectral bands will include blue (0.45–0.52 m), green (0.52–0.60 μm), red (0.63–0.69 μm), near

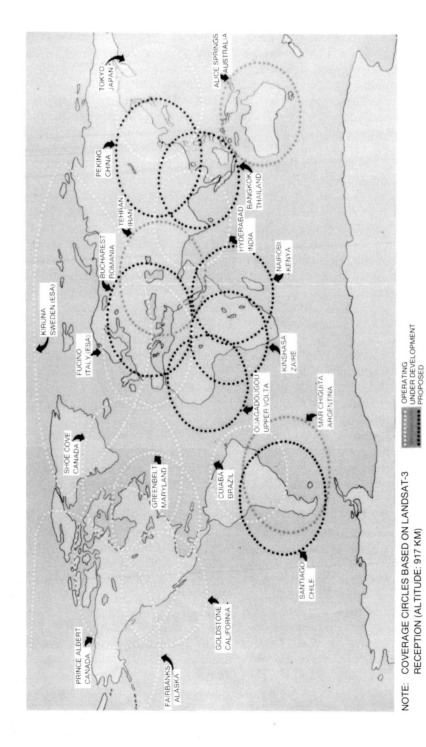

NOTE: COVERAGE CIRCLES BASED ON LANDSAT-3
RECEPTION (ALTITUDE: 917 KM)

OPERATING
UNDER DEVELOPMENT
PROPOSED

Fig. 5.7 Current and probable Landsat ground stations. (Source: NASA.)

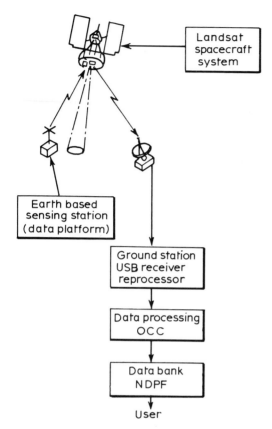

Fig. 5.8 Elements of the Landsat DCS (Data Collection Service) System. (Source: NASA.)

infrared (0.76–0.90 μm), mid infrared (1.55–1.75 μm) and thermal infrared (10.4–12.5 μm).

The Landsat D system equipped with the Thematic Mapper presents a number of opportunities together with some problems. The picture element size will diminish from about 80×70 m in the present Landsat 1–3 programme to about 30×30 m. This should assist in many applications, particularly in land use studies. It will, however, bring a very large increase in the rate of data acquisition. This will go up from about 15 megabits s^{-1} for early Landsat systems to about 110 megabits s^{-1} with the Thematic Mapper. Likewise the number of detectable grey levels will increase from 64 in early Landsat missions to 256 in Landsat D. This increased radiometric resolution will be extremely important in detecting differences in vegetation.

5.8 Spacelab and the Space Shuttle

The Space Laboratory (Spacelab) will be carried into orbit in the payload bay of the Shuttle orbiter (see Chapter 19) and remain attached to the orbiter throughout the mission (Plate 5.4). At the end of each mission the orbiters will make runway landings similar to those made by conventional aircraft (Fig. 5.11). The Spacelab will then be prepared for its next mission. An important feature of the system will be its quick turn-round capability of approximately two weeks.

The Spacelab concept provides a considerable degree of flexibility in relation to a variety of applications. First, the Space Shuttle flight characteristics may be varied so that the orbit inclination, orbit altitude (200–900 km) and resulting ground coverage may be selected for mission compatibility. In the early stages of the programme the East Coast launch site at Kennedy Space Centre will be used so that the possible range of inclinations is 28.5–57°. Later in the Spacelab programme the West Coast site at Vandenberg is planned to become available, thereby ensuring orbit inclinations up to 104°. Also the Orbiter orientation (all directions with an accuracy up to 0.5° per axis) and flight duration time (up to 30 days) can be adjusted as required.

The second type of flexibility results from the modular approach adopted in the design. The sensor platform consists essentially of two parts. First, a pressurized laboratory (module) offering a 'shirt sleeve' environment for the space crew, scientists and equipment. Second, a pallet area outside the laboratory which is unpressurized but can be used to mount certain items of equipment that can be remotely controlled. The module and pallet can be varied in size by selecting from the available Spacelab elements (see Table 5.2) so that the configuration fits the mission requirements. The module may be composed of one or two identical cylindrical shells (2.7 m long, 4.1 m in diameter) whereas the pallet is composed of up to 5 segments, each 2.9 m long. Three basic configurations are apparent – module only, module plus pallet and pallet only. Further flexibility is provided by the availability of the common payload support equipment. Also certain subsystem components such as cold plates, equipment racks, and recorder can be

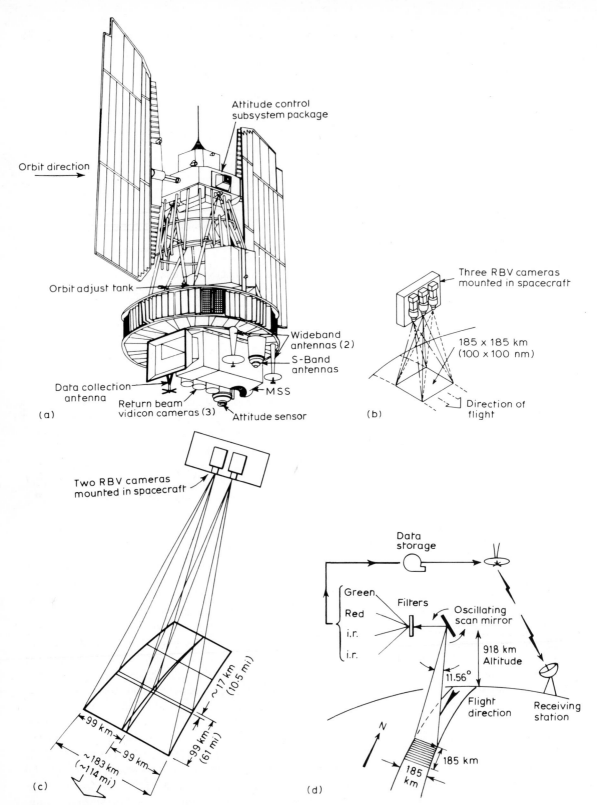

Fig. 5.9 (a) General configurations of Landsats 1 and 2. (b) Return Beam Vidicon System for Landsats 1 and 2. (c) Return Beam Vidicon System for Landsat 3. (d) Landsat Multispectral Scanner. For each terrain scene, four images are transmitted to a receiving station. (Source: NASA.)

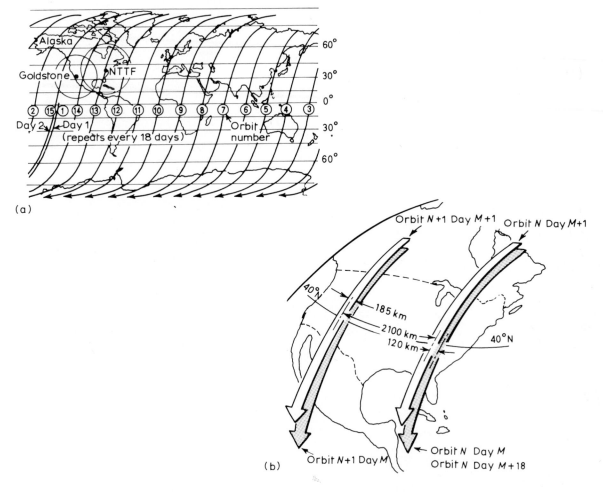

Fig. 5.10 Landsat orbits (a) providing world coverage and (b) over the United States on successive days. Note the 62 km sidelap of successive image swaths at 40°N latitude. (Source: Sabins, 1978.)

used or removed as required. This provides for flexibility in loading.

The purpose of Spacelab is to provide a ready access to space for a broad spectrum of experimenters in many fields and from many nations. It may be noted that there are two principal objectives associated with any experimental Earth-observation programme using advanced sensing methods:

1. To understand the capabilities and limitations of the various sensor systems proposed and to develop measurement techniques that can be applied to the different areas of application under investigation;

2. Using the proven remote sensing instruments and methods derived above, to conduct experiments in the Earth observation disciplines aimed either at purely scientific objectives or at perfecting measurement methods for later applications oriented missions.

The role of Spacelab in meeting the two objectives is shown in Fig. 5.12. It will be seen that the manned space laboratory can act as a bridge between the initial ground and airborne measurements and final systems based on long-life, automatic satellites.

Spacelab is essentially a European manned space laboratory. It is planned to fly two remote sensing

Plate 5.4 The European Space Agency (ESA) builds Spacelab, a habitable laboratory that will be carried in the Space Shuttle orbiter cargo bay for Earth-orbital periods up to 30 days. Spacelab will be used to conduct experiments in the physical sciences, health sciences, and manufacturing processes. (Source: ESA.)

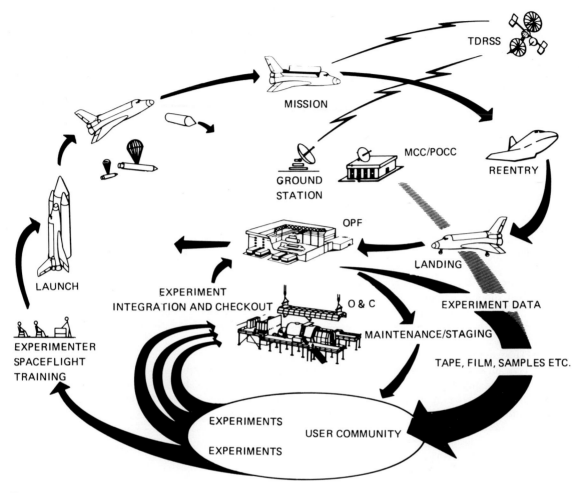

Fig. 5.11 Shuttle – Spacelab operations cycle. (Source: NASA.)

experiments, the Metric Camera and the Microwave Remote Sensing Experiment (MRSE) as part of the First Spacelab Payload (FSLP). Both experiments are provided to the European Space Agency by the Federal Republic of Germany, as general instrumentation for the FSLP. The Metric Camera has been discussed in Section 4.2.1. The Microwave Remote Sensing experiment embodies a radar facility which in the active modes transmits microwave energy in the 9.65 GHz band (X-band) to Earth targets. A sensitive low-noise receiver detects the backscattered radar signals. The microwave instrument will, however, operate in three modes:

1. A high-resolution mode as a Synthetic Aperture Radar (SAR);

2. A two-frequency Scatterometer (2FS) mode; and
3. A passive mode as a passive microwave radiometer.

In the SAR mode the Earth's surface will be imaged and it is anticipated that imagery with a ground resolution of 25×25 m will be obtained. The system parameters for the SAR mode are given in Table 5.3.

In the two-frequency scatterometer mode (2FS) the instrument will be directed towards measurement of ocean surface-wave spectra by using the complex backscattering of the ocean surface at two adjacent microwave frequencies. It is anticipated that the long ocean wave spectrum components for waves between 10 and 500 m will be detected together with wave direction.

Table 5.2 Spacelab facilities. (Source: ESA)

Available to users	Spacelab configuration			
	Short module +9 m pallet	Long module	15 m Pallet	Independ. susp. pallet
Payload weight (kg)	5500	5500	8000	9100
Volume for experiment equipment				
Inside module (m³)	8	22	—	—
On pallet (m³)	100	—	160	100
Pallet mounting area (m²)	51	—	85	51
Electrical power (28 V DC 115/200 V at 400 Hz AC)				
Average (kW)	3–4	3–4	4–5	4–5
Peak (kW)	8	8	9	9
Energy* (kWh)	300	300	500	500
Exp. Support computer with central processing unit and data acquisition system	←———64 K core memory of 16-bit words———→ ←————320 000 operations s⁻¹————→			
Data handling				
Transmission through orbiter	←————Up to 50 Megabits s⁻¹————→			
Storage digital data	←————Up to 30 Megabits s⁻¹————→			
Instrument pointing sub-system IPS	Mounted on pallet, will provide arc second pointing for payloads up to 5000 kg			

*Energy can be increased by the addition of payload-chargeable kits, each providing 840 kWh and weighing approx. 350 kg (at landing).

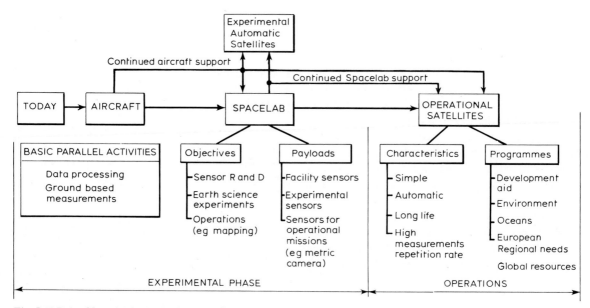

Fig. 5.12 Role of Spacelab in the development of remote-sensing instruments and techniques. (Source: ESA.)

Table 5.3 *System parameters for the SAR mode.* (Source: ESA)

Frequency	9655 MHz
Bandwidth	>10 MHz
S/N for $\sigma_0 = -25$ dB	6 dB
RF power	240 average, 3000 W peak
Doppler bandwidth	3 kHz
Depression angle	45°
Swath width	8.5 km
Resolution (3 dB)	25 m (az), 25 m (el)
PRF	10 ± 0.3 kHz
Pulse length	8 μs
Data rate	32 Mbps
Encoding	4 bits 1 and Q

Table 5.4 *Optimum sensor groupings about a single major sensor.* (Source: ESRO, 1974)

Payload cores or modules	Core sensors	Support sensors
Multispectral scanner core	Multispectral scanner	Spectrometer Cameras i.r. radiometer i.r. scanner* Laser profiler
Synthetic aperture radar core	Side looking radar	Cameras Scatterometer
Passive microwave radiometer core (multifrequency)	Passive microwave radiometer	Cameras i.r. radiometer i.r. scanner Scatterometer

*If not included in the multispectral scanner.

Future payloads of Spacelab are likely to consist of three main groups of sensors. These are facility, standard support and experimental sensors. *Facility sensors* are considered as basic sensors common to a wide range of mission types and users. Typical examples of facility sensors are multispectral scanners and the active and passive microwave sensors. *Standard support sensors* and equipment are any additional sensors required to support the main sensor payload. Examples of these are metric camera, multispectral cameras, tracking telescope, precision altimeter, precision non-scanning spectro-radiometer, data receiving and display equipment and consoles allowing the crew to operate and monitor sensors during the mission. *Experimental sensors* could be extensions or modifications to the basic facility payload, or new sensors specially developed for space application. New sensors might include laser radar, imaging spectrometer, and polypanchromatic side-looking radar.

It is possible to identify optimum groupings of sensors in which each module would comprise one facility/support sensor plus the necessary supporting sensors. Three sensor complements using different core sensors are shown in Table 5.4.

5.9 The relationships between Landsat, Skylab and the Space Shuttle

NASA's Goddard Space Flight Center (GSFC) began a conceptual study of the feasibility of Earth resource satellites in 1967. The results of the Landsat programme have certainly exceeded most expectations.

In parallel to the Landsat programme NASA had the idea of developing an Orbital Research Laboratory and this concept was subsequently built into the Apollo programme. This project was named the Skylab project and the general configuration of Skylab is shown in Fig. 5.13. The Skylab mission of 1973 carried a three-man crew and the observations were coordinated with the collection of ground data on test sites and aircraft underflights. Skylab can be regarded as a step in the development of the Space Shuttle and its successful completion was marked by the acquisition of detailed imagery of many parts of the world. It should be noted, however, that due to its lower orbital angle its coverage was limited to 50°N to 50°S only.

In this discussion examples have been taken from the American (NASA) and European (ESA) programmes. It is noteworthy that in Russia a similar progression from Spacecraft (Soyuz) to the modular build up of an orbital space station (Salyut) has taken place as a major part of the Soviet space effort.

5.10 Acquisition of Landsat images

At the outset of any piece of work the user is confronted by practical questions such as:

(a) Where can data be obtained?
(b) What forms of data are available?
(c) Are there any changes in the state of operation of the satellite?

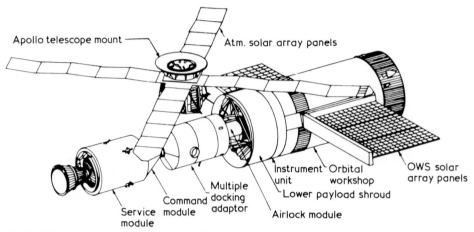

Apollo telescope mount

Atm. solar array panels

Instrument unit

Orbital workshop

OWS solar array panels

Lower payload shroud

Multiple docking adaptor

Command module

Airlock module

Service module

Fig. 5.13 General configuration of Skylab platforms.

The user should bear in mind that the forms of data and the operating efficiency of particular satellite missions may vary. The most convenient point of contact should, therefore, be determined and close liaison maintained with the distribution centre so that any changes in the availability of data are known at an early stage.

The United States has taken steps to operate Landsat in the *international public domain,* and as a result images of any part of the world can be acquired by Landsat and these are available for purchase by any person. The user wishing to acquire data should first of all establish which is the best point of contact. In the USA the EROS Data Center (EDC), Sioux Falls, South Dakota, 57198, is the appropriate principal source although other suppliers include the US Department of Agriculture, Aerial Photography Field Office, Administrative Services Division, PO Box 30610, Salt Lake City, Utah 84115, and General Electric Space Systems, Photographic Engineering Laboratory, 5030, Herzel Place, Beltsville, Maryland, 20705.

In Europe Landsat data is available through the *Earthnet* programme set up by the European Space Agency. The Earthnet Programme office is located at Frascati, Italy, and is responsible for monitoring a

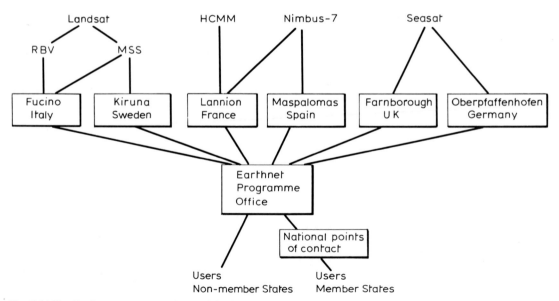

Fig. 5.14 The Earthnet programme. (Source: ESA.)

network of stations receiving data from Landsat, HCMM (Heat Capacity Mapping Mission), Nimbus 7 and Seasat systems (Fig. 5.14). It is also responsible for interfacing with the users and distributing data. Users in Europe should, however, apply to their National Point of Contact for remote sensing data and these (as 1980) are listed in Table 5.5.

Images of Canada may be obtained from the Center for Remote Sensing, 2464 Sheffield Road, Ottawa, Canada or from Integrated Satellite Infor-mation Services Ltd., PO Box 1630, Prince Albert, Saskatchewan, Canada. Likewise, South American images recorded at Cuiaba, Brazil can be obtained from Institute de Pesquisas, Especiais (INPE) a/c Direcao, Av. dos Astronautas, 1758, Caixa Postal 515 12 200 Sao Jose dos Campos, S.P. Brazil.

5.10.1 Indexing of Landsat images

The EROS Data Center publishes monthly

Table 5.5 National points of contact in Europe

Belgium	**Italy**
Services de Programmation	Telespazio
de la Politique Scientifique	Corso d'Italia 43
Rue de la Science 8	Roma
1040 Bruxelles	Tel 06–8497306 Twx 610654 TELEDIRO
Tel 02–2304100 Twx 24501 PROSCI B	
	Spain
Denmark	CONIE
National Technological Library	Pintor Rosales 34
Centre for Documentation	Madrid 8
Anker Engelundsvej 1	Tel 34 1 2479800 Twx 23495 INVES E
2800 Lyngby	
Tel 02–883088 Twx 37148 DTBC	**Sweden**
	Swedish Space Corporation
Netherlands	Tritonvagen 27
National Aerospace Laboratory	S–17154 Solna
NLR	Tel 08 980200 Twx 17128 SPACECO S
Anthony Fokkerweg 2	
1059 CM Amsterdam	**Switzerland**
Tel 020–5113113 Twx 11118 NLRAA NL	Bundesamt fur Landestopographie
	Seftigenstrasse 264
France	CH–3084 Wabern
GDTA, Centre Spatial de Toulouse	Tel 031–541331 Twx 32498 UEM CH
18 Avenue Edouard Belin	
31055 Toulouse	**UK**
Tel 61–531112 Twx 531081 F CNEST	National Remote Sensing Centre
	Space Department
Germany	Royal Aircraft Establishment
DFVLR Hauptabteilung Raumflugbetrieb	Farnborough
8031 Oberpfaffenhofen	Hants GU14 6TD
Post Wessling	Tel 252 24461 Twx 858442 PE MOD G
Tel 08153–28740 Twx 526401	
Ireland	
National Board for Science & Technology	
Shelbourne House	
Shelbourne Road	
Dublin 4	
Tel Dublin 683311 Twx 30327 NBST EI	

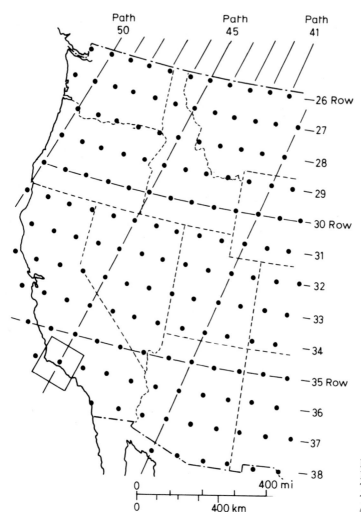

Fig. 5.15 Landsat index map prepared by EROS Data Center based on nominal image centrepoints. The square shows coverage of a Landsat image. The location of this image is path 45, row 36. (Source: Sabins, 1978.)

catalogues listing, by latitude and longitude, the newly acquired Landsat images. As well as listing images by location and date of acquisition, the image quality is classified. For black and white images, categories of 2 = poor, 5 = fair and 8 = good, are identified. Colour images are graded from 0 to 9 on a continuous scale of increasing quality. Cloud cover is given in tens of percent.

Similar lists of imagery can be obtained from National Points of Contact in Europe and elsewhere. Furthermore the user can, on supplying the latitude and longitude of the boundary points of an area of interest, obtain lists of images covering that area.

The EROS Data Center has prepared Landsat Index Maps for the world. The maps show the nominal image centre points. These are defined by the intersection of north-to-south orbit paths and east-to-west rows. Thus the centre point of the image outlined in Southern California as shown in Fig. 5.15 is defined by path 45, row 36. Browse file facilities consisting of microfilm viewers and images are available at most of the National Points of Contact. All remote sensing data acquired by Earthnet for European users is catalogued on the LEDA files (Line Earthnet Data Availability) in the ESA Information Retrieval Service at Frascati, Italy. Users who have access to a DIALTECH terminal can interrogate the LEDA files themselves.

5.10.2 *Landsat data products*

The formats of Landsat satellite images available

Table 5.6 Landsat 1, 2, 3 and D images available from EROS Data Center, and prices effective from October 1982

Archival products	Price ($)
Photographic products	
70 mm film positive (B/W)	26
70 mm film negative (B/W)	32
10 in film positive (B/W)	30
10 in film negative (B/W)	35
10 in paper (B/W)	30
20 in paper (B/W)	58
40 in paper (B/W)	95
10 in film positive (colour)	74
10 in paper (colour)	45
20 in paper (colour)	90
40 in paper (colour)	175
16 mm microfilm (B/W)	60
35 mm slide (colour), from existing collection	4
16 mm microfilm (colour, 100 ft roll)	150
Digital products	
9 track, 800 bpi CCT (MSS scene—all available bands)	650
9 track, 1600 bpi CCT (MSS scene—all available bands)	650
9 track, 6250 bpi CCT (MSS scene—all available bands)	650
9 track, 800 bpi CCT (RBV single subscene)	650
9 track, 1600 bpi CCT (RBV single subscene)	650
9 track, 800 bpi CCT (RBV set of 4 subscenes)	1300
9 track, 1600 bpi CCT (RBV set of 4 subscenes)	1300
14 track, high density tape (variable content)	Variable
Generation of colour composite (false colour infrared)	
Surcharge on product price	$ 195
Special acquisitions. (Special acquisitions signify Landsat D MSS scene data that are not scheduled for routine collection, but which are provided upon user request).	
Delivery of preprocessed digital data to the requestor's site via communication satellite; per MSS scene collected at a time and place specified by the requester	$ 790

from National Points of Contact or from EROS Data Center are listed in Table 5.6. In addition computer compatible tapes (CCTs), data for each Landsat scene can be acquired. All CCTs contain one set of MSS data for a Standard 185×185 km scene based on the Landsat worldwide reference system (path/row or track/scene), or one 98×98 km subscene of RBV data. The tape standard is 9-track, 1600 bpi, phase encoded, industry-compatible. As an option 9-track, 800 bpi, industry-compatible tapes are also usually available but in this case one scene occupies two tapes.

Each image has descriptive annotations, the type of annotation varying according to the source of the data. For EROS data, the annotation is as shown in Table 5.7.

In other countries the annotation may vary somewhat. For example, images obtained from the United Kingdom National Point of Contact at the Royal Aircraft Establishment (RAE, Farnborough), have the annotations shown in Table 5.8 and Plate 5.5.

Table 5.7 Style of annotation at bottom of EROS Data Center image

Sequence left to right	Interpretation
21 October 1972	Date image was acquired
C N34–33/W 118–24	Geographic centrepoint of image (degrees min^{-1})
N N34–31/W 118–19	Nadir of spacecraft (degrees min^{-1})
MSS	Multispectral scanner image
4 (or 5, 6, 7)	MSS Spectral band – one specified
SUN EL39	Sun elevation – degrees above horizon
Az 148	Sun azimuth – degrees clockwise from North
190	Spacecraft Heading, in degrees
1255	Orbit revolution number
G (or A or N)	Ground Recording Station
	G = Goldstone; A = Alaska
	N = Network Test and Training Facility at GSFC
1090–18012	The unique frame idenification number composed as follows:
1 or 5	Landsat 1 (5 denotes Landsat 1 for days greater than 999 since launch).
090	Days since launch. This was October 21, 1972. (Beginning in early 1977 this part of the identification number is expanded to a four-digit number, such as 1104, to accommodate 1000 and higher days)
18	Hour at time of observation, GMT
01	Minutes
2	Tens of seconds

Table 5.8 Style of annotation for UK Landsat product

Sequence left to right	
TIPS/Landsat-IPS	Telespazio Image Processing System – Landsat Image Processing System
Landsat 2	Satellite identification
Centre 5.56 N 2.88 W	Latitude and longitude of image centre
219/24 02 July 77	Path/Row and acquisition date.
7 (or 5, 6, 7)	Band identification – one specified
SUN Az 133 EL 50	Sun angle azimuth and elevation-degrees.
RAE Farnborough	Image source
Process 22	Process identification for applied corrections or enhancement.
22 January 1978	Master generation date
E-2	Earth Resources 2
12434	Orbit number
10072	Greenwich Mean Time at image centre in hours, minutes and tenths of minutes

NB. On the left of the image two grey scales are produced
    LIN — Linear scale with 18 density wedges
    ENH – Contrast enhancement wedge

5.11 Conclusion

In this chapter we have concentrated upon the availability of Landsat and Spacelab missions for land and ocean monitoring. Of course, other satellites have provided valuable experimental data and particular interest has been shown in data from Seasat 1, devoted to monitoring sea conditions but also of interest in land applications.

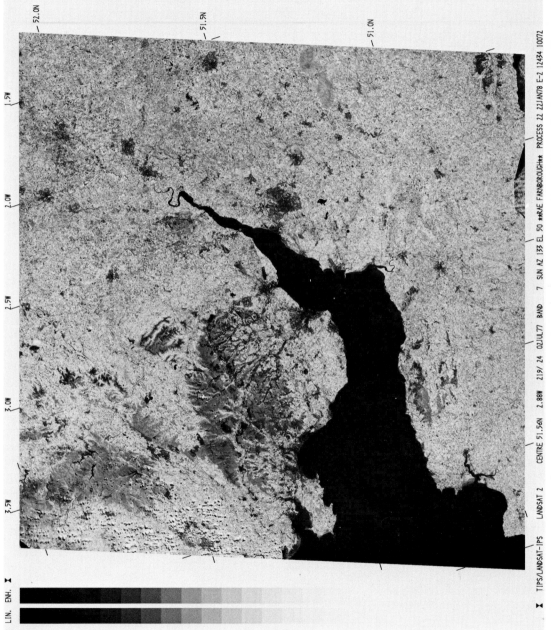

Plate 5.5 Landsat 2 image of part of South West England and South Wales bordering the Bristol Channel. Note style of annotation of the scene used in the UK. Compare with Plate 15.2 for land use and vegetation identification. (Courtesy, Space Department, RAE, Farnborough; Crown Copyright Reserved.)

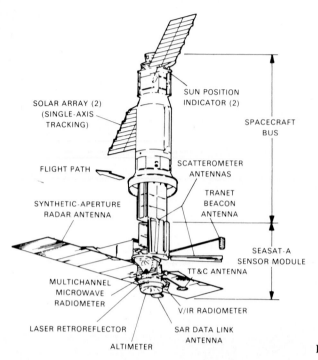

SOLAR ARRAY (2)
(SINGLE-AXIS
TRACKING)

SUN POSITION
INDICATOR (2)

SPACECRAFT
BUS

FLIGHT PATH

SCATTEROMETER
ANTENNAS

TRANET
BEACON
ANTENNA

SYNTHETIC-APERTURE
RADAR ANTENNA

SEASAT-A
SENSOR MODULE

TT&C ANTENNA

MULTICHANNEL
MICROWAVE
RADIOMETER

V/IR RADIOMETER

LASER RETROREFLECTOR

SAR DATA LINK
ANTENNA

ALTIMETER

Fig. 5.16 General configuration of Seasat 1 platform.

This satellite, specially developed for ocean monitoring (see Fig. 5.16) completed 106 days of successful operations before a catastrophic failure occurred on 9 October 1978 and NASA officially declared the satellite lost on 21 November 1978. Detailed discussion of the instrumentation and capabilities of Seasat will be found in Section 12.5.

Another satellite of considerable interest for land monitoring was that provided by the Heat Capacity Mapping Mission (HCMM). This satellite, launched on 26 April 1978, was operative until October 1980. Its principal characteristics and applications are also discussed in Chapter 12.

In the meantime, the Russians orbited their first Landsat equivalent on 18 June 1980 under the designation Meteor 30. This was equipped with two multispectral scanners, one a high-resolution system resolving 30 m elements on the ground, the other a medium-resolution system with a ground resolution of 170 m. Such systems are designed primarily to meet the needs of Eastern Bloc countries for Earth resource data.

It is clear that satellite missions are now beginning to emerge from the experimental stage and progress towards the operational systems capable of satisfying the needs of users. The speed of technological advance is such that new concepts and new platforms are being rapidly realized. These new developments are considered further in Chapter 19.

6 Collecting in situ *data for remote sensing data interpretation*

6.1 Introduction: The need for *in situ* data

The use of remote sensing techniques demands that there should be some method of calibrating the sensors in use or checking the accuracy of the data obtained. For example where infrared thermal line-scan equipment is being used it is useful to have some temperatures measured on the surface to check the accuracy of the data. Also some equipment is designed to detect relative differences in temperature and display them to advantage; i.e. they are self regulating systems which alter the scale over which measurement is made according to the dynamic range in the target area. In such cases it is essential to measure temperature at points within the target area in order to obtain some absolute values from the sensor data.

There is also a need for ground truth to check the accuracy of interpretations made on the basis of sensor data. For example if sensor data is being used to identify agricultural land use it is necessary to know the actual condition of a sample population of fields in order that the per cent accuracy of the identification can be determined.

It is common practice to use the term 'ground truth' for observations made on the surface of the Earth in relation to remote sensing studies. In many ways, however, the term is unsatisfactory because the measured variable is often in water, ice or air rather than on a ground surface. For this reason some workers prefer to use the term 'surface truth'. Even this term is not wholly satisfactory because it is sometimes difficult to define the measured 'surface' very precisely. As an example, it is often necessary to monitor meteorological factors such as wind-speed, solar radiation, rainfall, atmospheric humidity and cloud cover in order to assess the effects of atmospheric conditions on sensor performance. Some of these measurements may be made close to the ground (microclimatic observations), others may be obtained from standard meteorological screens (mesoclimatic observations), but some may be recorded at considerable heights in the atmosphere by radiosonde techniques.

Similarly a wide variety of methods are employed for measuring *in situ* data in water bodies. Oceanographic observations may include measurements of sea temperature, salinity, wave motion (height and wavelength), and biological content as well as meteorological conditions. These measurements are made from a variety of platforms including weather ships, coastal protection vessels, automatic data collection buoys and coastguard stations.

In estuaries and rivers additional factors such as suspended sediment load, biological oxygen demand, water reaction and pollution are often recorded. These observations may be made from boats or by automatic samplers used from the banks.

In this chapter we shall draw examples from ground and water surface studies but the general problems of data collection such as sampling and selection of sites for observations are relevant to other types of surfaces also. The data transmission methods are always essentially the same in each case.

It should be noted that atmospheric, hydrologic and oceanographic studies commonly depend on data from extant stations or station networks. Special efforts to collect *in situ* data in these contexts are often unnecessary.

6.2 Selection of ground truth sites

The location of areas for ground truth collection in support of particular remote sensing 'campaigns' may be decided on the basis of a number of criteria. These include study objectives, sample size satisfactory for statistical purposes, repeatability and continuity of the experimental study, access to the study area, availability of existing ground truth data for the area, personnel, equipment resources and the orbit characteristics of the space platform.

The shape of the ground data collection area is dependent on statistical requirements and speed of access. The allocation of sample plots in a data collection area is made more statistically efficient if the area can be stratified into relatively homogeneous areas. For stratification to be useful, strata boundaries should separate areas where within class variance is less than between class variance. The number of survey plots can then be estimated by the method of proportional allocation. Thus, the shapes of homogeneous areas may influence the shapes of ground survey areas. In addition, the sampling technique used in the ground observations (grid, area, line) may influence the shape of the area, for example line sampling along existing road networks may be preferred to block area sampling.

The size of the ground data collection area will be affected by study objectives, statistical considerations, scale of the ground phenomena, angle of view of the sensors and time factors. Where sensor testing is the objective, it is desirable to select the smallest ground truth area which allows detailed ground monitoring by equipment and personnel over a wide range of ground/atmospheric conditions. Many small sites may be necessary if it is desired to check on the validity of a spectral signature for a particular surface condition. An example of such a site is that used by one of the authors at Long Ashton near Bristol (Plate 6.1). The total size of the site is 120 hectares within which smaller areas of approximately 2 hectares were repeatedly monitored.

Where relationships are sought between the sensor response and particular surface conditions (e.g. a particular crop such as grass) it is necessary to obtain sufficient samples of the crop (generally more than 30) to allow statistical tests to be carried out. The size of area which will provide sufficient samples must be determined.

Time constraints affect the size of ground data collection areas through (a) quantity of data required, (b) resources available for collection, and (c) the rate of change in ground environmental conditions. For example, where evaporation rates are high it may be necessary to monitor changes in soil moisture content frequently (several times a day) whereas in conditions of low evaporation loss the soil moisture change may be slow and fewer observations are required in a given time.

The prime objective of ground data collection is to provide a contemporaneous record of ground conditions at the time of imagery. In practice it is difficult to obtain synchronous data for more than a small area or selected sample sites. The aim, however, is to obtain sample ground truth data within a short time of the acquisition sensor data. In planning ground data collection, special attention should be given to the rate of change of the variables to be observed. These variables can be categorized as transient or non-transient. Data recording of transient features (e.g. crop stage, leaf cover, windspeed, surface moisture) must be near synchronous. Recording of non-transient features (e.g. slope, aspect, soil texture) can be carried out prior to, or after, the sensing mission.

As an example of rate change in a transient surface condition one may note that data for soil temperature variation on slopes of an Exmoor Valley (Fig. 6.1) show that frequent observations are necessary on south facing slopes, but the rate of change on north facing slopes is much less.

The nature of the ground data required varies according to the type of investigation being made. Geological surveys often demand that rock and soil samples be taken for analysis and description (see Plate 6.2). Hydrological studies require a range of information including stream gauging, suspended sediment contents, water-table measurements and local climatic data. In studies of soil conditions the ground truth data normally include records of soil phase (or soil series), soil moisture, soil temperature, soil texture, structure, stoniness, organic matter content, soil colour and bulk density.

Some data can only be obtained a short time before the remote sensing mission takes place because the phenomena under investigation are continually changing. For example, in crop studies it is necessary to observe transient features such as the type, stage, height and colour of crop, disease types, weed species, effects of husbandry (ploughed, harrowed, drilled, rolled, wheeling marks), the grazing method, number of livestock and per cent crop cover.

Other types of data are more permanent and

Plate 6.1 Ground truth site, Long Ashton. Note the soil moisture measuring equipment in foreground, including neutron moisture probe and tensiometers. (Photo: L. F. Curtis.)

features such as the morphology of the terrain can normally be recorded by field survey and analysis of contour maps before the sensing missions take place. The morphology of the ground over which the mission is carried out is an important element in respect of data interpretation. Gradient, slope form and aspect often have significant effects on sensor data and this is particularly the case where radar data is being used. Terrain classification and evaluation techniques are now well developed and they can be used for recording site morphology at various levels of detail.

In the following sections selected aspects of *in situ* data collection for crop and sea monitoring are reviewed. It must be emphasized, however, that each combination of sensors and user applications requires careful consideration before the mission takes place. Frequently, the nature of the data to be collected will be affected by the geographic location of the area of investigation.

6.3 Ground truth for crop studies

The range of ground truth data required varies

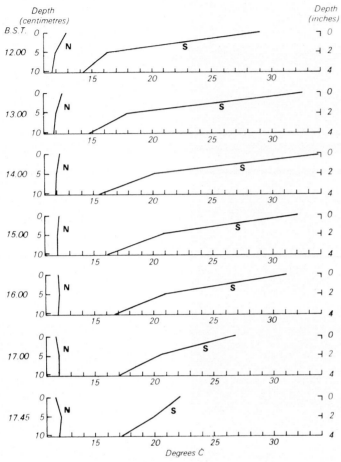

N - *North facing site (Exmoor complex)*

S - *South facing site (Exmoor complex)*

Fig. 6.1 Contrasts in soil temperatures on north and south facing slopes on Exmoor, England. (Source: Curtis, L. F., 1971.)

Plate 6.2 Field examination of rock and soil samples in relation to air survey. Adequate labelling and field description are essential elements of camp work. (Courtesy, Hunting Technical Services Ltd., Elstree.)

according to the farming region and its soil, relief and climatic conditions. Furthermore, the cost of acquisition of remote sensing imagery often determines that it be used for more than one purpose, e.g. in the agricultural context for crop recognition, soil drainage mapping and land quality evaluation. Generally, four categories of data are required for multipurpose land use studies in rural areas as follows:

1. Site morphology.
2. Crop/vegetation cover characteristics.
3. Cultivation/husbandry features.
4. Soil surface conditions.

The range of data to be collected should also be related to the organizational structure and personnel resources. A summary table (Table 6.1) gives manpower requirements for ground data collection in respect of remote sensing studies in three English counties. The prime objective of ground data collection is to provide a contemporaneous record of ground conditions at the time of imagery. In practice it is rarely possible to obtain detailed synchronous agricultural data for more than a small area or selected sample sites. In planning ground data collection, special attention should be given to the rate of change of the variables to be observed. These variables can be categorized as transient or nontransient. Data recording of transient features (e.g. crop stage, leaf cover) must be nearly synchronous. For example, data for spring barley in Nottinghamshire showed that mean per cent leaf cover increased from 18 to 40 in a period of 8–10 days in the first half of May. These changes are of sufficient magnitude to necessitate repetition of ground data collection since the proportion of bare soil exposed beneath a growing crop will have a major effect on image response. Quantitative observations of soil exposure using quadrat sample methods are time consuming. A single observer measuring crop cover using a 50×50 cm, 100 point quadrat for 500 observations per field could cover approximately 25 fields per day. The most successful method of

Table 6.1(a) Assessments of manpower requirements for ground data collection. (Source: Curtis and Hooper, 1974)

Area observed (km²)	Total fields	Total observers	Date	Sampling method	Sampling density	Prior training	Progress (km h⁻¹)	Fields per hour per man
635	933	10 (Working in pairs)	June/July	Line traverse (road)	1 field in 4 along traverse	Agricultural officers familiar with area	8	22
70	341	1	May	Line traverse (road)	Continuous – all fields on traverse	Agricultural officers familiar with area	4.4	15
70	341	1	August	Line traverse (road)	Continuous – all fields on traverse	Agricultural officers familiar with area	3.8	14

(b) Sample study of the time allocation in ground truth data collection

Task	Percentage of total time	
	May (45 h)	August (46 h)
Ancillary data collection – soil samples and farming operations	12	5
Data collection and recognition on traverses	48	52
Data checking and office compilation of data interpretation of imagery	40	43

DATE . . . . LAND USE * . . . . FIELD REF . . . .

CROP CONDITIONS   *SOIL CONDITIONS*

Stage   * [   ]   % Soil exposed   [   ]

Height: average   [   ]   Surface   general   [   ]
         range   colour: pattern   *   [   ]
         pattern   * [   ]   extent   *   [   ]
         extent   comment   [   ]
         comment

Colour: general   * [   ]   Roughness: furrowed   [   ]
         pattern   * [   ]   normal tilth
         extent   * [   ]   cloddy
         comment   panned

Disease: type   [   ]   Surface   abundance   [   ]
         extent   * [   ]   stones: %
         comment   size
                                                    type

Weeds: species   [   ]   Surface   *   [   ]
         density   * [   ]   moisture
         comment

Husbandry   *Site*
                                          Morphology:
         Ploughed   [   ]
         Harrowed   Gradient   *   [   ]
         Drilled   Slope type   *   [   ]
         Rolled
         Wheelings   Microrelief   [   ]
                                            type
         Grazing   [   ]   extent   *
           method   Field   *   [   ]
         Livestock   Boundary

                          * Codes available (e.g. extent
                             1: < 5%; 2: 5-50%; 3: > 50%)

                    General Comments

[   ]

Fig. 6.2 Sample data collection form for crop and land use studies. (Source: Curtis and Hooper, 1974.)

estimating leaf cover in a cereal crop such as barley may be to establish a relationship between crop stage and leaf cover for a sample of fields by quadrat measurements. Data for crop stage/leaf cover relationships are shown in Fig. 6.3, leaf cover in each field having been determined by 500 point observations.

Where large quantities of ground data are to be collected and handled it is necessary to develop a system of computer storage. In these circumstances coding of data in a computer compatible form becomes desirable. Such coding is relatively straightforward where a limited range of data are to be recorded and where class intervals are known or

can be predicted. In our experience, however, it is difficult to provide unambiguous codes which will cover all, or even most, land use conditions of possible significance to imagery evaluation. There is a real risk that different ground surveyors will code the same conditions differently. Attempts to devise comprehensive coding systems can, therefore, lead to complex systems which impair speed and efficiency in field surveys. Thus, where coding is employed, experimentation is necessary to test the coding system and to train field staff.

6.4 Sea truth observations

Sea truth observations may be obtained from a wide variety of sea-borne platforms. These may include weather ships, oil rig platforms, data collection buoys and specially commissioned ships. As in the case of crop studies, the monitoring of sea conditions must be carried out at the time of the remote sensing mission. One of the chief problems facing the mission planner is that of coordinating and positioning ships and buoys so that the transient state of the sea can be most accurately assessed. Frequently, the most satisfactory method is to combine continuous sampling along a ships traverse with discrete samples at selected points along the traverse.

A good example of this approach is provided by the successful collection of data for the EURASEP programme organized by the Joint Research Centre, Ispra, in 1977. In this investigation sea truth was required for analysis of the performance of the Ocean Colour Scanner (OCS). A number of ships were used to obtain coastal data off Belgium, France and Holland. In a ship's traverse, continuous samples were taken of chlorophyll *a*, turbidity, salinity, and temperatures from the top 10–15 cm of water. At particular positions discrete samples were taken from the surface and at depths of 1 m and 5 m in a vertical profile. These samples provided a 5 l quantity sample which was then subdivided while still being agitated into subsamples of 2 l each for chlorophyll and sediment analysis, 250 ml for plankton analysis and 100 ml for yellow substance analysis (see Fig. 6.4). A standard format for sea station identification and sample numbering is essential in such work. Also positioning must be accurate and this is often achieved by a DECCA

Quadrat samples 6th 8th & 16th May 1973

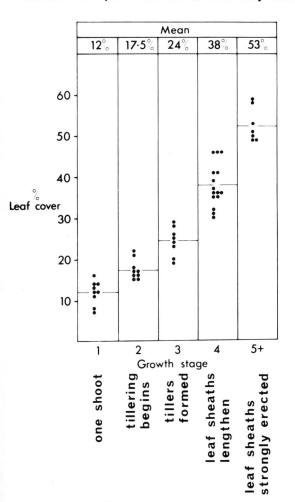

Fig. 6.3 Relationship between growth stage and leaf cover in Spring Barley. (Source: Curtis and Hooper, 1974.)

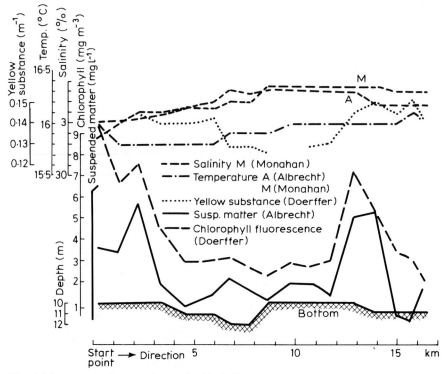

Fig. 6.4 An example of traverse data obtained in the Eurasep Programme on Track Z, 29 June 1977, off the coast of Belgium. Eurasep (European Association of Scientists in Environmental Pollution) was jointly programmed by the EEC (General Directorate XII) and the Joint Research Centre, Ispra. (Source: Sorensen, 1979.)

navigation system. A traverse speed of 5 knots has been found to be acceptable in such sea truth observations.

6.5 Data collection systems for transmissions to satellites

In modern satellite surveys a data collection system (DCS) provides the capability to collect, transmit, and disseminate data from Earth based sensors. Normally such a system involves data collection platforms, satellite relay equipment, gound receiving site equipment and a ground data handling system (Fig. 6.5). The data collection platform (DCP) collects, encodes and transmits ground sensor data to the space platform (e.g. Landsat observatory). The general characteristics of the data collection platform are shown in Fig. 6.6 (a) and (b). Such a platform will accept analog, serial-digital, or parallel-digital input data as well as combinations of those. Eight analog bits of 64 bits of digital input can be accepted. The format of a DCP message prior to decoding consists of 95 bits.

In the case of the Landsat system the spacecraft acts as a simple relay unit. It receives, translates the frequency and then retransmits the burst messages from the DCPs. No onboard recording, processing or decoding of the data is performed in the Landsat system. In the design of the Spacelab, however, it is anticipated that some on board processing of DCP data from Earth sensors may be carried out.

When the DCP data is retransmitted from the spacecraft it is put on a subcarrier of the Unified S-band (USB) which allows for narrow band telemetry to three primary receiving sites (Fig. 6.6 (c)).

The role of *in situ* data collection platforms can be well illustrated by reference to the Meteosat programme (Fig. 6.7). In addition to the transmission of WEFAX (Weather Facsimile) compatible data the satellites can interrogate observational platforms on ships, buoys, hydrological stations and ground stations. An Automatic Collection and Telemetry (ACT) series has been developed for operation with the Worldwide Meteorological Satellite network (Meteosat, Goes). The ACT unit samples data from

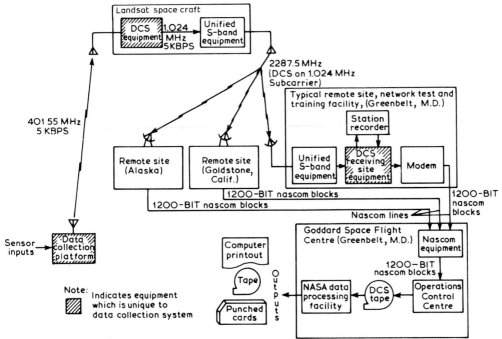

Fig. 6.5 Data collection system for the Landsat Satellite. (Source: NASA.)

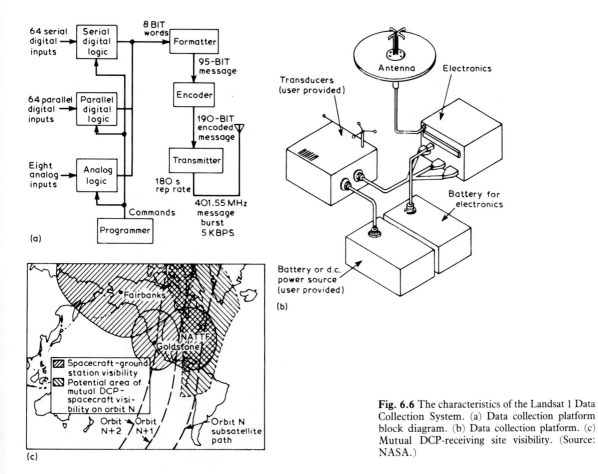

Fig. 6.6 The characteristics of the Landsat 1 Data Collection System. (a) Data collection platform block diagram. (b) Data collection platform. (c) Mutual DCP-receiving site visibility. (Source: NASA.)

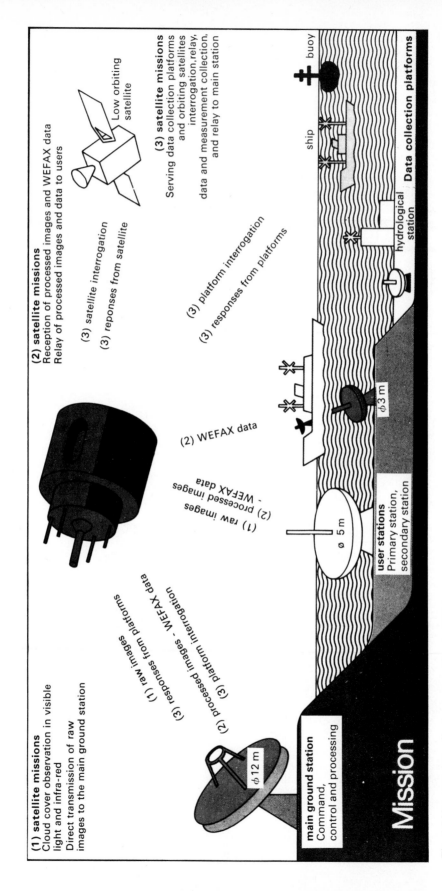

Fig. 6.7 Meteostat data collecting system. (Source: ESA.)

the sensors installed on the ground/sea truth station at preset times, checks data limits, and correctly formats the data, which it then stores (up to 5 K bits) until the allotted transmission time. If the data is outside the preset limits (e.g. above a dangerous flood level) the unit initiates an 'alert' transmission, giving an immediate warning of dangerous conditions. The specification for the standard Meteosat version of the data collection platform provides for operation on three channels (402.0 to 402.2 MHz) and a data storage capacity of 5192 bits.

Data collection platforms have also been used very successfully as aids in transatlantic yacht races. For example, five yachts participating in the transatlantic race for two-man crews held in 1979 carried Argos meteorological keypads. These keypads were used for twice daily transmission at synoptic time 0 and 12 hours of data for the following: atmospheric pressure, water temperature, wind strength and direction, and wave height and direction. The remainder of the yachts in the race were equipped with Argos system position tracking transmitters which also carried data from a sea surface temperature probe. Between 26 May and 15 August, 1979, 13 400 location calculations were made using data gathered from 22 300 passes. The winning yacht was named 'VSD' and its position was determined 213 times over 34 days using data from just one satellite (Tiros-N). The track of the winning yacht is shown in Fig. 6.8 but it may be even more important to note that accurate location of four yachts in distress and their resultant rescue was one of the results of the Argos facility.

It will be evident that a monitoring system using surface based sensors installed at a number of truth sites on ice, sea and land could be very effective if linked by satellite transmissions. Such ground based systems would provide not only calibration data for the overhead space and airborne sensors but would also allow overhead sensors to be directed to points of interest. Many of the future sensing systems on satellites will be capable of being steered to observe targets of opportunity by means of ground control. One may envisage that in the future thermal sensors will be activated and steered to monitor forest fires, radar sensors will be directed to monitor floods and camera systems will capture records of crop damage. The control of these overhead sensors will depend upon information from ground truth sites. In many respects one can expect the ground truth stations of the future to provide the datum points around which the space and airborne sensors will draw the contours of the environmental conditions at particular moments of time.

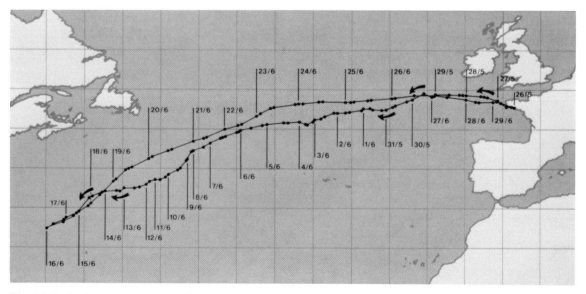

Fig. 6.8 Track of the winning yacht 'VSD' as recorded by the Argos system. (Source: ESA.)

7 Remote sensing data forms: their enhancement and interchangeability

7.1 General considerations

It is essential that automated techniques should be developed for the analysis of the vast amount of remote sensing data now being produced. Nevertheless the reader will appreciate that the simplest techniques are usually the most cost-effective.

Before embarking on automated analysis the user should remember that, given training and education, an experienced and skilled interpreter can extract much valuable information from data presented in image form. The eye-brain system of the human analyst is often the most efficient and flexible means of making practical use of remote sensing data. Frequently, it is quite sufficient to analyse the image qualitatively in terms of point, line or area features, and to interpret the results in terms of known phenomena or classes of phenomena (see Chapter 17).

Where measurements and map-making of an accurate kind are needed it is necessary to apply photogrammetric techniques. Here mechanical and optical devices are normally employed. Whichever objective is in view, it is usually necessary to 'pre-process' or enhance the initial raw data before photo-interpretation or photogrammetry can be carried out.

Most remote sensing data we see in the media or on display in our departments and laboratories, are in 'hardcopy' form. Hardcopy pictures fall into two categories, 'photographs' and 'images'. *Photographs* are obtained directly by cameras and the photographic process, the whole scene within the field of view of the camera being registered instantaneously on a film base which is then developed to give a photographic print or transparency. *Images* are recorded more indirectly, usually by scanning devices which provide variable

electrical ('analog') signals. These may then be converted into a picture, often built up line by line, on paper or screen, which can then be photographed to record the composite image more permanently.

It should be understood that the data first observed and recorded by remote sensing apparatus, whether it be a latent photographic image or a sensor signal, cannot be used for analysis and interpretation until it has been transformed into a different, more permanent, format. The initial record made by a sensor may be in one of the following forms:

(a) Analog data, continuously variable voltages recorded on magnetic tape representing, for example, the variation of radiation from a target traversed by a scanning radiometer line by line.
(b) Digital data, representing observations from small areas of the target, and recorded on magnetic or punched paper tape.
(c) Latent photographic images.

In order to proceed with further work the data usually have to be converted into other formats, including:

(a) Photographic transparencies or positive prints.
(b) Computer-compatible digital magnetic tapes.
(c) Computer print-outs, plots, diagrams.
(d) Analog magnetic tapes.

The three basic data formats, image, analog and digital are *all completely interchangeable*. However, it is possible to identify three main tasks at the beginning of the data handling process, which are fundamental to the provision of data to the environmental scientist who wishes to use its contents for research or operations. These involve:
(a) Processing of photographic material.
(b) Analog-to-digital conversion of signals recorded

on magnetic tape and production of computer-compatible tape (CCT).

(c) Digitization of images and, conversely, production of images stored in digital form on magnetic tape.

Let us now proceed to examine in greater detail the key considerations related to such tasks.

7.2 Photographic processing

Photographic processing can consist of either black and white technology or colour technology. In black and white processing attention has to be given to a number of variables which affect the quality of the product:

(a) The *characteristic curve* shows density as a function of log exposure. Density is a measure of the degree of blackening of the exposed film, plate or paper after development and it can be seen in Fig. 7.1(a) that density is nearly proportional to the log of exposure in the central part of the S-shaped curve. The *proportionality factor* as measured by the ratio a/b is referred to as the film *gamma(γ)*. Where density measurements are to be made on a number of films it is necessary to process each film to the same gamma.

(b) *Spectral sensitivity* of the film describes the sensitivity of the film in a given region of the spectrum. Panchromatic films, as their name suggests, are sensitive to all parts of the visible spectrum up to wavelengths of 700 nm and must be developed in complete darkness whereas certain duplicating films are sensitive only up to 500 nm and can be developed in red light.

(c) The *modulation transfer function* (MTF) is the ratio of intensity variations in the image to those occurring in the original. In other words it is an expression of the resolving power of the film and the achievement of maximum resolution depends on the chemistry and duration of the photographic process.

(d) The *dimensional stability* of the film used in the photographic process is important in order to limit distortion in the image. Also some information concerning the *granularity* of the film is desirable since it affects the amount of detail that is registered.

Stability improves with thickness of the film but in space applications films must be as thin as possible in order to conserve weight (see p. 46). It is also necessary to control the film environment if film stability is to be maintained – especially the temperature and humidity. Polyester base material is usually superior to other materials in dimensional stability. For example its expansion coefficient is normally 0.001–0.01 per cent per °F whereas cellulose triacetate is less good by a factor of 2–3 times. The *granularity* of a film becomes important because it represents unevenness in the emulsion which creates 'noise' in the photograph image. A microdensitometer trace on a uniformly exposed and processed film can be used to detect irregularities or discontinuities and thus assess its granularity. Root mean square graininess can be determined from the standard deviation of the density measurements and from the diameter of the

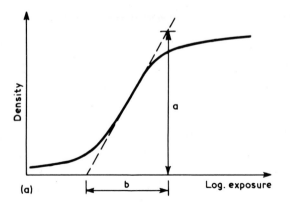

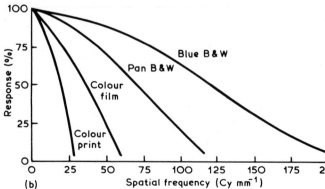

Fig. 7.1 (a) Shape of typical characteristic curve for photographic film. (b) Film transfer function curves.

scanning aperture:

$$G = K\sigma_K(D) \qquad (7.1)$$

where G = Rms graininess
K = diameter of the scanning aperture used
$\sigma_K(D)$ = standard deviation of density measurements.

In processing photographic film it is usually necessary to control the density of the product. This is done by a method which removes silver from the negative and is termed *reduction*. Ammonia and potassium persulphates are reducers which are commonly used but their action is liable to be uneven and hard to control. Other methods include 'dodging' which may be carried out manually by inserting tissue paper or a mask to decrease the density range. Automatic dodging printers are available and are reliable and efficient. The *printing papers* used in the photographic process can accommodate a wide range of density ranges and chloride, chlorobromide and bromide papers are the major types used.

In colour technology the chief areas of interest are in dyeing and reversal processes, resolution, colorimetry, colour balance and reproducibility.

The colour films used are primarily natural colour and infrared colour. The use of aerial colour only became significant in the 1960s following the breakthrough achieved by film manufacturers who succeeded in producing a range of colour emulsions with speed and resolution characteristics comparable to panchromatic films. Aerial colour films are available as either reversal colour films or negative colour films. Generally, the reversal films are used for medium to high altitude photography and negative colour films for work at low altitudes.

It has been appreciated for some time that the visual contrast presented by colour to the human eye is much greater than that of panchromatic photography. The eye will distinguish about 200 gradations on a neutral or grey scale, whereas it is capable of differentiating some 20000 different spectral hues and chromas (intensities of colour). Therefore colour photography immediately offers the opportunity of more detailed study of surface objects.

Infrared colour is a false-colour reversal film. It differs from ordinary colour film in that the three sensitized layers are sensitive to green, red and infrared radiation instead of having the usual blue, green and red sensitivities. The green sensitive layer is developed to a yellow positive image, the red sensitive to a magenta image and the infrared to a cyan positive image; thus the colours are false for most natural objects.

False colour film is highly sensitive to the green–red wavelengths of light as well as the near infrared. As a result it has particular characteristics such as that water and wet surfaces image in blue and blue–grey tones. An important feature of this type of film is that it records healthy green vegetation in various shades of red in the positive images (see Tables 4.1 and 4.2).

Colour films are generally characterized by lower spatial resolution and higher contrast than black and white (Fig. 7.1(b)). Colour balance is particularly important in treatment of false colour films because unlike in true colour photography the user has no reference available for comparison. The main aim of colour balancing is to ensure that brightness differences occurring in the target area are reproduced faithfully in the film image without affecting hues and colour response. In order to achieve this adjustments of the exposure for each layer are required. Sequential printers are available which allow independent layer corrections. Problems sometimes arise, however, due to variations in the characteristics of stored false colour film. These can be largely overcome by storing the film at $-20°C$ and making some densitometric tests before processing.

The photographic equipment necessary for photographic processing includes enlargers, contact printers, film and paper print processors, mixers, drier/cutters, microfilm processors and light tables.

7.3 Photographic enhancement techniques

If the analyst is to recognize information in the image, it must be reproduced in a way that permits visual identification. Unfortunately, in some environmental scenes important information is characterized by very low spectral reflectance levels, or by low inherent contrast between the subject and its background, or both. There are also atmospheric effects which further complicate the situation. Every photo-interpreter is familiar with the feeling that

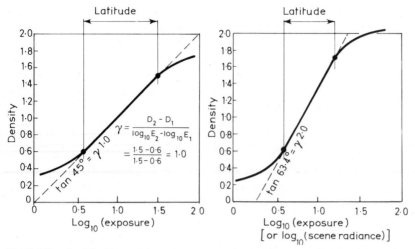

Fig. 7.2 Density (D), Gamma (γ), and Exposure (E) relationships. Two films of different gammas (contrasts) are compared. At γ 1.0, scene brightnesses ($\log_{10} E$) on the straight part of the response curve produce densities at the same contrast ratios as viewed by the sensor; i.e. $\Delta \log_{10} E = \Delta D$. When γ 2.0 film is used the apparent contrast is amplified by a factor of 2: i.e. $\log_{10} E = 2(\Delta D)$. Note that increased gamma leads to reduction in exposure latitude. (Source: Ross, 1976.)

more information is buried in the image than he is able to see, or interpret with confidence and accuracy. As a result various enhancement techniques have been employed to emphasize the tonal differences between objects.

The enhancement techniques most commonly employed consist of contrast stretching, photographic masking and density slicing. Increasing or manipulating the contrast of data is a powerful means of raising such data to levels where it can be recognized by visual interpretation. Gamma (γ) and contrast are often used as interchangeable terms for describing an image characteristic but gamma is a measurable value whereas assessment of 'contrast' is highly subjective and varies among observers. Nevertheless, in photographic image enhancement processes, films of different gammas from 1.0 to over 6.0 may be employed to achieve higher contrast. For example, in Fig. 7.2 two films of different gammas are compared where the apparent contrast of the scene is amplified by a factor of 2.

In the case of photographic masking a positive and negative image of the scene are registered together and a new image, the mask, is printed from the combination. The mask may be either positive or negative, and will add to, or subtract from, the density range of the image being masked. For example, where the data used consists of multispectral photography or scanning images three frames might be selected from a nine frame set because they show the most marked differences. These frames N1, N6, N7 can then be treated as shown below (Table 7.1).

The positives and negatives are printed subtractively (superimposed) in registration on to pre-punched film to form intermediates. Each intermediate is made from superimposition of one positive and one negative each from a different band. Each intermediate is then printed additively

Table 7.1 Enhancement processing

Camera negatives	Film positives	Duplicate negatives	Intermediates	Additive printing filters	Integral colour print
N1	→ P1	→ N1'	1_1 (N1' + P6) → G	→ Colour
N6	→ P6		1_2 (N7' + P1) → R	→ derivative
N7	→ P7	→ N7'	1_3 (N1' + P7) → B	→

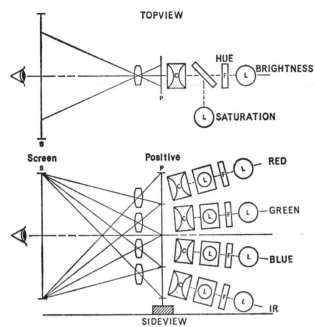

Fig. 7.3 Additive colour viewing. The positive transparencies (P) are illuminated by light from lamps (L) passing through filters (F), inner saturation lamps controlling colour saturation and condensing lenses (C). They are then viewed on the screen (S). (Source: Curtis, 1973.)

with a filter so that the image of each one is transferred to only one of the colour layers in the colour film (i.e. I_1 = magenta layer; I_2 = yellow layer; I_3 = cyan layer). The resulting images when developed from a single tri-pack film provide the colour derivative transparency for viewing. The intermediates represent differences of reflectance between bands. Where no differences occur, the negatives and positives cancel out each other, so that no colour shows on the final enhancement. It has been found that the relative brightness and colour differences of objects appearing in the colour derivatives can be used to detect such changes as moisture content, soil density and vegetative conditions (see Colour Plate 9.2).

From the above descripton it will be evident that the making of colour enhancement prints is complex and time-consuming. It is, therefore, somewhat expensive and it is not always easy to ascertain the best selection of negatives which will give the maximum amount of information at the interpretation phase. In these circumstances, it is an advantage to have the capability of experimenting with different combinations of the negative frames by means of a visual display. In order to achieve this a method is required for projecting spectral positive transparencies, one superimposed upon the other in accurate registration while illuminating each with a

different coloured light. If the transparencies are illuminated by the primary colours, a full colour reproduction of the scene is produced. It is also possible to make reconditioned false colour (infrared colour) and artificial derivatives of various kinds by suitable combinations of transparencies and colour addition. Additive colour viewers (Fig. 7.3) are important aids to photo-interpretation. The brightness and saturation controls of each spectral transparency permit alterations of the final composite screen presentations. In this way subtle differences in the scene can be detected which would not be readily distinguished by examination of normal panchromatic or colour prints.

Photographic density slicing converts the analog, continuous-tone image into one displaying a series of steps, each representing a separate, different increment of density, completely isolated from lesser or greater densities. It is also called 'equidensitometry' or 'isodensity contouring'. Agfa-Gevaert makes Contour Film, a special emulsion for this purpose. Very precise and narrow density slices can be made with conventional lithographic films and developers. For example, lithographic films such as Kodak Ortho Film 2556 can reach very high gammas when processed as recommended and can prove to be ideal for density slicing.

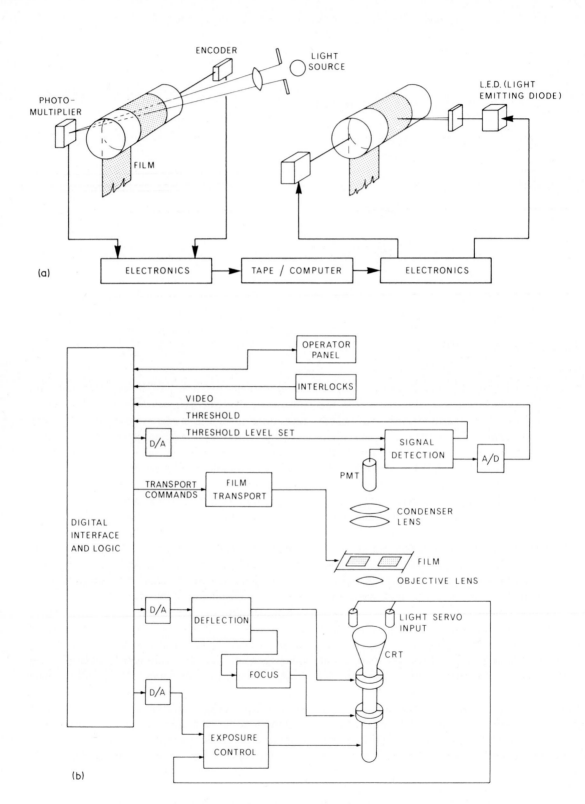

Fig. 7.4 Image recording systems. (a) Drum scanner principles. (b) Flying spot scanner schematic diagram.

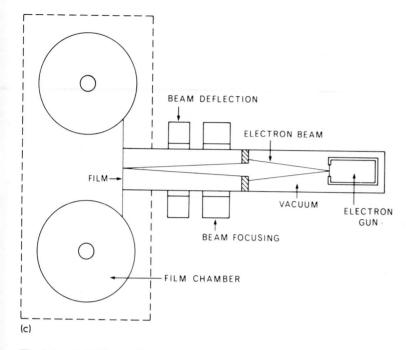

FILM→

BEAM DEFLECTION

ELECTRON BEAM

VACUUM

ELECTRON GUN

BEAM FOCUSING

FILM CHAMBER

(c)

Fig. 7.4 cont. (c) Electron beam recorder.

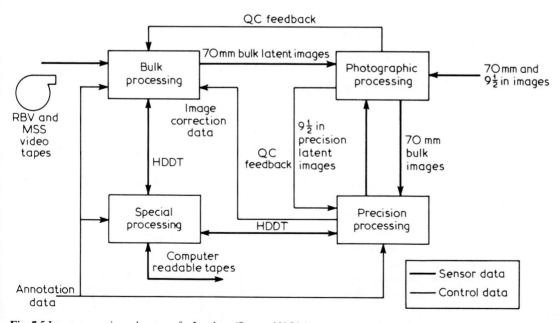

QC feedback

RBV and MSS video tapes

Bulk processing

70 mm bulk latent images

Photographic processing

70 mm and 9½ in images

Image correction data

HDDT

QC feedback

9½ in precision latent images

70 mm bulk images

Special processing

HDDT

Precision processing

Computer readable tapes

Annotation data

Sensor data

Control data

Fig. 7.5 Image processing subsystems for Landsat. (Source: NASA.)

7.4 Conversion of optical and digital data

A fundamental requirement in Earth sensing studies is the ability to convert images between the optical and digital domains. Therefore there are 'converters' which are optical/digital (O/D) or digital/optical (D/O). The task to be fulfilled is *either* the production of computer or analog data from an image *or* the forming of an image from computer or analog data.

7.4.1 Optical to digital conversion

One of the most common types of instrument for the conversion of film data into analog, and thence computer, data is the microdensitometer. This instrument allows the film to be traversed by a beam of transmitted light. The light source is accurately controlled so that the difference between the light beam received and the light beam emitted by the film can be measured by means of a photomultiplier. The spot size of the transmitted light can be varied, so the amount of detail in the original film can be sampled according to the nature of the study to be undertaken. For example, in order to examine variations in the density of reflectance from a field containing different crops the spot size might be 50 μm. On the other hand scanning an image for highly contrasting and coarse detail such as rivers may be carried out with a raster (sample frame) size of 200 μm.

Some microdensitometers use a flat bed system in which the film is laid flat on a moving table mechanism and light is transmitted vertically through the image. However, the most common type of densitometer in use in remote sensing studies is the drum microdensitometer. Drum type machines are usually faster than flat bed machines and are high resolution microdensitometers where the film is fastened on to a rotary drum and scanned by a light source when in rotary motion. A typical high speed digital microdensitometer which has been widely used is the System P-1000 Photoscan, which provides the following scanning times for a 12.5×12.5 cm film (Table 7.2). A schematic diagram showing the basic characteristics of the drum scanning microdensitometer is shown in Fig. 7.4(a).

Another method of converting film data into

Table 7.2 Scanning times for the P–1000 Photoscan

Drum speed (rev s^{-1})	Raster (μm)	Data rate (Hz)	Scantime (min)
2	50	14.4	20
4	100	14.4	5
8	200	14.4	1.25

digital data is by scanning the film by means of flying spot CRT (Cathode Ray Tube) and vidicon systems (see Fig. 7.4(b)). This machine offers high speed of conversion, easy interaction with computers, possibility for data reduction by selective access to images for spectral separation, and for filtering and image improvement techniques. Flying spot scanners can read 64 levels on a grey scale and record 60 levels. They require only 20 μs for reading and 27 μs or less for recording. Spot sizes of 0.03–0.05 mm can be selected for scanning.

Traditional light sources are normally used for these scanning machines. However, exceptionally fast machines have been developed using laser beams. Laser scanners consist essentially of three components, namely an optical system, a film transport system and a rotating scanner system. The laser provides a high power, collimated beam of monochromatic light which scans the film image. Typical specifications of commercially available equipment are given in Table 7.3.

Another form of converter is the electron beam recorder (EBR). The basic EBR consists of a high resolution electron gun, an electron optical system for controlling the electron beam, a film transportation mechanism, an automatic vacuum system and regulators and electronic circuits which operate the recorder. The electron gun provides an electron spot (3–10 μm diameter) which is focused by coils on the sides of the vacuum tube. A schematic diagram of an electron beam recorder is given in Fig. 7.4(c).

7.4.2 Digital to optical conversion

With the development of multispectral scanning devices it is now necessary to have a means whereby digital (D) data can be converted into analog (A) data. Such a system enables photographic imagery to be generated from digital data.

Table 7.3 Laser scanner characteristics (RCA, Model LR-70 and LR-71). (Source: ESRO CR-295, 1973)

Model	Parameter	Performance
LR-70	Resolution (at 50% MTF)	20 000 pixels per scan
	Video bandwidth	DC to 75 MHz
	Geometric fidelity	1 part in 20 000
	Grey scale	$16\,(2)^{\frac{1}{2}}$
	Film format	5 in $\times$ 1500 ft
	Scan rate	To 5000 scan s^{-1}
	Film velocity	Variable from 0.25 in s^{-1}
	5 in frame scan time	4 s
LR-71	Resolution (at 50% MTF)	100 lp mm^{-1}
	Video bandwidth	DC to 30 MHz
	Geometric fidelity	1 part in 20 000
	Grey scale	$16\,(2)^{\frac{1}{2}}$
	Film format	5 in $\times$ 1500 ft
	Scan rate	Variable to 7500 scans s^{-1}
	Film velocity	Selected ranges between 0.15 to 40 in s^{-1}

Table 7.4 Data processing facility requirements for Landsat Return Beam Vidicon observations. (Source: NASA)

RBV input (3 bands)	Scenes per week (%)	Scenes per week	For each scene	Items per week
1316 scenes/wk 3948 images/wk	100% Bulk	1316	3 B–W Masters	3 948
			30 +	39 480
			30 −	39 480
			30 + Prints	39 480
	20% Bulk colour	263	1 C −	263
			10 C Prints	2 630
	5% Precision	66	3 B–W Masters	198
			30 −	1 980
			30 +	1 980
			30 + Prints	1 980
			1 C −	66
			10 C Prints	660
			Ability to digitize	—
	1% digitized	13.2	3 Copies computer readable	$\approx$ 158 tapes

Key:
B–W	Black and White
C	Colour
+	Positive transparency
−	Negative transparency
+ Prints	Positive paper prints
C Prints	Colour positive paper prints

Generally D-A conversion is achieved by reading digital data in the computer from computer compatible tape, transforming it into a form acceptable for digital-analog hardware equipment, and then passing the transformed data into the D-A device. Since analog tape recorders are not able to stop and start rapidly (unlike digital recorders) it is desirable to maintain a continuous flow of data to the analog hardware. This is often achieved by setting up an input and output queue. The system is arranged so that there is always a backlog of input data waiting for processing and also a reserve (buffer) amount of data in the output to the analog hardware. This enables corrections to the input data to be made without interruption of flow to the analog device.

In the case of the Landsat data production the facilities must be capable of handling very large quantities of items. The scale of the operation can be assessed by considering production requirements for RBV (Table 7.4) and MSS sensors (Table 7.5). The format of the image processing subsystems for Landsat is shown in Fig. 7.5.

7.4.3 Conversion steps for Landsat data

The Landsat data is obtained in analog form as MSS and RBV electrical signals. This is then converted into latent images for bulk photographic processing. Also the analog data is treated so that precision images and colour images can be made. The system also provides for the creation of computer compatible tapes (see Fig. 7.5).

The digitizing procedure is accomplished by an A-D converter in a number of steps. The area to be digitized (resolution element) is selected by 'gate' settings which can be set up manually on the A-D converter so that only certain regions are digitized. When the 'gated' section is read by the analog tape recorder the data value of each channel is carried into a system which shifts right the appropriate number of places dictated by a data resolution switch which incorporates the number of significant bits used for each data value.

When the computer word is completely filled a transfer takes place whereby the computer stores the word from the A-D converter in sequential

Table 7.5 Data processing facility requirements for Landsat multispectral scanner system observations. (Source: NASA)

RBV input (4 bands)	Scenes per week (%)	Scenes per week	For each scene	Items per week
1316 scenes/wk 5264 images/wk	100% Bulk	1316	4 B–W Masters	5 264
			40 +	52 640
			40 −	52 640
			40 + Prints	52 640
	20% Bulk colour	263	2 C −	526
			20 C Prints	5 260
	5% Precision	66	4 B–W Masters	264
			40 +	2 640
			40 −	2 640
			40 + Prints	2 640
			2 C −	132
			20 C Prints	1 320
			Ability to digitize	—
	5% Computer readable	66	3 Copies computer readable	≈ 713 tapes

Key: B–W   Black and White
      C   Colour
      +   Positive transparency
      −   Negative transparency
    + Prints   Positive paper prints
    C Prints   Colour positive paper prints

addresses in core. After the final scan line has been digitized and transferred to the computer a signal is sent to the computer that the block of data is complete. The computer then places the digitized information onto magnetic tape in the form of one scan line per digital record. Often there is overlap between scan lines made by the sensor, i.e. the same part of the ground region is sampled twice. This results in a superfluous amount of analog data (oversampling has occurred). In these circumstances a 'smoothing' or 'filtering' programme can be incorporated in the digital computer. This programme normally has two objectives. First, to reduce the number of scan lines by an integer factor which will give true 'aspect' to the ground sampling. Second, to reduce the instrumental 'noise' present in the recording process. This can be accomplished by averaging of data over a number of lines.

The advances in micro-electronics are such that converters have become progressively faster and more efficient. One of the main problems now to be overcome is that of developing less bulky items for the storage of data. In the next decade the traditional computer tape is likely to be supplanted by other, less space demanding, memory banks.

8 Manual data analysis and interpretation

8.1 The role of manual data analysis and interpretation

Although an increasing amount of satellite data is being treated by semi-automatic computer analysis it is still the case that most operational uses of remote sensing data involve manual procedures. Such analysis often entails traditional image inspection and interpretation. This is carried out by environmental scientists with various objectives and different types of background disciplines.

Likewise, most of the topographic maps now made by national and commercial agencies are produced by manual techniques of photogrammetry. These techniques are frequently supplemented by mechanical and computer aids.

In the following sections we shall deal with these relatively traditional remote sensing practices of photo-interpretation and photogrammetry. Attention will be focused on interpretation and photogrammetric evaluation of aerial photographs. However, the reader will appreciate that other types of images such as satellite pictures can be treated similarly.

Once the objectives of the interpretation have been selected the completeness and accuracy of the results depend on an interpreter's ability to integrate the elements of the image either consciously or subconsciously. It is for this reason that many would argue that image interpretation is as much an art as a science. Certainly interpretation of aerial photographs is a deductive process which proceeds in stages. The first stage consists of a general examination which enables the interpreter to take account of the general patterns of relief, vegetation and cultural development. The second stage involves concentration on specific areas or features and requires a systematic approach towards identifi-cation. The final stage of evaluation normally involves some consideration of classification of the phenomena in the scene.

8.2 The methods of photo-interpretation

It has been suggested that it is the *level of reference* which governs the performance in photo-interpretation. The level of reference can be regarded as the amount of knowledge which is stored in the mind of any person or group of persons interpreting photographs. Three levels of reference can be distinguished – general, local and specific. The general level is the interpreter's general knowledge of the phenomena and processes to be interpreted. Extra knowledge is added by the interpreter's intimacy with the relevant local environment. Finally, there is the specific level. At this level the interpreter employs highly specialized knowledge involving a much deeper understanding of the processes and phenomena he wishes to interpret.

One can identify two basic tasks in photo-interpretation as follows:

1. A process of establishing the identities of the objects and elements detected in the image, and
2. A process of searching for their meanings.

The first process uses photo-image characteristics such as shape, size, tone, shadow, pattern, texture, situation and resolution to identify the objects. The second process employs more sophisticated types of analysis and deduction to discover meaningful relationships. It is the second process which seeks to give order to the qualitative information obtained and thereby provide some form of classification. Classification in itself demands some form of theorization. As a result the skill of the interpreter is

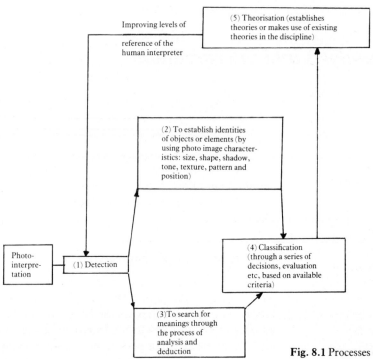

Fig. 8.1 Processes of photo-interpretation. (Source: Lo, 1976.)

raised and a feed-back of skill raises the level of reference of the human interpreter (Fig. 8.1).

Although we all 'interpret' photographs reproduced on television or in newspapers, special training is required for image or photo-interpretation. This is partly because of the unfamiliar viewpoint of the imagery and partly because of the special types of information which are usually demanded as the end products of their analysis. Nine elements of photographic interpretation (related to the image characteristics mentioned above) are regarded as being of general significance, largely irrespective of the precise nature of the imagery and the features it portrays. Selected examples are given in the list that follows. Most of these elements apply to imagery obtained by other sensors also, especially those of the multispectral scanner type:

(a) Shape. Numerous components of the environment can be identified with reasonable certainty merely by their shapes or forms. This is true of both natural features (e.g. geologic structures) and man-made objects (e.g. different types of industrial plant).

(b) Size. In many cases the lengths, breadths, heights, areas, and/or volumes of imaged objects are significant, whether these are surface features (e.g. different tree species) or atmospheric phenomena (e.g. cumulus vs. cumulonimbus clouds). The approximate scale of many objects can be judged by comparisons with familiar features (e.g. roads) in the scene.

(c) Tone. We have seen how different objects emit or reflect different wavelengths and intensities of radiant energy. Such differences may be recorded as variations of picture tone, colour, or density. They permit the discrimination of many spatial variables, for example on land (different crop types) or at sea (water bodies of contrasting depths or temperatures). The terms light, medium and dark are used to describe variations in tone.

(d) Shadow. Hidden profiles may be revealed in silhouette (e.g. the shapes of buildings or the forms of field boundaries). Shadows are especially useful in geomorphological studies where micro-relief features may be easier to detect under conditions of low-angled solar illumination than when the sun is high in the sky. Unfortunately, deep shadows in areas of complex detail may obscure significant features, for example the volume and distribution of traffic in a city street.

(e) Pattern. Repetitive arrangements of both natural and cultural features are quite common, which is fortunate because much photo-interpretation is aimed at the mapping and analysis of relatively complex features, rather than the more basic units of which they may be comprised. Such features include agricultural complexes (e.g. farms and orchards), and terrain features (e.g. alluvial river valleys and coastal plains).

(f) Texture. This is an important photographic characteristic closely associated with tone in the sense that it is a quality which permits two areas of the same overall tone to be differentiated on the basis of microtonal patterns. Common photographic textures include smooth, rippled, mottled, lineated and irregular. Unfortunately, texture analysis tends to be rather subjective since different interpreters may use the same terms in slightly different ways. Texture is rarely the only criterion of identification or correlation employed in interpretational procedures. More often it is invoked as the basis for a subdivision of categories already established using more fundamental criteria. For example two rock units may have the same tone, but different textures.

(g) Site. At an advanced stage in a photo-interpretation procedure the location of objects with respect to terrain features or other objects may be helpful in refining the identification and classification of certain picture contents. For example, some tree species are found more commonly in one topographic situation than in others, whilst in industrial areas the association of several clustered, identifiable structures may help us determine the precise nature of the local enterprise. For example, the combination of one or two tall chimneys, a large central building, conveyors, cooling towers, and solid fuel piles point to a correct identification of an installation as a thermal power station.

(h) Resolution. More than most other picture characteristics, resolution depends upon aspects of the remote sensing system itself, including its nature, design and performance, as well as the ambient conditions during the sensing programme, and subsequent processing of the acquired data. Resolution always limits the size and therefore in many cases, the nature, of features which might be recognized. Some objects will always be too small to be resolved (e.g. fair weather cumulus clouds on the average low-orbiting weather satellite photograph), while others lack sharpness or clarity of outline (e.g. the exact position of a shoreline is often difficult to deduce from an air-photo of average scale, say 1:10000).

(i) Stereoscopic appearance. When the same feature is photographed from two different positions with overlap between successive images, an apparently solid model of the feature can be seen under a stereoscope (Fig. 8.2). Such a model is termed a stereo model and the three dimensional view it provides can aid interpretation. This valuable information cannot be obtained from a single print and is usually less easily obtained from scanner images. Further discussion of stereoscopic images can be found on p. 123.

In practice these nine elements assume a variety of ranks of importance. Consequently, the order in which they may be examined varies from one type of imagery to another, and from one type of study to another. Sometimes they can lead to assessments of conditions not directly visible in the images, in addition to the identification of features or conditions which are explicitly revealed. The process by which related invisible conditions are established by inference is termed 'convergence of evidence'. It is useful, for example, in assessing social class and/or income group occupying a particular neighbourhood to note features such as detached or tenement types of building forms, proportion of open space, trees, size of gardens, road patterns, traffic and proximity to industrial plants or railroad facilities (see Chapter 17). Likewise in estimating soil moisture conditions in agricultural areas the tonal or colour registration of vegetation, species identification, existence of field drains, riverine features, saline encrustations, and land use provide contributory evidence for interpretation of the soil moisture regime (see Chapter 13).

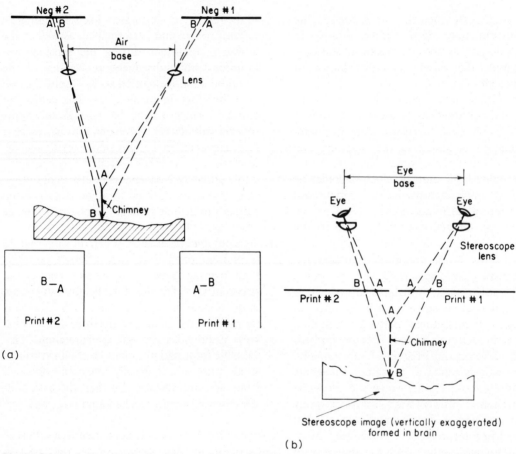

Fig. 8.2 Schematic diagrams showing (a) the appearance of a vertical object, e.g. a chimney on both negative and positive. Note the lean of the chimney top on the prints due to height displacement. (b) The stereo image formed in the brain by viewing the prints through a stereoscope.

Photo-interpretation may be *regional* or *spatial* in its approach and objectives, as in the case of terrain evaluation or land classification. Fig. 8.3 is a sample photo-interpretation key for land use and vegetation mapping. On the other hand, photo-interpretation may be highly *site-specific*, and related to very precise goals. In the field of transport studies, for example, features such as car parks, railway yards or shipping berths are the subject of special study.

It is important to bear in mind that the degree of accuracy achieved in photo-interpretation may vary considerably depending on the nature of the subject, the type of photography and the skill of the interpreter. Consideration must be given, therefore, to the question of how much ground checking will be required to produce a result comparable in objective accuracy to a ground survey.

Whereas a map offers classification and symbolization of the features it records the photograph does not. This lack of symbolization makes a photograph more difficult to interpret but it also makes it potentially more flexible and informative for those prepared to set up their own classifications on the basis of the image characteristics.

8.3 Measurement and plotting techniques: selected aspects of photogrammetry

Various aspects of plotting from scanner data and scanner images are discussed elsewhere in relation to particular types of image data (e.g. radar, infrared linescan). This section will, therefore, concentrate on measurement and plotting from aerial photographs. The science of photogrammetry deals with the techniques involved and a detailed consideration of photogrammetric techniques cannot be given in the space of this volume. The

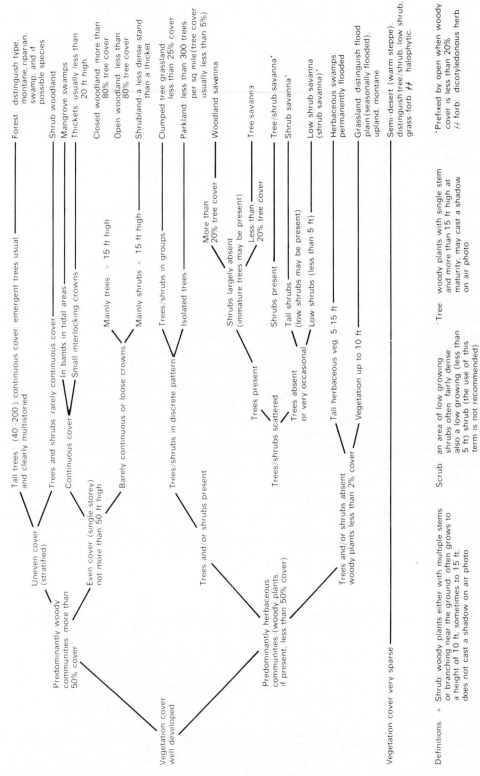

Fig. 8.3 A land use/vegetation key for mapping from aerial photographs in north-east Nigeria. (Source: Alford *et al.*, 1974.)

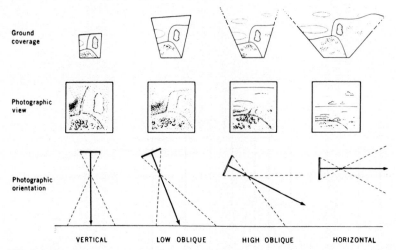

Fig. 8.4 The characteristics of vertical and oblique photographs taken from the air. The high oblique view is similar to that obtained from high ground or a tall building. The horizontal view is like that obtained from a slight eminence. (Source: Spurr, 1960.)

reader is, therefore, directed towards further reading (see end of book); only selected points of background knowledge are presented here.

Aerial photographs may be either *vertical* or *oblique* according to whether the camera axis is vertical or not. A 'high oblique' includes the visible horizon whereas a 'low oblique' does not (Fig. 8.4). Oblique photography is most often obtained by single cameras but it is sometimes obtained by multiple camera arrangements of the trimetrogon type. In this a single, vertically orientated, camera is flanked on either side by oblique viewing cameras. These systems are cheaper than one employing a single vertical camera, since this requires more flight lines to cover a given area. However the lateral distortions of scale are greater in obliques and less easy to correct.

It is important to recognize that all air photographs are perspective pictures and various types of planimetric inaccuracy are contained within them. These inaccuracies chiefly arise from scale distortions due to height differences in the objects viewed and to tilt distortions (Fig. 8.4). In addition the planimetric accuracy of objects is affected by the displacement of the image due to height variation (Fig. 8.2).

In order to overcome these problems of scale distortion and height displacement it is necessary to

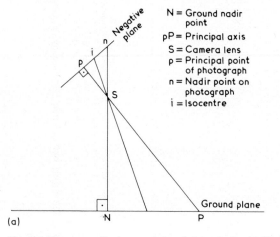

N = Ground nadir point
pP = Principal axis
S = Camera lens
p = Principal point of photograph
n = Nadir point on photograph
i = Isocentre

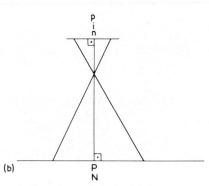

Fig. 8.5 The geometric characteristics of photographs. (a) Tilted photograph to show principal point, isocentre and nadir point. (b) Vertical case when principal point, isocentre and nadir point coincide.

use overlapping photography capable of providing a stereoscopic image. Before considering the techniques employed attention must be focused on the principal geometric characteristics of air photographs which are fundamental to the photogrammetric processes. The geometrical properties of the principal point, isocentre and nadir point are particularly important (Fig. 8.5). Briefly stated they are as follows:

(a) A photograph is angle true with respect to its isocentre for all points in the plane of the isocentre.
(b) Straight lines drawn through the photographic nadir point pass through features which would lie on a surveyed straight line on the ground.
(c) When a photograph is vertical the isocentre and nadir point are coincident with the principal point which then assumes their properties.

In these circumstances angles measured from the principal point are as truly representative of angles between features as if they had been measured with a theodolite on the ground. Straight lines measured from the principal point also cross detail which would lie on a surveyed straight line on the ground.

These properties are only completely true when the camera axis is perfectly vertical. However, providing the tilt of the camera does not exceed 2° and the height variation is small in relation to the flying height (less than 10%) the properties can be held to apply.

Thus it is possible to use the principal points of a number of photos as though they were plane table stations. Rays can be drawn from the principal point on each photograph through points of detail. Intersections can then be made by tracing off the rays from each photograph as they are successively placed under a transparent overlay along a base line. This method is termed the radial line plot and it provides a manual method of plotting the true positions of objects from air-photos.

Certain mechanical aids are used in this work by commercial firms. The rays are scribed on transparent film templets and then slots are cut to correspond to the rays. The technique is termed the 'Slotted Templet' method. The rays can also be scribed mechanically by the Radial Line Plotter apparatus.

Any method requiring measurement of the height of objects from air photographs is also dependent on stereoscopic cover between successive images. For this reason most vertical photographs are taken overlapping along the flight line (vertical line overlap). In this system features are photographed from successive camera positions (S_1, S_2, Fig. 8.6). Surface features (e.g. A, B, Fig. 8.6) will occupy different positions in relation to the principal points P_1, P_2 according to the heights of the features above a given datum. Thus the features A, B are imaged at the same photo point a, b on S_1 but at different positions on S_2 (a', b'). If we add the distances P_1a and P_2a' on the two positives this gives us a measure of the parallax of A. Similarly if we add the distances P_1b and P_2b' we can obtain a measure of the parallax of B. The difference between these two parallax measurements arises from the height difference

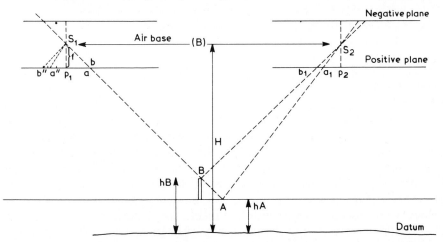

Fig. 8.6 Diagram to show the parallax of points A and B on two photographs taken at two positions S_1 and S_2.

between A and B, therefore differences in parallax can be used to calculate height differences from air-photos.

It is perfectly feasible to measure parallax on photographic prints with a ruler but generally speaking the differences are small so that a more accurate method is demanded. Therefore the Parallax Bar (or stereometer) has been developed which allows small differences in spacing to be measured with a micrometer gauge.

For contouring work it is necessary to employ optical–mechanical devices which will allow an operator to view the stereomodel (Fig. 8.2) together with movable markers which can be adjusted in turn to positions A or B. When viewed stereoscopically such markers appear as a single image placed directly on the feature in the stereomodel. If the markers are moved closer together (i.e. parallax increased) the 'fused' mark will appear to float above the feature in the stereomodel. If they are separated (i.e. parallax reduced) the marks normally fail to be held in fusion by the brain and they separate on the model.

A skilled operator can manipulate a 'floating mark' of this kind so that it rises and falls on the model, just resting on the surface all the time. By doing so an automatic trace can be made which shows the true plan position of the features traversed. Alternatively the operator can place the floating mark at a known height and then move it across the stereomodel following the contours of the surface. By this means a contour map of the stereo image can be made.

Various types of 'floating' mark equipment are used in photogrammetry. Some instruments depend on the principle of anaglyph projection (i.e. one image red, the other blue, viewed through spectacles with one red and one blue lens). An example of such equipment is the multiplex projector. In this system the floating mark normally consists of a spot of light in the centre of a white disc on a movable tracing table.

Larger, and more expensive, manually operated instruments are used for precision work. Many of these have 'space rods' incorporated into the design which accurately reconstruct the angular relationships within the stereo model by mechanical rather than optical means (Fig. 8.7). In this equipment the operator manipulates handwheels and foot controls

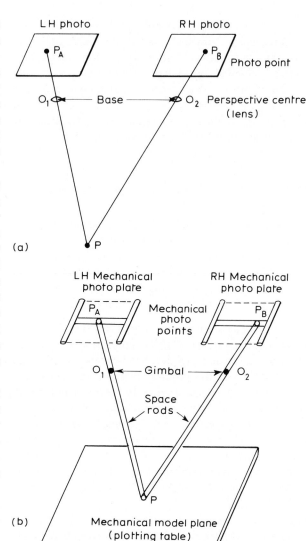

Fig. 8.7 Geometric and mechanical relationships in a stereo model and stereo plotter. (a) Geometric relationship between a point on the photograph and the corresponding point on the ground. (b) Mechanical representation of the image–ground relationship. (Source: Lo, 1976.)

in order to move and adjust the floating mark on the stereomodel. As he does so the scribing pencil draws on the table beside the stereoplotter (Plate 8.1).

Very high accuracies can be achieved by high order plotting machines. The limitations in plotting accuracy are set by the nature of the stereomodel produced by the photography and by the ground control available for setting the model into its correct spatial orientation. Thus it should be borne in mind that accurate photogrammetric work also

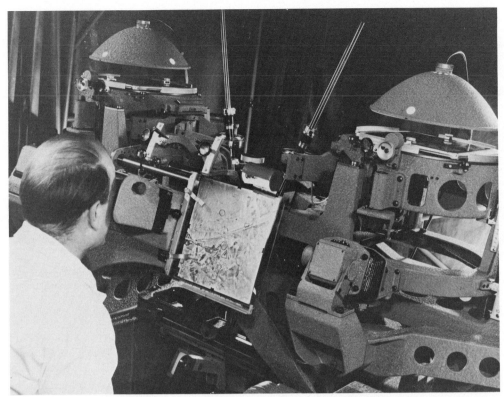

Plate 8.1 A Wild A.8 Stereo Plotter in which diapositives are viewed stereoscopically. The diapositives are mounted beneath the illumination lamps on either side of the apparatus. Note the 'space rods' projecting above the centre of the plotter. (Courtesy, Clyde Surveys Ltd., Maidenhead, Berks.)

requires accurate ground control. This is normally achieved by establishing the heights of selected well defined features on the photographs. The operator can then orientate the stereomodel in relation to a true datum and heights or positions are then maintained in true relationships within the model.

These manual techniques of photo-interpretation and photogrammetry are widely employed by commercial air survey companies and many of our existing maps have been constructed by these means (Plate 8.2).

Manual operation of a stereoplotter requires the employment of a highly skilled operator to control the work. As work loads increase there is a tendency to look for increasing automation in the photogrammetric process. Such automation is essentially a technical improvement on machine systems (Plate 8.2). The goal has been to achieve automation of the three phases of machine operation i.e. measurement, orientation and recording. There have been four main lines of development consisting of:

1. Photogrammetric digitizing of graphical outputs;
2. Provision of analytical or computer controlled stereoplotters;
3. Orthophotoprojection systems;
4. Automatic image correlation.

Although many semi-automatic systems are available in large commercial and national mapping agencies the individual user may find that the scale of his investigations does not warrant investment in such expensive equipment. The way forward may be that existing national and commercial facilities will be made available in a routine manner providing standard products for both large and small contracts.

It is important that the user should have a clear idea of the requirements for accuracy. In many cases the application of remote sensing data to environmental problems does not demand photogrammetric accuracy. In such cases the use of a device such as a Sketchmaster or a Transfer Scope (Plate

Plate 8.2 Wild A.5 Autograph plotters in operation with plotting tables seen in the foreground. A scribing pencil is automatically controlled as the operator manipulates the stereoplotter, placing the floating mark over the points of detail to be mapped. (Courtesy, Clyde Surveys Ltd., Maidenhead, Berks.)

Plate 8.3 Bausch and Lomb Transfer Scope. A graphical data transfer instrument which combines optical views of stereoscopic image and the base map. Used widely to transfer detail from photographs to maps. (Courtesy, Survey and General Instrument Co., Edenbridge, Kent.)

8.3) will be quite adequate providing suitable base maps are already available on to which the photo or image detail can be transferred.

When the basic procedures of photo-interpretation and photogrammetry were developing some sixty years ago, the amount and range of image data were limited. The explosive development of remote sensing techniques has necessitated the evolution of a wide range of preprocessing methods. These modern methods are designed to improve the precision of the data to be analysed and to allow surplus data to be discarded.

The new generations of satellites offer greater resolution capability, with corresponding increases in data flow. Thus data compression and filtering techniques are likely to become more and more important. In the following chapter numerical data preprocessing, processing and analysis are reviewed. It will be evident that in the processing stages large volumes of data can be manipulated or discarded in such a way that economic use of the data set can be achieved.

9 *Numerical data preprocessing, processing and analysis*

9.1 The need for numerical data manipulation

The analysis and interpretation of remote sensing data by manual means has much to recommend it. For example, the trained human eye can identify a very wide range of imaged features with both ease and accuracy.Consequently the interpretation of aerial photographs, space photographs, and other forms of remote sensing imagery is, perhaps, best carried out by hand if the data sets are small. But the volume of data generated by new remote sensing systems is growing rapidly, and already far exceeds the capability of trained interpreters in some environmental fields. It has been estimated that a single low-altitude Earth-orbiting satellite such as any of the Noaa (National Oceanic and Atmospheric Administration) weather satellites can yield something of the order of 10^{10}–10^{12} points of new data every day. Members of the Landsat family planned for the future may be capable of even greater outputs, yielding as much as 250 million data bits per second. Still further increases may be expected as bandwidth restrictions are eased. Computer-based interpretation techniques seem to afford the only possible solution to the problem of routine extraction of useful facts on a near-real time basis.

Another stimulus to the increasing use of computer analysis is the fact that numbers are required for environmental hypotheses to be tested rigorously. Although manual analyses can provide some such information and are the better suited to some types of tasks, data arrays of suitable magnitude for statistical processing can be compiled much more quickly by automatic analytical techniques. In particular, the eye is good at interpretation, but relatively poor at assessing basic image characteristics such as brightness and texture. So, the choice at present lies between range of features and accuracy of identification on the one hand and numbers of cases and speed of operation on the other. To a great extent the success of 'objective' or automated techniques of image analysis is dependent upon the type of question which is posed: increasingly, acceptable answers are being obtained in response to the posing of reasonable questions. For example, numerical methods are especially useful for the development of probabilistic statements about points, lines or areas. In feature recognition and/or classification, complete accuracy is not to be expected whether manual or objective methods of image analysis are deployed. In these contexts quantitative techniques may be preferred to qualitative methods in that they can be used to generate accuracy levels for the determinations which have been achieved.

9.2 General definitions

Mainly in the last two decades a proliferation of numerical methods has grown up for use in remote sensing, stimulated by the satellite-led information explosion and facilitated by the rapid rise of electronic computing. For present purposes it is useful to distinguish between three broad groups of numerical methods of widespread use in remote sensing. These groups are interrelated, or even overlap, and furthermore, the boundaries between them are variable and subject to change due to the differing and changing needs of different groups of users. However, the groups differ markedly with respect to the purposes to which they are put. Given the already vast and often highly technical literature which has grown up in this field, our concern here is to obtain a structural overview of the nature and use of numerical methods in remote sensing, to serve as a springboard both into the journal literature itself, and more immediately, into Part Two of this book

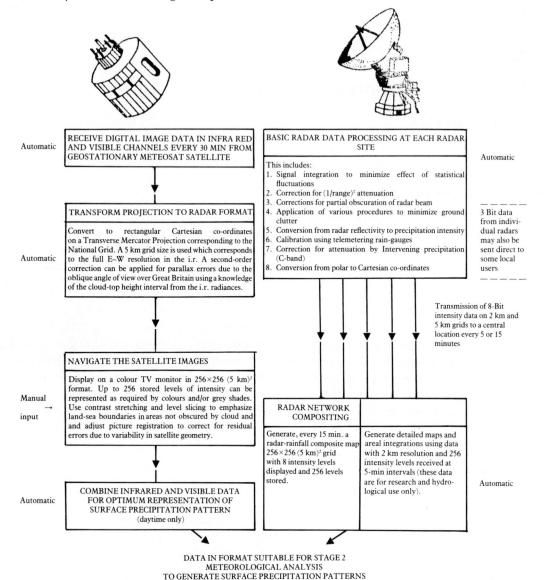

Fig. 9.1 Preprocessing of satellite and radar data for the 'FRONTIERS' plan to use radar and satellite imagery for very short-range precipitation forecasting (see also p. 328). (From Browning, 1979.)

where the applications of environmental remote sensing will be reviewed, exemplified and discussed.

For present purposes it is convenient to distinguish between three major groups of numerical data manipulations. We will consider them in the order in which they are usually invoked in connection with a given set of remotely sensed data.

9.2.1 Data preprocessing

This includes any process which has to take place before remote sensing data can be processed and analysed by the scientific user. Ideally, therefore, it should be undertaken before the environmental scientist acquires the data from a central reception facility or archive. It commonly involves such things as data navigation and registration, rectification, cleaning and primary calibration (see Fig. 9.1). It may be expected that such procedures will be carried out increasingly on satellite remote sensing platforms themselves, or at central reception and/or archiving facilities.

9.2.2 Data processing

This includes any data manipulation necessary or helpful for later scientific analysis and interpretation, but which the analyst or interpreter does not want to have to perform himself. Increasingly data processing is being undertaken or offered by central facilities, although in the simpler cases it is usually carried out by the user. Its chief object is to present remote sensing data in manageable forms and quantities. It often involves some element of selection from, and/or compression of the pre-processed data, and/or enhancement of selected features of particular significance to the customer.

9.2.3 Data analysis

This covers the assessment of remote sensing data implications through comparison with the nature and condition of the target, wherever possible assessed also by *in situ* observational information. Appropriate relationships, models and algorithms are often required for meaningful intercomparison of the two different types of data. Inter-relations ('training sets') observed in selected test areas for which both remote sensing and ground data are available are often used as the bases for dependent *supervised classifications* of remote sensing data from conventional data-sparse areas. In the absence of analyst-specified training data, *unsupervised classifications* are used instead. Here the search is for natural groupings inherent in the remote sensing data themselves, which may then be interpretable in terms of recognizable features or classes of features in the target area.

These three broad groups of numerical procedures support much modern data interpretation and application. Therefore it is necessary to consider the salient points in greater detail before we can go on to see how remote sensing is serving the community of environmental scientists today.

9.3 The quality and quantity of remote sensing data

Data from remote sensing missions are not always optimal in quality. Each remote sensing system has its own characteristic capability depending on such things as the instrument design, altitude of the plat-

form, and performance of the recording equipment. Other less predictable, often extraneous, influences may adversely affect the quality of the retrieved data, as exemplified by Figs. 9.2(a) and (b) and Table 9.1. Preprocessing functions are necessary to reduce or eliminate such inadequacies. Some such methods involve photographic film or deploy optical or electronic analog systems (see Chapter 7). Here we must concentrate our attention upon the digital approach.

9.3.1 Geometric corrections

Remote sensing data, commonly obtained either in snapshot or line-scan forms, come in a variety of scales and geometries (Fig. 9.3) Often these must be rectified before they can be compared in detail with existing data. The problem arises, for example, in studies designed to investigate the relative merits of imagery from aircraft or satellite platforms for the identification and mapping of features at a selected level of detail in a particular region. The problem is most intransigent where:

(a) Imaged areas are topographically rough.
(b) The images are obtained from systems with broad fields of view.
(c) The images are obtained from low-altitude platforms, where image displacement due to quite modest topographic features may be considerable.
(d) Systems other than framing cameras are used.

Rectification can be performed by optical and electronic means (of which the former is often the better and cheaper), or by digital techniques. Basically the problem involves bringing separate data into congruence, that is to say ensuring that point-to-point correspondences are achieved. These are necessary, within certain limits, if point decisions or point comparisons are intended using statistical pattern recognition techniques. The fast Fourier transform algorithm has been employed quite widely to provide the two-dimensional Fourier transform of an image, and then to facilitate rapid cross-correlation between pairs of images of data arrays.

Since the variations in the performance of an orbiting sensor system are proportionately less than those associated with lower-altitude systems, time-

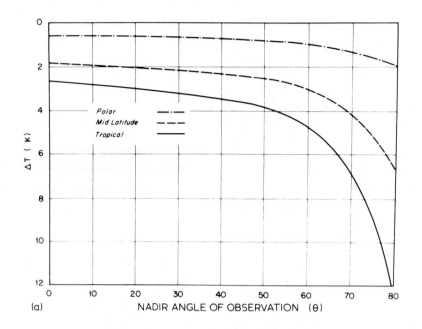

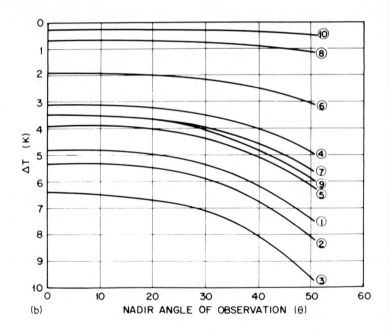

Fig. 9.2 (a) Estimated departures (ΔT) of Nimbus 2 HRIR equivalent blackbody temperatures as function of local zenith angle for three different air masses. Unfortunately this represents a gross over-generalization of the influence of the Earth's atmosphere on infrared transmissions toward space. (b) Calculated departures of Nimbus 4 THIR 11.5 μm channel temperatures from ground surface temperatures as a function of local zenith angle for ten different atmospheres (see Table 9.1). Unfortunately the structure of the atmosphere is complex and rapidly changing and even these more detailed correction functions leave residual errors in the satellite data after processing. (Source: Sabatini *et al.*, 1971.)

Table 9.1 Computations of atmospheric effects on temperatures derived from the THIR radiometer (11.5 μm channel) on the weather satellite, Nimbus 4. Though detailed, the table represents a broad generalization of a medium capable of large and rapid variations in temperature and moisture content

Atmosphere	Temperature surface (K)	Precipitable water (cm)	'Observed' equivalent black-body temperatures at nadir angles of:			
			0°	15°	30°	50°
(1) Standard	288.15	1.96	283.38	283.24	282.77	280.70
(2) Tropical	299.65	3.63	294.37	294.23	293.74	291.56
(3) Subtropical summer	301.15	3.94	294.77	294.60	294.02	291.45
(4) Subtropical winter	287.15	1.93	284.02	283.93	283.61	282.18
(5) Mid-latitude summer	294.15	2.64	290.29	290.19	289.79	287.95
(6) Mid-latitude winter	272.59	0.74	270.64	270.58	270.41	269.53
(7) Sub-arctic summer	287.15	1.84	283.59	283.49	283.13	281.55
(8) Sub-arctic winter-cold	257.28	0.36	256.58	256.56	256.48	256.13
(9) Arctic summer	278.15	1.65	274.51	274.39	274.00	272.24
(10) Arctic winter (mean)	249.22	0.18	248.95	248.95	248.91	248.74

differing images from satellites may be brought into acceptable congruence more easily and more cheaply per unit area than those from aircraft. Herein is one of the principal advantages of a satellite remote sensing system. Landsat is of sufficient significance today to merit special mention.

Although Landsat images are relatively free from panoramic distortions and relief displacements, there are image distortions of both *systematic* (or predictable) and *random* (or unpredictable) types for which compensations must be made before really accurate analyses can be made of these data. On the ground, each Landsat MSS pixel is basically trapezoidal (not square or rectangular) in form, as a result of the general viewing angle of the sensor system. Minor deviations from the norm stem from factors such as variations in the altitude, attitude and velocity of the satellite. Corrections for systematic distortions, like the skewed-parallelogram effect of Earth rotation beneath the satellite during

imaging, can be made relatively easily, in this case by offsetting each scan line by an appropriate amount.

Corrections for random distortions are more difficult, and are usually accomplished through reference to carefully-selected *ground control points*, such as highway intersections, small water bodies etc. For such points both image (column, row) and ground (latitude, longitude) coordinates are established, and the values then submitted to a least-squares regression analysis to determine coefficients for two 'transformation equations' which interrelate the two sets of coordinates. The full process by which the Landsat MSS geometric transformations are applied to the original data is termed 're-sampling'. It involves:

(a) The definition of a geometrically uniform 'output' matrix in terms of ground coordinates.
(b) The transformation by computer of the coordinates of each output cell into corresponding coordinates in the image data set.

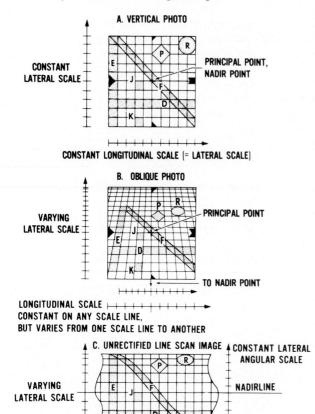

Fig. 9.3 To exemplify the differences in the scales and geometries of different types of remote sensing data. The letters reference points or areas to show effects of different geometric distortions. (Source: Taylor and Stingelin, 1969.)

(c) The transferral of each pixel value from the image data set to its appropriate place in the output matrix. The result should be a matrix of digital image data which is geometrically correct in respect of a set of ground coordinates.

9.3.2 Radiometric corrections

These are necessary to lessen the effects of a number of variations in radiation or image density. The aim is to achieve a product with a high level of invariance with respect to the radiance of the scene, whether the radiance is reflected or emitted energy. Common image defects are associated with:

(a) A decline in light intensity away from the centre of a photographic lens.
(b) The relation between the angle of view and the angle of solar radiation.

(c) Variations in illumination along a sensor flight path. These are especially significant in connection with polar-orbiting satellites.
(d) Variations in viewing angles of sensors scanning across the flight path of the sensor platform.
(e) Image variations arising from the performance of sensor systems.
(f) Variations in photo-processing and developing procedures.

Numerical methods of combatting such effects include several 'normalization' techniques. Most involve the assessment of the nature and degree of the effect caused by one or more of the sources of variation in recorded radiation density, and the development of appropriate physical remedies (e.g. filters or films) or numerical weighting functions (e.g. calibration curves and algorithms) to improve the data.

9.3.3 The reduction of noise

Noise may be defined as any unwanted (either random or periodic) fluctuation of a signal which may obscure it and make its analysis and interpretation more difficult. Random noise may be caused by the performance of remote sensing systems during recording, storing, transmission, and especially in ground reception of the data. Periodic noise may be caused by radio interference, and by certain components of the sensor/platform complexes themselves. It is possible to dampen random noise in some cases through the application of statistical smoothing functions to the data presented in numerical form, but such noise can rarely be eliminated. Rather, efforts are necessary to keep it within acceptable bounds. Periodic noise is easier to eliminate because its pattern in both one-dimensional and two dimensional products is more systematic. Fig. 9.4 illustrates the noise outline for the tape-recorder on

Noaa 2. The rotating mirror in the Noaa 2 Scanning Radiometer, and the satellite's tape recorder drive motor, were controlled by a common power source oscillator. Consequently the noise from this source associated with the tape recorder – inherent in the data transmitted to the ground – tended to be periodic. Cloud pictures contained regular 'streaky' patterns, which, if untreated, would have been passed on to secondary products. When the pattern of noise is known, appropriate modifications can be made to ground station hardware to lessen its effect.

In the case of Landsat 3 an unwanted image characteristic resulted from the improper functioning of one of the six identical detectors in the Multispectral Scanner: a noticeable banding appeared with a six-line interval. Several techniques were developed to reduce the impact of this 'striping' effect, which could not be fully removed by nominal radiometric corrections (Plate 9.1). The simplest method was to multiply the pixel values in the de-

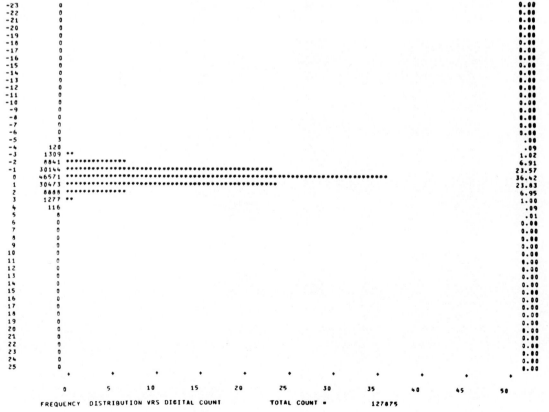

Fig. 9.4 Noise pattern, Noaa 2 tape recorder (absolute response) shown for a supposedly inert recording of night time data from the visual channel scan sensor. Such effects may create systematic streaking in display products, and produce systematic contamination in the generation of more quantitative applications. (Source: Conlan, 1973.)

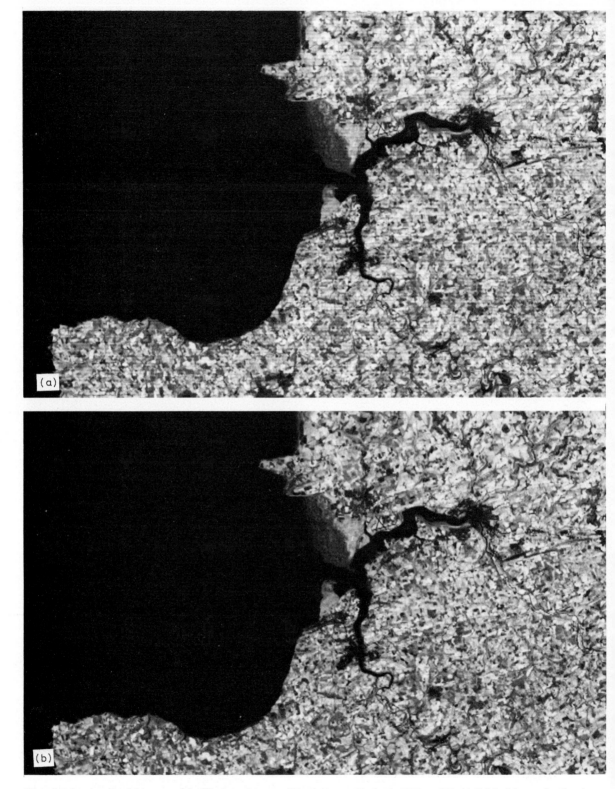

Plate 9.1 Landsat Band 7 images of the Woolacombe area of North Devon, England, 27 May 1977. (a) Original image showing the Landsat 3 striping effects. (b) Digitally destriped image. (Courtesy, Space Department, RAE, Farnborough; Crown Copyright Reserved.)

ficient lines by normalizing factors to give adjusted values with mean and standard deviations equal to those in the rest. However, all such methods yield improvements which are more apparent than real: there is no way in which fully appropriate adjustments can be made for fundamentally defective data.

9.3.4 Data calibration

Substantial sections of this volume are concerned with the analysis and interpretation of remote sensing data in terms of familiar characteristics of the human environments. Since our funds of environmental data from conventional sources involving *in situ* sensors are much greater than those of remote sensing data, having been built up over a much longer period of time, it is common practice to calibrate the remote sensing data in terms of conventional parameters. Whether or not this practice will continue after the imbalance in the two data types has been reduced, redressed or even reversed remains to be seen. Already some environmental patterns mapped from satellite altitudes have been expressed simply in terms of the measured radiances (e.g. stratospheric radiances in terms of voltages), with no attempt to calibrate the results in terms of commonly accepted environmental units. At present this approach is restricted mainly to situations in which conventional ('ground truth') data are few or unavailable.

Often prelaunch calibration procedures are carried out to provide the numerical voltage-to-temperature or voltage-to-brightness response relations for the sensor systems. For example, the calibration programme for operational weather satellites involves the preparation of families of calibration tables for each sensor for converting normalized raw counts into effective blackbody radiative temperature responses in the case of infrared sensors, or into calibrated brightness responses for the visual channel. These values are then corrected for any bias as a function of sensor temperature, a measurement regularly available in the telemetry data.

The final step involves those corrections necessary for the interpretation of measurements of the natural Earth scene. For example, allowances must be made for atmospheric attenuation in the case of infrared sensors. This 'limb darkening' correction is generally taken as a function of local zenith angle, with the assumption that water vapour is the primary absorbing constituent (see Table 9.1). For the visual channel on current Noaa satellites the algorithm in use at present is a simple cosine function of the solar zenith angle.

9.3.5 Data compression

The reduction of the raw data obtained from a remote sensing sytem into a volume no larger than that required for a particular programme of analysis and interpretation may be effected by one or more of a wide range of standard statistical techniques. These include smoothing and averaging processes, sampling techniques, and feature extraction methods. Smoothing or averaging processes are commonly applied by numerical means to data in either one-dimensional (e.g. line scan) or two-dimensional (data array) forms. Supplementary processing may be carried out by the computer to ensure that the assumptions of subsequent numerical processes might be met; for example, weighting factors may be applied to normalize the frequency distribution of the new, reduced population. If the original data are to be reduced by selection instead, the principles and practices of statistical sampling theory are invoked. The problems of feature extraction are not difficult to resolve providing that the features to be removed for further study are related to simple characteristics of the raw data themselves, for example, regions of radiation temperatures above a given threshold in infrared studies, or areas with selected ranges of brightness in densitometric analyses of conventional photographs. Further data-reduction processes may then be applied to the information selected for careful scrutiny. However, the problems of selective feature extraction are very difficult when combinations of data characteristics are invoked simultaneously. Some of the attendant difficulties of pattern analysis and pattern recognition are discussed in Section 9.4.

9.3.6 Image enhancement

Specific users of remote sensing data often require that the features of special interest to themselves be emphasized or enhanced at the expense of other, background, features. Such enhancement can be carried out photographically (see Chapter 7) or digitally. The commonest procedures include density slicing, edge enhancement (image sharpening), contrast stretching (increasing greyscale differences), and change detection by image addition, subtraction or averaging. The digital approach has the advantage of flexibility, the photographic approach has the advantage of cheapness (see Chapter 8). The digital approaches may be summarized as follows:

(a) Density slicing. Given image data presented in a digital form it is possible to simplify the information by reducing the number of classes through the selection of appropriate threshold levels. Sometimes a very simple two-class classification is adequate, for example, where a categoric yes/no type of decision is to be made. Numerous digital systems have been developed for rapid density slicing. In such systems any greyscale level or levels may be selected on a photographic or electrical imprint, with output via a line printer on a flatbed or drum plotter, or directly onto a photographic film writer. The procedure involves the development of suitable algorithms and computer programmes. 'Quantization noise', or spurious contouring, may result if the slices are not chosen carefully.

(b) Colour enhancement. Here preselected colours are accorded to chosen categories of contents in visible or infrared images. In this way, more obvious distinctions can be made between different features of each scene (Plate 9.2, see colour plates).

(c) Edge enhancement. This can be used to sharpen an image by restoring high-frequency components through the removal of scanline noise, or to emphasize certain edges to aid interpretation by taking the first derivative in a given direction. Unfortunately the computer cannot easily distinguish desired lineaments from others of similar greyscale change, and these will be enhanced as well. A common method of edge enhancement of Landsat MSS images involves the doubling of the deviation of each, or selected, pixels, from the local averages. These averages can be computed from 'pixel neighbourhoods' whose shapes and sizes can be varied to suit the

requirements of the data user. Plate 9.3 illustrates one edge-enhancement process and its results.

(d) Contrast stretching. Any image of low contrast or any portion of the greyscale of an image may be enhanced in digital processing through suitable adjustments to the array of picture points. Many satellite sensor systems (e.g. Landsat MSS) are designed to accommodate a wide range of scene illumination conditions. Consequently the pixel values in many scenes spread over relatively narrow portions of the full ranges which are possible. Contrast stretch enhancements expand the ranges of the original pixel values to increase the contrast between features with similar spectral responses, and thereby facilitate feature recognition and image interpretation. Fig. 9.5 illustrates some of the available contrast stretching processes. The *linear stretch* (Fig. 9.5(c)) expands uniformly the observed histogram of brightness levels in a selected scene to cover the full range of available values. The *histogram stretch* (Fig. 9.5(d)) is better still, for image values are assigned here on the basis of their frequency of occurrence: more display values (and hence more radiometric details) are assigned to the more frequently-observed levels in the histogram. For more specialized analysis and interpretation use may be made of a 'special stretch' (Fig. 9.5(e)) in which features represented by a narrow range of brightness levels are greatly enhanced through the expansion of this narrow range to occupy the full range of display values, to the exclusion of other, perhaps previously dominant, features of the scene.

(e) Spectral contrast stretching. In the previous category we considered procedures designed to accentuate intensity contrasts in individual waveband data sets. Sometimes it is valuable to accentuate the contrasts between different bands of image data. One common example is the *ratio image* which is generated by computing the ratio of digital values for two spectral bands for each pixel in an image. A useful characteristic of the products of spectral contrast stretching of this type is that they are relatively free of extraneous effects such as differential illumination (e.g. relief light and shadow) across a scene. However, it is often difficult to determine which

Plate 9.3 The Landsat 3 image shown in Plate 9.1, after destriping and edge enhancement. (Courtesy, Space Department, RAE, Farnborough; Crown Copyright Reserved.)

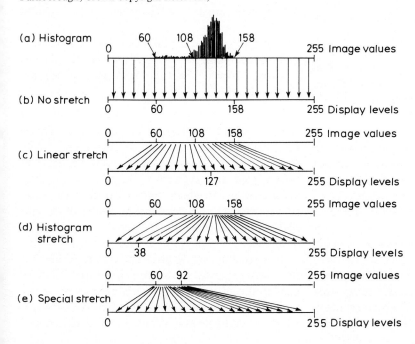

Fig. 9.5 Principle of contrast stretch enhancement. (From Lillesand and Kiefer, 1979.)

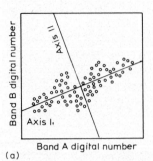
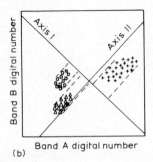

(a) Band A digital number (b) Band A digital number

Fig. 9.6 Rotated coordinate axes used in multispectral transformations. (a) Principal components analysis. (b) Canonical analysis. (From Lillesand and Kiefer, 1979.)

is the most appropriate choice of wavebands for a particular application. So, statistical techniques have been developed to transform pixel values on to alternative sets of measurement axes so that they may be described – and later reproduced in image form – in the most efficient manner possible for the task in hand. Such techniques are illustrated by Fig. 9.6, which relates to a simple, two-channel data set. In Fig. 9.6(a) a random sample of pixels from a selected image have been plotted according to their observed values. New axes (I and II) are introduced to provide a more efficient description of the plotted data. Such axes are characteristic of *principal component analyses* which can be undertaken for multidimensional space: each axis or component accounts for successively smaller portions of the observed variance in the data set. In Fig. 9.6(a), Axis I accounts for much more of the variance than Axis II. A principal-components enhancement can be generated by displaying the image in terms of the new pixel values, expressed in relation to Axes I and II, not A and B as before. Where more prior information is available for a region, training set pixel values (from sites of known feature type) can be used as the basis for *canonical analysis*. In this case new axes are positioned so that the discrimination of given feature types can be maximized. In Fig. 9.6(b) three feature types can be discounted on the evidence of Axis I alone.

(f) Change detection. Change detection is generally carried out by hand, using interactive man-machine methods, for example the flicker procedure whereby the first and second images are presented alternatively to the observer many times a second. Apart from simple techniques

like the detection of greyscale change from one photograph to another, digital procedures have not yet been developed with notable success in this interesting and important, but rather intransigent field.

9.4 The analysis and interpretation of selected features

9.4.1 Quantitative feature extraction

Quantitative feature extraction from remote sensing data involves four steps, namely:

1. The development of methods whereby significant features or variables may be isolated.
2. The preparation of appropriate quantitative data arrays.
3. The determination of the essence of the structure of the data.
4. The reduction of the information so that only the essential point in, or structure of, the data is carried forward into the final stage of decision classification.

The first step involves the logical or *syntactical* structuring of the design of the experiment. This is the planning stage, in which consideration is given to what will be measured and why; how the data will be sampled and aggregated; how many variables should be used; which methods of data retrieval, preprocessing, and data compaction will be employed; and how the final analytical stages will be organized.

The second step involves some form of line-by-line data scanning to quantify the variables to be used in the analysis. These variables may be uncorrected greyscale (from a photograph, colour waveband or multispectral channel), or a normalized

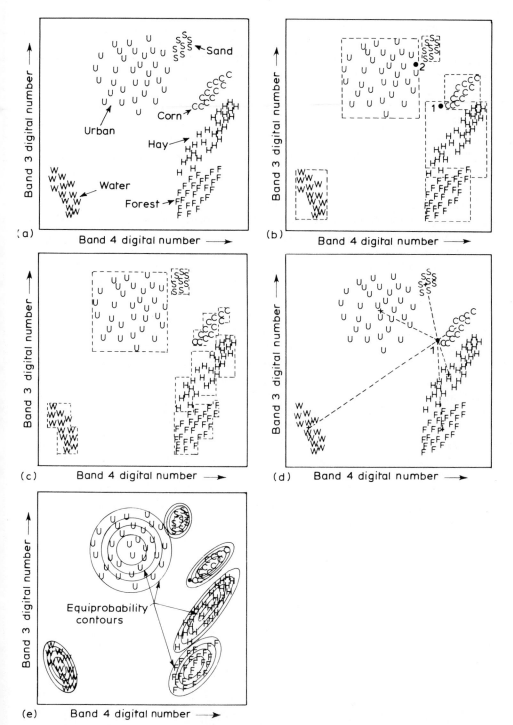

Fig. 9.7 Some common Landsat data classification strategies. (a) Pixel values from two wavebands are plotted on a bispectral scatter diagram. (b) A 'parallelepiped' classification where pixels are grouped on the basis of the highest and least class values in each band. (c) A 'parallelepiped' classification whose stepped boundaries represent each group of pixels more precisely. (d) A 'minimum distance to means' classifier where an unclassified pixel (1) is compared with established class means by computing distances between it and their centres. (e) An 'equiprobability contour' classifier to accomodate new data like (1) using maximum likelihood rules. (After Lillesand and Kiefer, 1979.)

variable. In most remote sensing applications the principal variables used in each channel will be tone, texture or height measurements. These relate to greyscale, greyscale organization and variability, and stereoscopic characteristics, respectively. Resolution on the ground, and resolution in density, or greyscale (the number of detectable shades of grey), may limit the subsequent stages. On other occasions limitations are imposed by the analyst himself, for example with the number of final categories required for successful completion of his task.

The third stage involves the search for key indications of the phenomena or relationships under scrutiny. Often it is sufficient to select a few aerial photographs from a large set, or the indications from some rather than all the multispectral channels available. Extraneous or highly cross-correlated data can be eliminated, reducing the measurements or variables to the minimum which is substantially orthogonal in feature space. Thus the fourth step of the extraction process is reached.

In many programmes of data analysis and interpretation attention is focussed either upon point features, or upon non-point features having length, or length and breadth. In order that image contents may be correctly interpreted and classified, it is common practice for 'training sets' of data to be compiled (see also p. 131). These are subsamples whose identification is completely known. Where data from two or more wavebands are available it is now commonplace to consider the relations between and amongst digital values (or 'vectors') in multi-dimensional space. Some approaches frequently adopted in Landsat MSS image analysis are illustrated by Fig. 9.7. Here, for the sake of simplicity, pixel values from only two bands are considered; similar processes can be applied readily to all four or five data bands.

In these and other ways classifications of data from known regions can be used as stereotypes of points, lines, or areas, by comparison with which unknown areas can be assessed. Thus new data may be interpreted through their resemblance to sets of templates or known 'fingerprints' in a 'supervised' approach. The statistical parameters of each training set are calculated by analog or digital computer and can involve many variables which may have a bearing on the characteristics of the key site, ground truth station, or ground survey area. These include such factors as topography, vegetation and soil content, farming practices and environmental management as well as time of day, season of the year, and prevalent weather conditions. In calculating the statistical parameters of the training set, the greater the number of characteristics considered at a point, or the greater the number of resolution elements within an area, the better the classification accuracy will be.

Training sets are commonly used where multispectral scanner data are available. The data-display techniques which are used to facilitate feature extraction are numerous and include histograms for each category and by each channel (see Fig. 9.8).

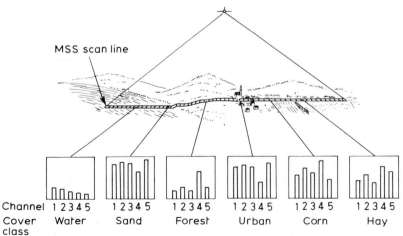

Fig. 9.8 Selected MSS measurements made along one scan line. Channels cover the following spectral bands: 1, Blue; 2, Green; 3, Red; 4, Reflected i.r.; and, 5, Thermal i.r. (From Lillesand and Kiefer, 1979.)

From such data manipulations matrices of correlation and covariance may be compiled. The training set data usually involve parametric discriminants, using Bayesian probabilities and maximum likelihood functions. An extra advantage of these methods is that they commonly employ thresholding procedures in which items unlike the sample are assigned to a discard class, which may then be analysed in greater detail.

Many methods are being used for comparing new or unclassified data with established training sets. These include nonparametric procedures such as *linear discriminant functions* which are much faster in terms of computational time than the maximum likelihood decision rule, which, in practice often becomes a quadratic rule requiring a large number of multiplications for each decision. Some workers have used *composite sequential algorithms* based on clustering techniques. Others have developed models based on boundary conditions. Still others have developed 'table look-up procedures', much faster than maximum likelihood functions in practice and nearly as accurate. Operations in the spatial frequency domain have been investigated for areal analysis and classification.

Analog computer techniques can be operated rapidly to provide image comparisons based on brightness levels, areas above threshold brightness levels, and the numbers and shapes of selected picture features. Unfortunately such systems lack the flexibility of digital computers, and each new capacity or operation demands new hard-wired components.

Conventional training-set techniques are, however, not without disadvantages; in particular the ground truth sites or areas must be carefully chosen, presupposing detailed knowledge of the region. Often such knowledge is not sufficiently detailed to ensure the representativeness of the training set, and further problems are caused by variability in time and space within the training area. Consequently efforts are being made to develop techniques which are described as *unsupervised*, for application to areas for which training sets are unavailable. Such techniques involve *clustering algorithms*, which may take the form of manual routines whereby selected features are identified by eye and mapped. The technique relies on the fact that the multi-channel radiances of different features tend to cluster in different places in the corresponding multi-dimensional space. This allows the computer to calculate the classes present and the resolution elements which belong to them. At a later stage the classes can be identified by reference to new ground truth, or through comparisons of sample data with a bank of spectral signatures. The particular advantages of such a method are that it is quick and cheap, no prior knowledge of the site is required, human bias is minimized, and future ground control stations can be located more objectively than in the 'supervised' techniques.

9.5 Operational, automatic decision/classification techniques

Often in the early stages of any remote sensing programme some retracing of one's steps is unavoidable in the march towards an operational feature recognition system. In the early, research, stages most attention is accorded to picture analysis rather than feature recognition. Often the best methods of data processing and analysis can be established only by trial and error, by the comparison of results obtained by different means. The expenditure of time, money and effort, and the complexity of the computer hardware and software to complete each method of approach must be assessed in terms of the resources which would be available operationally, and in terms of the requirements of the consumer judged by the additional criteria of accuracy, frequency and resolution.

Once the appropriate method of data analysis has been established the chief aim of most operational programmes is the correct recognition, identification, or classification of the contents of fresh imagery (Plates 9.4 and 9.5, see colour plates). Numerical techniques are superior to manual or photographic techniques in that the decision and classification rules are, by definition, quantitative or statistical, whereas those for the human interpreter are judgmental and substantially qualitative. Against this there is the disadvantage of automatic techniques: that allowances are difficult to make for phenomena or conditions which are unevenly distributed or are

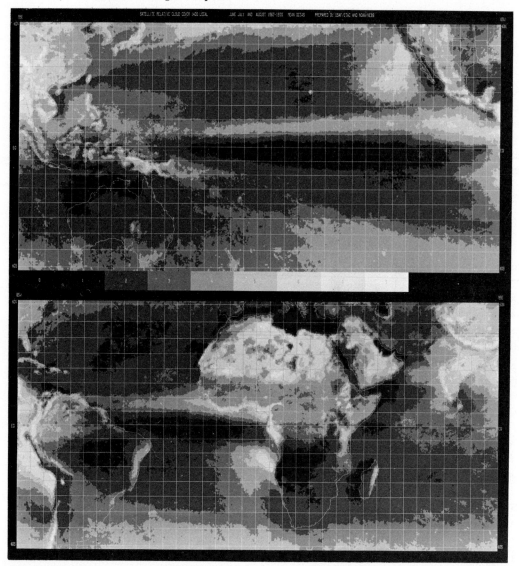

Plate 9.6 Photographic computer maps from ESSA Satellites, averaged for the season of June, July and August for a four year period (1967–1970). The use of the Mercator projection facilitates studies of tropical brightness patterns in general and trans-equatorial links in particular. Background brightness has been retained, hence, for example, the brightly-reflective desert areas of the Old World. The south Asian summer monsoon cloud over the Indian sub-continent contrasts strongly with the more typical east-west orientated ITCB clouds across the Atlantic and Pacific Oceans. (Courtesy, NOAA.)

present only from time to time. Such 'environmental noise' can be edited out by manual interpretation techniques, but may yield spurious patterns or results in automatic procedures.

Two cases in point are *background brightness*, which appears as an unwanted feature in climatological cloud imagery from weather satellites (see Plate 9.6) or *foreground brightness* related to clouds in Earth surface investigations. In satellite meteorology cloud cover maps (nephanalyses, see Section 10.3.1) can be prepared by hand to show, among other things, the percentage cloud cover categorized by broad classes. These can be summed and averaged to yield mean cloud cover maps exclusive of most background brightness effects. However, digital techniques (Fig. 9.9) can now provide acceptable results through the inclusion of a stage at which the minimum brightness level for each picture through the period in question is considered to be due to the

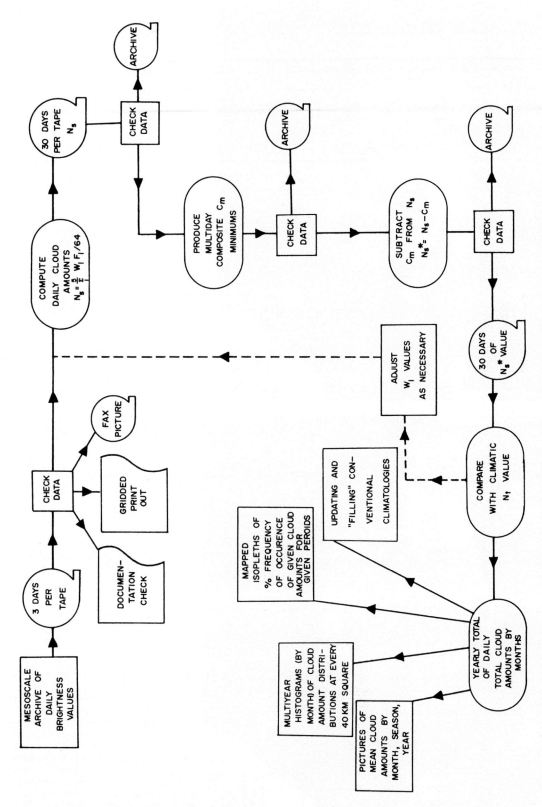

Fig. 9.9 Flow diagram for the production of a climatology of total-cloud amount from the mesoscale archive of the weather satellite data. (Source: Miller, 1971.)

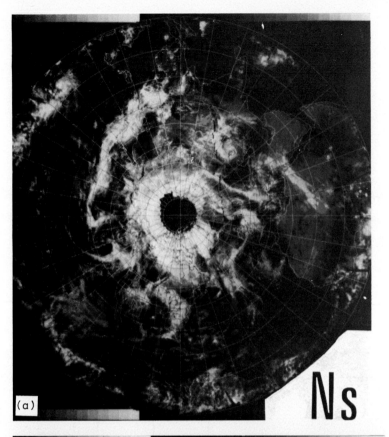

(a)

Ns

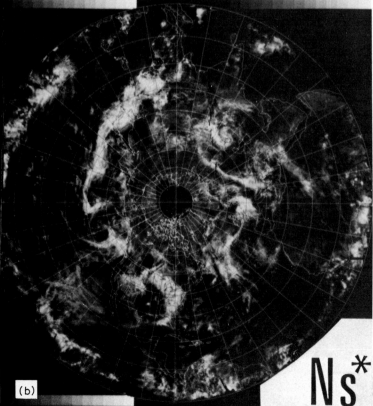

(b)

Ns*

Plate 9.7 (a) The total Earth-cloud scene represented in the mesoscale resolution of computer-rectified, mapped imagery from operational weather satellites. Note the high brightness of the Arctic ice and snow and the Saharan sands as well as the clouds. (15 April 1969.) (b) The cloud scene for 15 April 1969, remaining when the 30-day composite minimum cloud amounts were subtracted from the total Earth-cloud scene. (Source: Miller, 1971 Courtesy, US Air Weather Service, Scott Air Force Base, Illinois.)

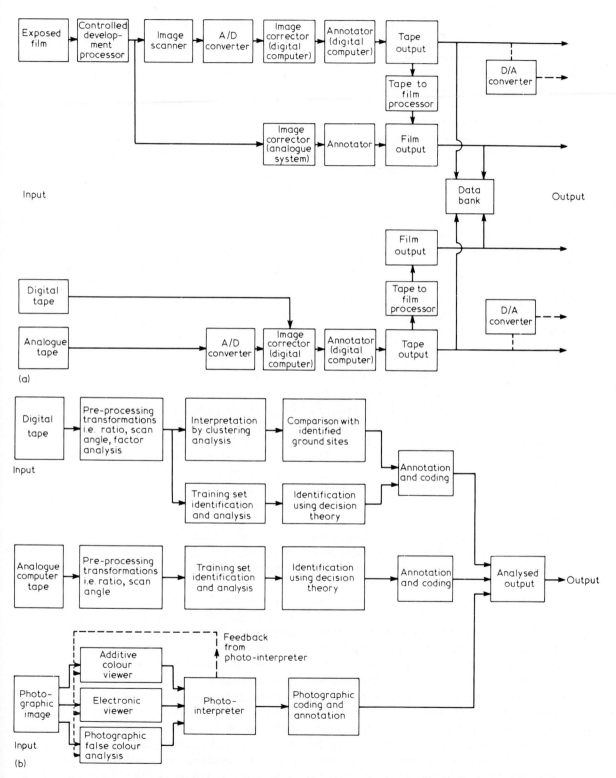

Fig. 9.10 Aspects of an automatic feature recognition facility for centrally processing data from Landsat satellites. (a) A data pre-processing facility. (b) A feature recognition facility. (Source: EMI, 1973.)

albedo of the surface of the Earth. This level is especially high in frozen regions, over sandy deserts, and where sun glint from water surfaces is strong. The brightness minima are subtracted from the daily mean brightness levels for every picture point to leave a value more representative of mean cloudiness itself (Plates 9.7 (a) and (b)). Unfortunately it is obviously more problematic to apply a technique like this to mapping of cloud cover on a short-term (e.g. daily) basis, so manual methods prevail for meteorological as opposed to climatological use.

At the end of any decision-making classification procedure, an output is produced; this may be in one of a variety of forms, including graphical, cartographic and tabular modes.

(a) Graphical outputs. These usually consist of histograms of the frequency of occurrence of identified categories of features, or related categories.

(b) Map outputs. Maps are common, popular, long-hand or computer products in environmental remote sensing programmes. They may be qualitative (showing types of features, e.g. crops), quantitative (showing values of features or conditions, e.g. surface brightness) or selected combinations of the two. The mapped patterns may be isoplethed or choroplethed, they may be single value or multiple category presentations, they may be based on absolute values, ratios, percentages, and changes through time; they may depict only point, line or area features, or all three simultaneously. In short, maps are highly versatile forms of digital programme outputs, and are justly significant in remote sensing as a whole.

(c) Tabular outputs. Once again a variety of forms and contents are possible. For example, the accuracy of identification may be summarized in matrix form, giving correct identifications on the diagonals, and errors of omission and commission on the x- and y-axes. Such tabular outputs can be prepared easily by computer. This may form part of the checking ('verification') procedure by which the programme results are assessed against those obtained by other, especially conventional, methods. Any remote sensing programme for the analysis and interpretation of the target area stands or falls finally on its ability to provide worthwhile, especially cost effective, results. Verification techniques often lead to the improvement of the procedures of analysis and interpretation by highlighting the errors.

We may conclude this necessarily brief and superficial introduction to the practices and problems of numerical processing by reference to an automatic feature recognition facility suitable for central processing of Landsat satellite data. Figs 9.10 (a) and (b) illustrate the data flow patterns through each part of the total facility, not only for digital and analog inputs, but also for inputs in a photographic form. After the initial simple image corrections have been made, and geometric and radiometric images have been reduced to a minimum, the data may be annotated, and a central store (the data bank) compiled (see Fig. 9.10(a)).

In the feature recognition facility (Fig. 9.10(b)) the flow pattern is basically the same. Variations occur in the digital format due to the different forms of recognition techniques in use – basically the supervised and the unsupervised approaches. For photographic image inputs three forms of interpretation methods are shown, all of which require a photo-interpreter to analyse the results, and, if necessary change the processor parameters. This consequent feedback is represented by the dotted line.

Although much time and money has been expended in the development of feature recognition facilities, no truly automatic feature recognition complex has yet been constructed. Many recognition problems are very involved, on account of the variety and variability of components of the natural environment, and the factors which have a bearing upon them. The problem is exacerbated by the fact that many natural features – unlike many man-made features – have diffuse outlines, and irregular shapes. These features are especially difficult to identify and classify by objective methods, and human, subjective, and relatively qualitative methods will play an important part alongside quantitative techniques in remote sensing for some time to come: *interactive* systems and techniques are likely to be dominant through the 1980s.

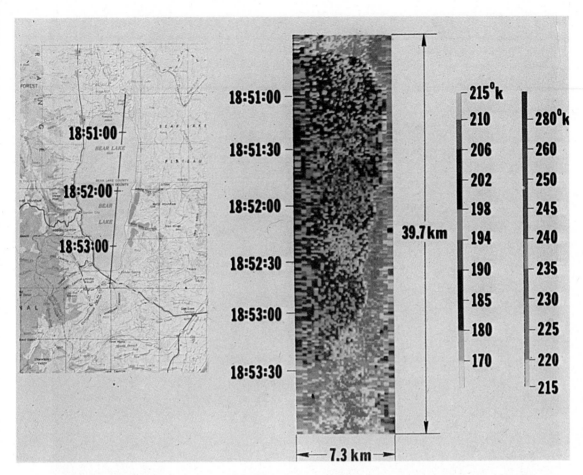

Plate 3.3 The flight path and microwave image (1.55 cm waveband) for 3 March 1971, Bear Lake, Utah/Idaho, from an altitude of 3400 m. (From Schmugge *et al.*, 1973.)

Plate 9.2 Tiros-N image of Great Britain, 12 July 1979.

Plate 9.4 Composite image made by computer from bands 4, 5 and 7 of 43 Landsat scenes.

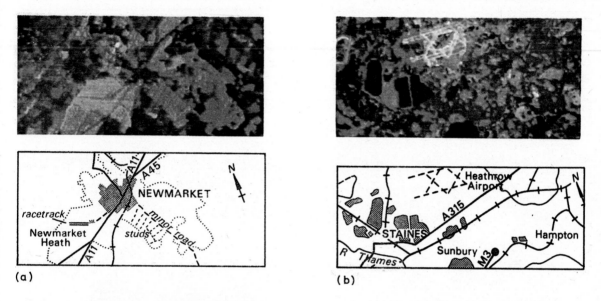

(a)

(b)

Plate 9.5 Examples of various land use categories obtained by computer processing of Landsat 1 data for the Department of Environment. The processing minimizes degradation of the picture detail. Scale of reproduction 1:250 000. See key maps for selected features. (a) Urban area of Newmarket. Rural areas image in red and green. Newmarket Heath images in yellow. (b) Heathrow airport and adjoining areas. Rural areas are imaged in red and green, reservoirs in black and construction works in white. (Source: Smith, 1975.)

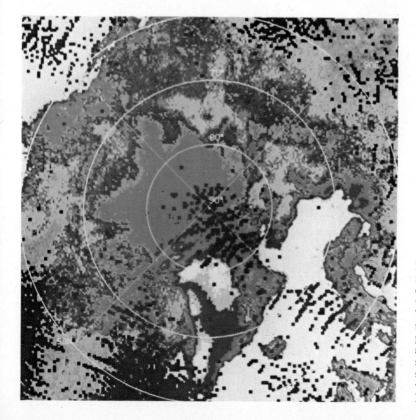

Plate 11.2 An ESMR thermographic map of the world for 16 January 1973, obtained from Nimbus 5 in the 1.55 cm waveband, from an altitude of 1100 km. The maritime pattern is influenced by the general poleward gradient of temperature, patterns of ocean currents, and active rain areas. Overland the stronger microwave signals from the solid surface generally obliterate variations due to rain. (Courtesy, NASA.)

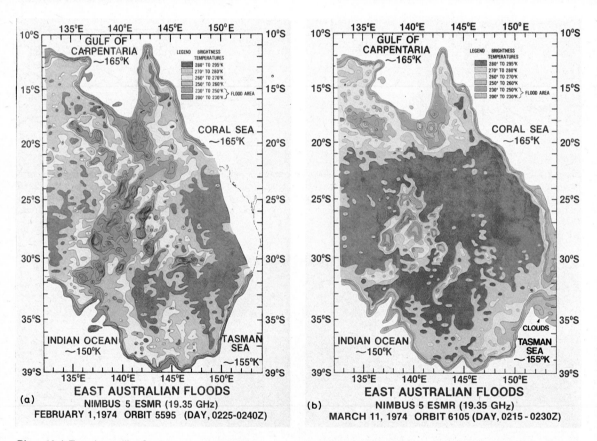

Plate 12.4 East Australian floods as revealed by Nimbus 5 Electrically Scanning Microwave Radiometer (19.35 GHz): (a) day, 0225–0240 GMT, 1 February, 1974; (b) day, 0215–0230 GMT, 11 March 1974. Blue (250–230 K) and purple (230–200 K) surface brightness values indicate flooded surfaces. (Source: Allison and Schmugge, 1979.)

Plate 15.1 Infrared colour image of part of Exmoor National Park, England. Dark brown and greenish grey colours show moorland of heather, bilberry, gorse and grasses. Red tones show reclaimed fields of pasture. Pale grey fields have been cut for hay. Deciduous trees appear pink in tone. (Source: Exmoor National Park.)

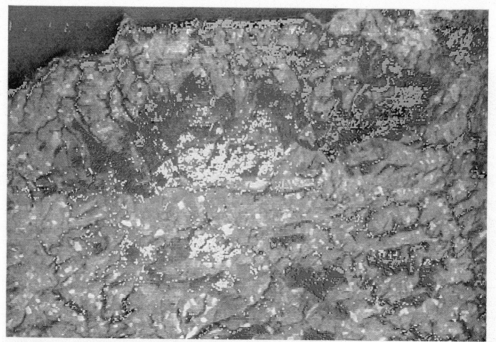

Plate 15.3 Colour image from display screen of IDP 3000 processor showing Exmoor National Park as analysed using supervised classificatory systems. Yellow – Grass moorland, Purple – Heather moorland, Green – Woodland, Red – Farmland. (Source: Exmoor National Park.)

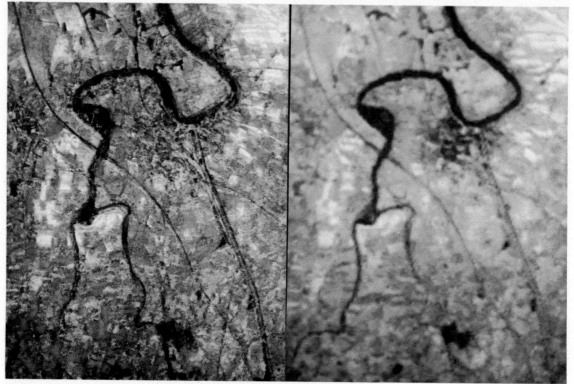

Plate 16.2 Example of improved resolution spacecraft imagery. The right of these two Landsat pictures of Goole on the River Ouse, taken on 1 February 1980, appears blurred, or out of focus. This is because a very small part of the multispectral scanner (MSS) image (the width represents about 12 km on the ground) has been 'blown-up'. The image produced by the other sensor on Landsat 3, the high resolution monochrome return beam vidicon (RBV) camera, has been registered digitally with the MSS picture. By using colour information from the MSS bands and intensity information from the RBV, a high resolution colour picture can be achieved and is indicative of the spacecraft imagery which should be available in the mid-1980s. (Courtesy, Space Department, RAE, Farnborough; Crown Copyright Reserved.)

(a)

(b)

Plate 17.3 (a) This thermal picture shows the heat loss from a large industrial complex. (b) For comparison, this shows a conventional aerial black and white photograph of the same site. The colour image depicts the variation in grey tones on the black and white print. Each colour represents a discrete temperature banding, which varies between sites, from grey and light blue (the coolest) to purple and white (the warmest).

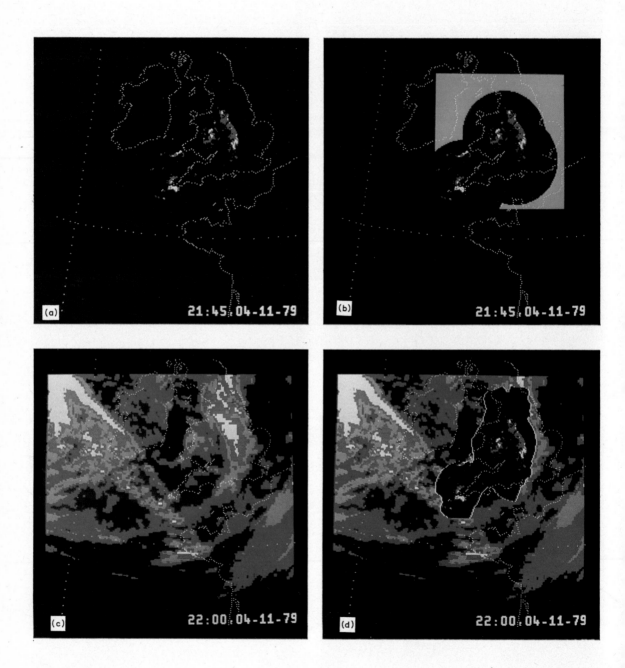

Plate 19.1 Satellite and radar composite images from the FRONTIERS project (see Fig. 9.1). (a) Echoes from 3 radars, and (b) Echoes from 3 radars, with limits of radar coverage in blue, both for 2145Z, 4 November 1979; (c) colour-enhanced Meteosat i.r. image for 2200Z, 4 November 1979 (high cloud tops in red or white); (d) radar echoes from (b) embedded within Meteosat image (c). Through such means relations between cloud and rainfall can be examined, modelled, and forecast. (Courtesy, K. A. Browning and the Ministry of Defence.)

Part Two: Remote Sensing Applications

10 *Weather analysis and forecasting*

10.1 Remote sensing of the atmosphere

10.1.1 Advantages of remote sensing systems
for weather studies

It is legitimate to ask why remote sensing should be welcome in support, or even in place, of conventional weather observation techniques which are well-developed, and long-established in many areas. This question is vital, especially since the total costs of remote sensing systems can be high in comparison with *in situ* sensor systems. The advantages of remote sensing systems for weather studies include the following, although, of course, not all may necessarily apply to a particular investigation:

(a) The sensor does not need to be carried into the medium which is to be measured.

(b) The measurement system does not modify the parameter being measured, since it is, by definition, remote from it.

(c) A high level of automation can be achieved, usually with minimal effort.

(d) It is often possible to scan the atmosphere by remote sensing means in two to three dimensions, unlike the single point measurement capability of most *in situ* sensor systems.

(e) Integration of a given parameter along a line, over an area, or through a volume, is often obtained readily in the direct output of a remote sensor.

(f) Sophisticated and elusive parameters, such as the spectrum of turbulence and momentum flux, may be available as direct outputs of certain remote sensing systems.

(g) High resolutions in time and space can be achieved with many types of remote sensing systems.

The three most common platforms for atmospheric remote sensors are the ground, aircraft (Plate 10.1) and satellites. The last have demonstrated great potential as vehicles both for operational meteorology and research. Much of the discussion that follows deals with the involvement of satellite systems in modern atmospheric science. Ground-based remote sensing systems are generally less well-developed, although they have a bright future in meso- and microscale investigations. Our account will dwell mostly on radar as an example of a ground-based system, in view of its relatively broad and varied development to operational or near operational levels. Aircraft-based systems are the least cost-effective; their almost entirely research-orientated applications have no significant place in an introductory review such as this book.

10.1.2 Ground-based remote sensing systems

Remote sensing of the atmosphere from the ground has been, and still is, a fragmentary affair. We shall see later that by far the greatest volume of atmospheric remote sensing data obtained today is gathered from a relatively small number of weather satellites. Two factors have restricted the development of ground-based techniques:

1. Most of the atmosphere is invisible within the light waveband of the electromagnetic spectrum. Consequently (acoustic sounding devices apart), the state and structure of the atmosphere could not be assessed by instrumental remote sensing methods until passive infrared and active microwave systems had been sufficiently developed.

2. The atmosphere is, from man's usual point of view, an enveloping, ubiquitous medium which, for many purposes, can be observed and

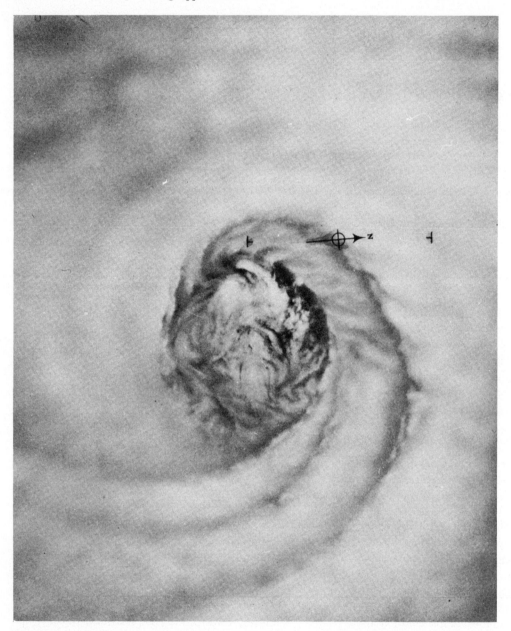

Plate 10.1 Typhoon Ida, 24 September 1968, photographed from the lower stratosphere by a U-1 reconnaissance aircraft. The wall clouds around the eye have a marked helical arrangement. (Courtesy, USWB.)

measured satisfactorily from within. Conventional weather observatories are immersed in the atmosphere, and, though subject to certain limitations, provide data of the types required by standard forecasting procedures. However, this compatibility has been, to a large extent, a child of necessity. Data from satellites are now adding a new dimension to well-established practices such as short-term weather forecasting even in areas generously supplied with conventional stations, and are completing the scene where such stations are sparse.

The earliest form of artificially assisted remote sensing of the atmosphere was cloud photography from the ground. Later, balloons and aircraft were employed to provide new views of these very significant aggregates of water droplets. The invention of films of different speeds and sensitivities, filters to eliminate specific wavelengths of radiation, and time-lapse techniques in photography have all made important contributions to the meteorology of clouds. However, there are limits to the value of the essentially local information such studies of the clouds can provide. Cloud photography today is dominated by the amateur enthusiast, not the professional weather observer. In synoptic weather forecasting three aspects of the clouds are routinely observed:

(a) The proportion of the sky which is cloud covered.
(b) The type or types of clouds present at different levels.
(c) The height(s) of the cloud base (or bases if the cloud is multilayered).

Of these three, the last is the only one which is sometimes evaluated by instrumental remote sensing from the ground. The height of the cloud base may be assessed either by searchlights (ceilingometers) or lasers.

Without doubt the most important ground-based system of atmospheric remote sensing – and that in most widespread use – is radar. This active microwave system was quickly developed for such purposes after the end of World War II. It may be used to identify and examine a wide range of atmospheric phenomena, including raindrops, cloud droplets, ice particles, snowflakes, atmospheric nuclei, and regions of large index-of-refraction gradients. Basically, the system consists of a transmitter, which produces power at the selected frequency, an antenna which radiates the energy and intercepts reflected signals, a receiver, which amplifies and transforms the received signals into video form, and a screen on which the returned signals can be displayed (see Chapter 4).

Often it is sufficient to know the strength of the echo from a target, and its distribution. No account need to be taken of the frequency of the returning radar wave by comparison with the transmitted wave. Hence 'non-coherent' radar systems are adequate. For other studies, it may be important to know the direction and rate of movement of a phenomenon with respect to the radar location. In such cases the *phase* of the received signal must be compared with the transmitted signal, in other words, a 'coherent' radar system is required. One class of radar system, commonly referred to as coherent (or Doppler) radars, is designed to exploit the 'Doppler shift' effect. This is the observed change in frequency of radiant energy reaching a receiver when the receiver and the target are in motion relative to each other. Doppler radar is used in meteorology for a wide range of purposes, including the measurement of turbulence, updraft velocities, and atmospheric particle-size distributions through tracking natural scatterers, and wind speed, through tracking artificial targets like rawinsonde balloons.

Some of the principal applications of non-coherent radars may be listed, and elucidated in conjunction with some common types of radar display units. These include:

(a) The examination of echo intensities in chosen directions. For such purposes the simple A-scope may be used (Fig. 10.1(a)). This presents the backscattered energy in profile form, and permits the operator to compare the echo with the transmitted pulse, which also appears upon the screen.
(b) The identification and distribution of rain areas, and the recognition of the types of rainfall included in them. Plan Position Indicator (PPI) display units (see Fig. 10.1(b)) are in quite widespread use for such purposes. By photographing the radar screen at intervals the development and movement of rain areas can be followed and predicted (see Colour Plate 19.1). PPI displays are widely used in operational meteorology for short-term rainfall forecasting and the monitoring of severe storms, and in research, for example in time-integrated studies of radar echo patterns and satellite cloud images and/or conventional data.
(c) The study of vertical profiles through the atmosphere. These may be obtained by a vertically-scanning radar, whose results may be displayed conveniently on a Range Height Indicator (RHI) (Fig. 10.1(c)). This, like the PPI, involves intensity modulation, i.e., the intensity of the bright spots on the visual display unit corresponds to the strengths of the returned signals.

The RHI system may be used to study such phenomena as cloud growth and development, or, in cloud-free conditions, atmospheric wave motions and turbulence, and even that most intriguing yet elusive phenomenon known as 'clear air turbulence' (CAT).

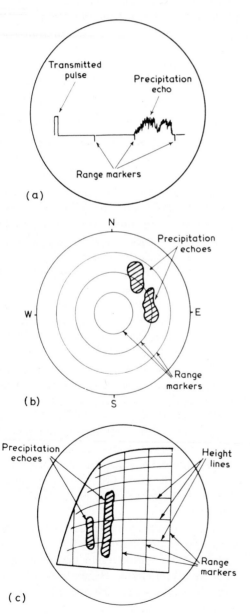

Fig. 10.1 The more commonly-used forms of radar displays, (a) the A-scope; (b) the plan-position indicator (PPI); (c) the range-height indicator (RHI). (Source, Battan, 1973.)

10.1.3 Aircraft-borne systems

For many years larger aircraft of both military and civilian varieties have been equipped with weather radar for detailed in-flight corrections to their flight paths. Such corrections may be made, for example, to avoid unnecessary encounters with powerful convective clouds of the cumulonimbus and cumulo-congestus types. Here strong thermal uplifts and pronounced atmospheric turbulence may occur.

In atmospheric research, airborne radars of the noncoherent type with vertically scanning antennae have been used to study the development and spread of precipitation with active mid-latitude fronts. The use of airborne Doppler radars for meteorological purposes is relatively new, an added complication accompanying the movement of the sensor platform as well as the target. We may look forward to successful operations employing such systems before the end of the 1980s.

10.1.4 Weather satellite systems

Remote sensing of the atmosphere received a tremendous fillip with the launching of the first specialized weather satellite, Tiros 1, in April 1960. It was not until the satellite era had opened that investigations of weather by remote sensing means became a practical possibility on anything other than a very piecemeal, localized scale. Prior to 1960 a number of spacecraft and satellites had made some preliminary observations of the Earth's atmosphere from orbital altitudes, e.g. high-altitude rockets of the Viking series, and members of the American Vanguard and Explorer satellite families, but the first Tiros (Televison and Infrared Observation Satellite) may be considered properly as the harbinger of satellite meteorology. By the time of writing over 100 satellites have been launched for the primary purpose of monitoring the Earth's atmosphere. Rather less than half of these have been 'operational' satellites, intended to back up and extend conventional weather observing facilities, and improve weather forecasts. Rather more than half have been 'research and development' (R and D) satellites, test-beds for new concepts, equipment and instruments. The principal atmospheric satellites concerned with observing weather patterns

Table 10.1 A summary of important atmospheric satellite famillies, 1959–1981

Family name	Country of origin	Number launched	Approx. period covered	Special remarks
Vanguard	USA	1*	Feb–Mar 1959	Early experimental satellites with primitive visible and infrared imaging systems
Explorer	USA	2*	Aug 1959–Aug 1961	
Television and infrared observation satellite (Tiros)	USA	10	Apr 1960–July 1966	First purpose-built weather satellites
Cosmos	USSR	22*	Apr 1963–Dec 1970	Some weather satellites in this large, cosmopolitan Russian satellite series
Nimbus	USA	7	Aug 1964–present	Principal American R and D weather satellite
Environmental survey satellite (Essa)	USA	9	Feb 1966–Jan 1972	First American operational weather satellite
Molniya	USSR	8	Apr 1966–May 1971	Dual purpose communication/ weather observation satellite
Applications technology satellite (ATS)	USA	4*	Dec 1966–June 1981	First meteorological geostationary satellite to test SMS concepts
Meteor	USSR	20	Mar 1969–present	Current Russian operational weather satellite series
Improved Tiros observational satellite (Itos)	USA	1	Jan 1970–June 1971	Noaa prototype
National Oceanic and Atmospheric Administration satellite (Noaa)	USA	5	Dec 1970–Dec 1979	Second generation American operational weather satellite series
Defense Meterological Satellite Program (DMSP)	USA	10	Feb 1973–present	Military weather satellites
Synchronous Meteorological Satellite (SMS)	USA	2	May 1974–Dec 1975	Testbeds for operational geostationary weather satellite systems
Global Operational Environmental Satellite (Goes)	USA	4	Oct 1975–present	First operational geostationary weather satellites
Geostationary Meteorological Satellite (GMS)	Japan	2	July 1977–present	First geostationary weather satellite cover of E. Asia/Pacific
Meteosat	Western Europe (ESA)	2	Nov 1977–Nov 1979	First geostationary weather satellites covering Africa/Europe
Tiros-N (prototype; series numbered Noaa 6 et seq)	USA	3	May 1979–present	Third generation American operational polar-orbiting weather satellites

*Denotes families which have included satellites designed for non-meteorological purposes also; numbers of weather satellites only listed here.

directly affecting man are listed in Table 10.1. Some other satellites have investigated solar phenomena and patterns in the upper atmosphere. In this account we need concern ourselves only with those which view man's immediate atmospheric environment – the weather in which he is normally immersed.

During the 1970s weather satellites from countries other than the 'big two' in space (the USA and the USSR) began to make significant contributions to the WWW. Of special importance have been the geostationary satellites Meteosat 1 and GMS 1 operated respectively by the European and Japanese Space Agencies. Since data from the Western weather satellites have been much more readily available than from their Russian

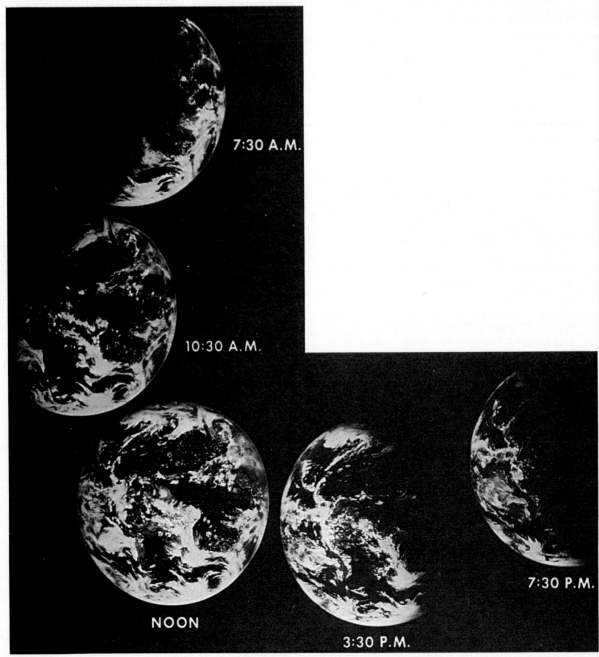

Plate 10.2 ATS geosynchronous satellite views of the illuminated portions of the visible disc of the Earth. ATS-III was in orbit above the mouths of the Amazon. These pictures show the passage of daylight from dawn to dusk around the globe. (Courtesy, NASA.)

counterparts we shall concentrate the remainder of our discussion in this chapter and the next on the Western weather satellite programmes and their outcomes, except in certain special instances.

Most of the weather satellites have occupied low-altitude, near-polar, sun-synchronous orbits. These are vital for coverage of polar and high-middle latitudes – which the growing group of geostationary satellites (ATS, SMS, Goes, GMS and Meteosat) in high-altitude equatorial orbits have been unable to monitor. Both types of satellite orbits are vital for adequate global weather coverage in the WWW. This requires four or five geostationary satellites to provide complete round-the-clock coverage of low and middle latitudes (effectively between about 50–55°N and S), imaging the Earth and its cloud cover at frequent intervals (generally every 30 minutes; Plate 10.2). Unfortunately, such a coverage has not been achieved fully operationally yet, largely because of problems of satellite design and operation. ATS, SMS and Goes satellites have given good, and mostly continuous coverage of the Americas, the western Atlantic and eastern Pacific Oceans for the best part of a decade, and the Japanese satellite GMS1 has fulfilled its design life. However, Meteosat 1 failed in November 1979 after only two years of operation, and launching and funding problems delayed its replacement, Meteosat 2 for nearly two years. The Russian satellite intended to cover the Indian Ocean region by late 1978 has still not appeared at the time this book went to press. Therefore, the requirements of the First GARP Global Experiment (FGGE) between November 1978 and November 1979 for complete circum-global geostationary satellite data could only be met by the temporary expedient of the movement of a stand-by Goes satellite from the Americas to the Indian Ocean for the FGGE period. This satellite was controlled temporarily by NASA and ESA jointly, under the same name of Goes IO (Indian Ocean). However, by 1980 only three geostationary satellites were operational, and significant gaps in the global coverage still exist today. Worse still, the completion of a global operational system of geostationary satellites as envisaged for the WWW is still in doubt for organizational and funding reasons.

In the meantime, many satellite data users outside the Americas continue to rely upon the Noaa polar-orbiters to meet their imagery requirements. With only brief exceptions, there have been two such satellites in continuous operation since 1966. Present Tiros-N type Noaa satellites provide 6 hourly imagery through two spacecraft orbiting at right angles to each other. For the future, complex and far-reaching plans are being drawn up for weather satellites and supporting research systems in the US Climate Program.

Through the long period of meteorological satellite operations there has been continuous development of the observing systems. Although it would be interesting to trace these developments, and of considerable value to relate them to the application and exploration of remote sensing principles we reviewed in Part One, space alone precludes such a study in the present context. The US National Space Science Data Center issues an annual *Report on Active and Planned Spacecraft and Experiments*. Payloads of any past, present and planned weather (and other Earth observation) satellites are all detailed in volumes in this series of NASA publications. For present purposes, summaries of the representative observing systems on current Noaa, Meteosat and Nimbus satellites will suffice to exemplify and clarify the value of meteorological satellites to weather and climate studies.

10.2 Representative weather satellites

10.2.1 Operational polar-orbiters: The Tiros-N payload

Tiros-N type satellites are 'third-generation' operational polar-orbiting environmental spacecraft designed to improve on and extend the functions of their predecessors, Essa 1–8 and Noaa 1–5. They are cooperative efforts of the USA, UK and France which capitalize on experience gained by these three nations in earlier satellite and sensor operations and experiments. These sun-synchronous satellites have been designed to operate in orbits inclined at 99° to the equator, at altitudes of about 850 km, giving southbound equator crossings between 0600–1000 Local Solar Time (LST), and northbound crossings between 1400–1800 LST. Instrumentation on Tiros-N and Noaa A (Noaa 6) is summarized in Table 10.2. The AVHRR (Advanced Very High Resolution Radiometer) is designed to provide swaths of imagery for central processing (after on-

Table 10.2 Tiros-N and Noaa A Instrumentation. (From Slater, 1980)

Instrument	Channels	Resolution (km)	Swath width (km)	Instrumental Observation	Climate Parameters
AVHRR/3	0.58–0.68 μm 0.725–1.0 1.53–1.73 3.55–3.93 10.3–11.3 11.5–12.5	1	3000	Visible and i.r. imaging	Sea surface temp., snow cover, albedo, clouds, vegetation cover, ice sheets
TOVS (3 instruments) HIRS	20 channels	30	1000	Multispectral i.r. radiation	Temp. profiles, humidity profiles
SSU	3 channels 668 cm^{-1}	147	1450	Selective absorption of CO_2	Stratosphere temperature profiles
MSU	4 channels 5.5 mm	100	1700	Passive microwave radiance	Temperature profiles
ERBSI⋆	2–50 μm 0.2–5 μm	1000 87	3000	Visible and i.r. wide field and scanner	Earth radiation budget
SBUV⋆	12 channels 0.255–0.344 μm	165	Nadir only	Ultraviolet backscatter	Ozone profile, total ozone

⋆These systems were proposed but not flown on Tiros-N and Noaa A; they are proposed for follow-on satellites in the series.

board storage) and Automatic Picture Transmission (APT) at 4 km resolution, and High Resolution Picture Transmission (HRPT) at 1.1 km resolution. APT and HRPT (see Plate 10.3) data are intended for local users, and stored data for global archiving. The TOVS (Tiros Operational Vertical Sounder) is comprised of three complementary sensors, each providing integrated radiances from narrow columns of the atmosphere for vertical profiling purposes (see Fig. 10.2). They are:

1. The Basic Sounding Unit (BSU), or High Resolution Infrared Sounder designed to profile temperatures from the surface to 10 mbar water vapour in three layers of the troposphere (both under cloud-free conditions), and total ozone content.
2. The Stratospheric Sounding Unit (SSU) designed to extend temperature profiles higher up the atmosphere.
3. The Microwave Sounding Unit (MSU) for broad temperature profile retrieval even in the presence of clouds.

The Tiros-N series also operate a Data Collection System (DCS) to acquire information from fixed and free-floating terrestrial and atmospheric platforms. Platform location is also possible by ground processing of the Doppler measurements of radio wave carrier frequencies. Thus, data can be collected from physically remote sites, and movement of bodies (e.g. transatlantic yachts) can be tracked using these facilities (see Fig. 6.8).

10.2.2 Geostationary satellites: the Meteosat 1 payload

An imaging radiometer, designed to give frequent (usually half-hourly) images of the visible disc of the Earth (about one-fifth of its surface area) constituted the principal payload on this satellite, and its successor, Meteosat 2. It consists of a large (400 mm aperture) telescope with a step-scan motor synchronized with the spin of the satellite to facilitate the construction of each separate image from 2400 adjacent scan-lines collected over 24 min at a satellite rotation rate of 100 rev min^{-1}. At the end of each individual imaging sequence the telescope is returned to its starting position in readiness for repetiton of the process. Optically-collected visible and infrared signals are converted into analog electrical signals in three wavebands, one each in the

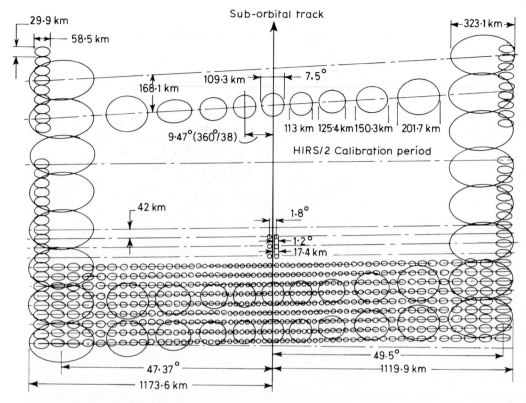

Fig. 10.2 Tiros Operational Vertical Sounder HIRS/2 (smaller element) and MSU (larger element) scan patterns projected on Earth: the microwave sensor integrates radiation over larger areas because naturally-emitted target radiation is weaker in the microwave than the infrared. (From Schwalb, 1978.)

visible (0.4–1.1 μm), infrared window (10.5–12.5 μm) and water vapour absorption (5.7–7.1 μm) regions giving resolutions of 2.25, 4.5 and 4.5 km respectively at the sub-point. A Data Collection System is also available for interrogation of Data Collection Platforms (DCPs).

10.2.3 Research and development satellites: the Nimbus 7 payload

Research and development satellites, of which there have been many, differ much more widely from one another than operational polar orbiters or geostationary satellites since the experimental ATS I, III and VI. Thus, the following summary of sensors carried by Nimbus 7 should be taken only as an example of the type and range of problems now thought significant for satellite research and development.

(a) The Scanning Multichannel Microwave Radiometer (SMMR). This measured radiances at five wavelengths in ten channels to provide data on sea surface temperature, cloud liquid-water content, precipitation (mean droplet size), soil moisture, snow cover, and sea ice.

(b) The Stratospheric and Mesospheric Sounder (SMS). This measured vertical concentrations of H_2O, N_2O, CH_4, CO, and NO, and the temperature of the stratosphere, to approximately 90 km.

(c) The Solar-Backscattered Ultraviolet/Total Ozone Mapper System (SBUV/TOMS). This measured direct and backscattered solar ultraviolet radiation, to provide data on global variations of solar irradiance, vertical distribution of ozone, and total ozone.

(d) The Earth Radiation Budget (ERB) sensor. This measured short and long wavelength upwelling radiances and direct solar irradiance to provide data on the solar constant, Earth reflectance, emitted thermal radiation, and anisotropy of the outgoing radiation.

Plate 10.3 Examples of Noaa 5 HRPT images: (a) i.r.

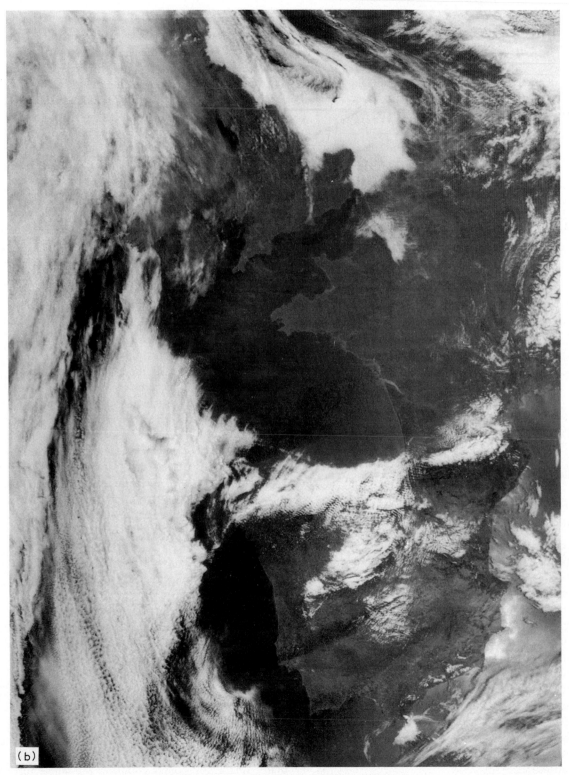

(b)

Plate 10.3 cont. (b) Visible, 10 May 1978. Note especially the greater sense of 'depth' in the i.r. cloud field due to the physical relationship between cloud top temperature and cloud top height, not found in visible cloud images. (Courtesy, Department of Electrical Engineering and Electronics, University of Dundee.)

(e) The Coastal Zone Color Scanner (CZCS). This measured chlorophyll concentration, sediment distribution, gelbstoffe ('Yellow stuff') concentration as a salinity indicator, and the temperature of coastal waters and open ocean.

(f) The Stratospheric Aerosol Measurements II (Sam II) experiment. This measured the concentration and optical properties of stratospheric aerosols as a function of altitude, latitude and longitude. Tropospheric aerosols are observable if no clouds are present in the field of view of the sensor.

(g) The Temperature-Humidity Infrared Radiometer (THIR). This measured the infrared radiation from Earth in two spectral bands (6.7 μm, and 11 μm), both day and night, to provide three-dimensional maps of cloud cover, temperature maps of clouds and of land and ocean surfaces, and maps of atmospheric moisture.

(h) The Limb Infrared Monitoring of the Stratosphere (LIMS). This surveys globally selected gases from the upper troposphere to the lower mesosphere. Inversion techniques are used to retrieve gas concentrations and temperature profiles.

10.3 Satellite data applications in meteorology

10.3.1 Routine data products and their uses

We have seen that, for each satellite system, there is a central ground facility which receives most or all of its observations. Here standard preprocessing, processing and analysis is carried out. The result is a wide range of products for subsequent use by the meteorological community. The breadth of such operations is admirably exemplified by Table 10.3. Of special significance are the vertical temperature profile retrievals, which augment radiosonde data in mapping the three-dimensional state of the atmosphere through reference to the relief on a recognized series of constant pressure surfaces ('significant levels') upwards through the atmosphere. Satellite-derived winds (see Fig. 11.4) are obtained by cross-correlating individual clouds in successive pairs of geostationary images.

More locally, in national and regional meteorological bureaux with APT (direct read-out) facilities, much reliance is placed on the analysis and

interpretation of cloud images in both the visible and the infrared. It is generally accepted that the analysis of images for conventionally classified cloud types is easier in the visible than the infrared; however, infrared images are advantageous in that they are available for night as well as day, and because their contents, relating to target emission rather than reflection, are physically more meaningful, revealing the temperatures of radiating surfaces. Where cloud is absent, land and sea surface temperatures can be evaluated; where cloud is present, evaluations can be made of the heights of the cloud tops above the ground (see Plate 10.4).

Table 10.4 summarizes the characteristics of clouds imaged by satellites whereby the principal cloud types may be identified. For many years it has been routine for forecasting offices to analyse the cloud contents of weather satellite images for the preparation of simplified clouds charts or *nephanalyses*. Fig. 10.3(a) summarizes the standard nephanalysis key, whilst Fig. 10.3(b) is an example of one such chart. It is important that each nephanalysis should be prepared in the same way, even when different analysts are involved. The information content of a nephanalysis includes cloud type, the degree of cloud cover, an indication of the structure of the cloud fields, interpreted features such as cyclonic vortex centres, and jet streams, and boundary lines for ice and snow as well as clouds. It has been estimated that the information content of a nephanalysis is two orders of magnitude less than that of the image on which it is based. However, it is usually sufficient for the purposes of short-term forecasters, who may compare the nephanalyses with conventional weather charts for the day, and correct these where necessary in the light of the satellite information. Especially where the forecasting area includes substantial sea areas and/or coastal zones the satellite evidence may be the salvation of a forecast through the correct or improved identification and positioning of, say, approaching fronts, or ridges of clear, anticyclonic weather. Current computer-based forecasting may provide good predictions of expected contour patterns for selected pressure surfaces, but much detail, e.g. frontal analysis, is still undertaken mostly by hand. In many smaller or poorer countries the whole forecasting procedure is still manual. In such respects and in such areas weather satellite imagery is of great value to the weather forecaster.

Table 10.3 Operational weather satellite products prepared by the US National Earth Satellite Service (see Hoppe and Ruiz, eds. 1974)

Manual and basic products	Unmapped image sectors	Man–machine combined products

1. Photographic imagery
  (a) Geostationary full disc frames
  (b) Geostationary mosaic loops
2. Facsimile
  (a) Great Lakes ice charts
  (b) US cloud cover depictions
  (c) Satellite Input to Numerical Analysis and Prediction (SINAP)
  (d) Northern hemisphere snow and ice charts
3. Alphanumeric
  (a) Satellite weather bulletins and tropical disturbance summary
  (b) Two-layer moisture analyses
  (c) Plume winds
  (d) Astrogeophysical Teletype Network (ATN) messages
	1. Facsimile
	  (a) Automatic Picture Transmissions (APT)
	  (b) Unmapped SMS Goes WEFAX sectors
		1. Alphanumeric
		  (a) Card deck, low and high level cloud motion vector field messages

Quantitative computer-derived products	Computer-derived image products	Archival products

1. Alphanumeric
  (a) Vertical Temperature Profile Radiometer (VTPR) soundings
  (b) Experimental Global Operational Sea Surface Temperature operations (GOSSTCOMP)
2. Facsimile
  (a) Unmapped SMS/Goes cloud motion vector facsimile formatted tape card deck for low level wind messages
	1. Photographic imagery
	  (a) Gridded, unmapped, pass-by-pass SR images
	  (b) SR hemispheric polar mosaics
	  (c) SR Mercator mosaics
	  (d) SR North America polar mosaic sectors
	  (e) Very High Resolution Radiometer basic images
	  (f) 5-day minimum brightness composites
	  (g) Augmented resolution map sectors
	2. Facsimile products
	  (a) Visual and Infrared Scanning Radiometer products for transmission on NWS facsimile networks
	  (b) Visual Scanning Radiometer products for transmissin via WEFAX
		1. Magnetic tapes
		  (a) Scanning Radiometer data tapes
		  (b) Sea surface temperature data tapes
		  (c) Vertical Temperature Profile Radiometer data tapes
		2. Photographic images
		  (a) Scanning Radiometer data
		  (b) Very High Resolution Radiometer data
		  (c) SMS/Goes data

Plate 10.4 Thermographic map of target radiomass in a Goes i.r. image, western USA and eastern N. Pacific, 22 June 1976. Complex enhancement curves are used in the preparation of such intricate 'three-dimensional nephanalyses' in which cloud tops are contoured objectively. (Courtesy, NOAA.)

10.3.2 Special forecasting applications

In many forecasting situations certain weather structures are of particular significance, usually because extreme events are associated with them. Over the years, research in many centres, but most especially in the National Environmental Satellite Service of the US Weather Bureau, has established that a very wide range of synoptic weather systems and structures can be identified in satellite imagery, either visible or infrared (see Fig. 10.4). Prominent amongst them are vorticity patterns. A wide range of tropical and extra-tropical cyclones of different types and at different stages of development have been recognized and modelled (see p. 169). Frontal forms, and the results of frontogenesis and frontolysis, can often be identified. Various cellular cloud patterns are known to be related to different instability conditions. In the upper troposphere, upper level troughs and jet streams can be identified

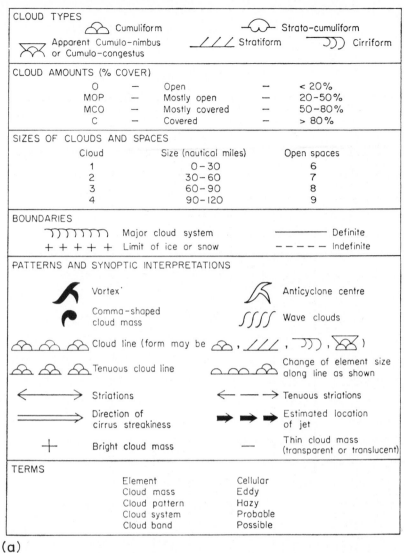

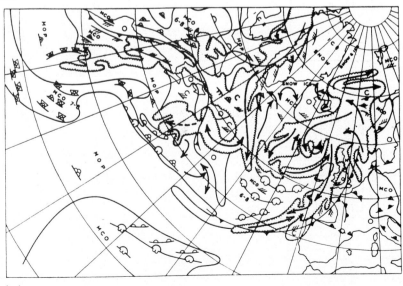

(a)

(b)

Fig. 10.3 (a) The standard nephanalysis key. (b) A sample standard nephanalysis summarizing cloudiness over the North Atlantic, on 17 July 1967. (Source: Barrett, 1970). An improved scheme has been suggested by Harris and Barrett, 1975.

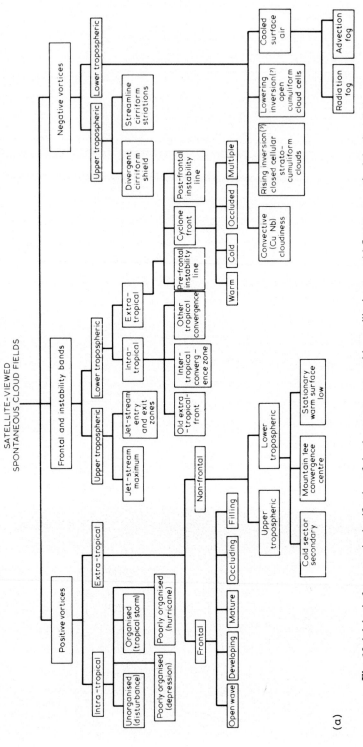

Fig. 10.4 A basis for a genetic classification of cloud systems portrayed by satellite imagery. (a) Systems organized dominantly by processes affected little by surface topography.

(a)

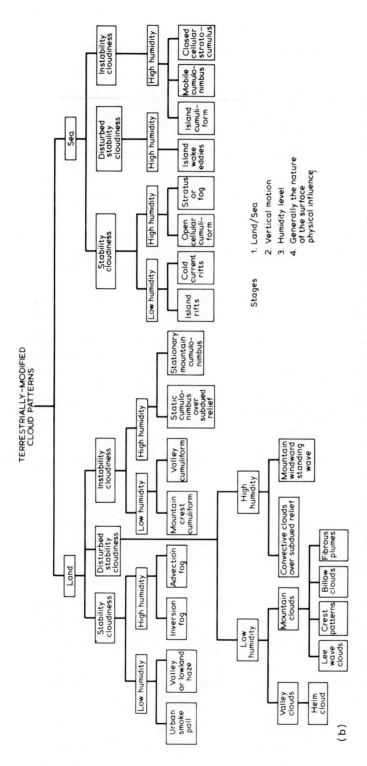

Fig. 10.4 cont. (b) Cloud systems and cloud rift patterns prompted mostly by surface characteristics. (Source: Barrett, 1970.)

Table 10.4 Characteristics of clouds portayed by satellite visible images

Cloud type	Size	Shape (organization)	Shadow	Tone (brightness)	Texture
Cirriform	Large sheets, or bands, hundreds of km long, tens of km wide.	Banded, streaky or amorphous with indistinct edges.	May cast linear shadows especially on underlying cloud.	Light grey to white, sometimes translucent.	Uniform or fibrous
Stratiform	Variable, from small to very large (thousands of square kms).	Variable, may be vertical, banded amorphous, or conforms to topography.	Rarely discernible except along fronts.	White or grey depending on sun angle and cloud thickness.	Uniform or very uniform.
Strato-cumuliform	Bands up to thousands of km long; bands or sheets with cells 3–15 km across.	Streets, bands, or patches with with well-defined margins.	May show striations along the wind.	Often grey over land, white over oceans, due to contrast in reflectivity.	Often irregular, with open or cellular variations.
Cumuliform	From lower limit of photo-resolution to cloud groups, 5–15 km across.	Linear streets, regular cells, or chaotic appearance.	Towering clouds may cast shadows down sunside.	Variable from broken dark grey to white depending mainly on degrees of development.	Non-uniform alternating patterns of white, grey and dark grey.
Cumulo-nimbus	Individual clouds tens of km across. Patches up to hundreds of km in diameter through merging of anvils.	Nearly circular and well-defined, or distorted, with one clear edge and one diffuse.	Usually present where clouds are well-developed.	Characteristically very white.	Uniform, though cirrus anvil extensions are often quite diffuse beyond main cells.

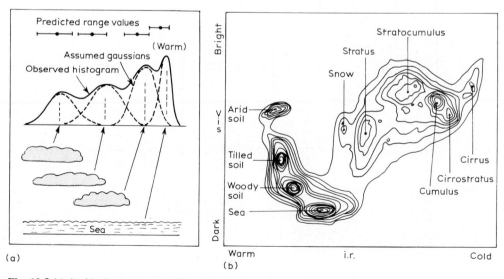

Fig. 10.5 (a) An idealized schematic of the Meteosat imagery one-dimensional histogram analysis concept. (b) A two-dimensional histogram for visible and infrared radiances observed by Meteosat. (From Fusco *et al.*, 1980, copyright, ESA.)

and classified. Anticyclones have always been more difficult to analyse, for they are characteristically cloud-free, but the shapes and forms of relatively clear areas, and their fringing clouds, have been used to differentiate between many different types of ridges and anticyclones. Last, but by no means least, many cloud features have been recognized as arising as a result of particular interactions between the atmosphere and underlying topography. Fig. 10.4 summarizes the richness of the cloud contents of satellite imagery through a *genetic classification* of systems or organizations of clouds: it represents such systems in relation to the atmospheric structures which prompt their growth and development.

Although every class of cloud systems represented in Fig. 10.4 has been accorded some attention in satellite meteorological research, some of these systems have been studied with special care because of their destructive capabilities. Weather satellites have almost certainly justified their costs since the mid-1960s through improvements in hurricane monitoring and forecasting alone. Probably no hurricane or tropical storm anywhere in the world has gone unnoticed since the first operational polar-orbiting satellite system was inaugurated by two Essa satellites in February 1966. In Chapter 18 (Hazards and Disasters) we will return to this important topic in much greater detail.

10.3.3 Research and development

There are two aspects of this matter which deserve attention here. One involves new sensors and new modes of data analyses, the other the search for an improved understanding of the structure and behaviour of the atmosphere. Continuing efforts are being made to improve existing satellite sensors, and to develop others which will provide more or better data. Most of the instruments flown by current operational satellites were tested first on Research and Development satellites like Nimbus. Others have been more purely research tools, intended to test seemingly logical propositions or possibilities.

Analytically, perhaps the most interesting and potentially useful research has been directed towards the development of an automatic ('objective') nephanalysis procedure. We have seen already how so-called 'three dimensional nephanalyses' are prepared from infrared data, but these classify cloud areas in terms of cloud top temperatures alone. Automatic identification and mapping of cloud type using single waveband data is probably impracticable, although some success has been achieved (where spatial resolution is not critical) by methods based on both *image brightness* (which is a point or pixel characteristic) and *texture* (a characteristic which can only be evaluated for groups of pixels). An approach using *bispectral* (visible and infrared) data has been developed at the European Space Operations Centre (ESOC) in West Germany for application to Meteosat imagery. This involves the production of both one and two-dimensional histograms of target radiances. In the first (Fig. 10.5(a)), a segment description provides predicted radiances, whilst a real segment yields an observed histogram which is processed to reveal the underlying Gaussians. From these, the contributions from high, middle and low cloud and the sea surface can be estimated. In the second case (Fig. 10.5(b)) a combination of the results of one-dimensional analyses of visible and infrared data yields a two-dimensional histogram in which the characteristics of several types of clouds and surfaces are typified. In this way new Meteosat data can be classified automatically in terms of the types of surfaces which are visible in the imagery.

A further possibility – perhaps the most promising of all – is that a suitable procedure could be developed using multispectral data, exploiting the known differences in radiant emissions from different types of clouds in several wavebands simultaneously (see Fig. 10.6). The recent arrival of a multispectral sensor on operational weather satellites (the AVHRR system on present Noaa satellites) should give the necessary fillip to this research and bring it to a successful conclusion. Certainly the way forward in the operational use of weather satellite data in meteorology must involve increased automation of image analysis and interpretation: the trend in instrument design and development is towards systems with higher bit-rates and more spatial and spectral information. To make best use of the data, more use must be made of computers to handle and process them.

Synoptically the biggest advances in our knowledge and understanding of the behaviour of the atmosphere are still being made within the

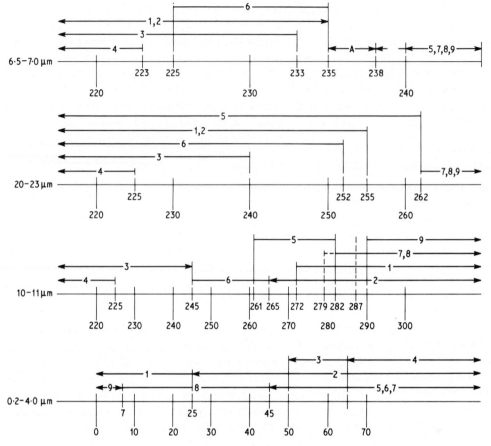

Key to cloud types:

1. Cirrus   2. Cirrus with lower clouds   3. Cumulonimbus and/or cirrostratus
4. Cumulonimbus   5. Middle clouds   6. Middle clouds with cirrus above
7. Stratus or stratocumulus   8. Cumulus   9. Clear   10. No decision

Fig. 10.6 A basis for nephanalysis using multispectral cloud data. This Cloud Type Decision Matrix was constructed for data retrieved from the Nimbus 3 instrument package. The horizontal graduations are K. Arrow heads indicate < or >. Broken lines show the thresholds of nadir angles > 40°. (Source: Shenk and Holub, 1973.)

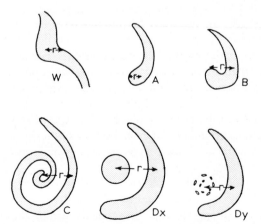

Fig. 10.7 Schematic cloud patterns associated with defined vortex types in mid-latitudes of the Southern Hemisphere. Cloudy areas stippled. r indicates distance taken as radius from vortex centre. For non-frontal vortices, r is distance to edge of cloudmass. (After Streten and Troup, 1972, from Barrett, 1974.)

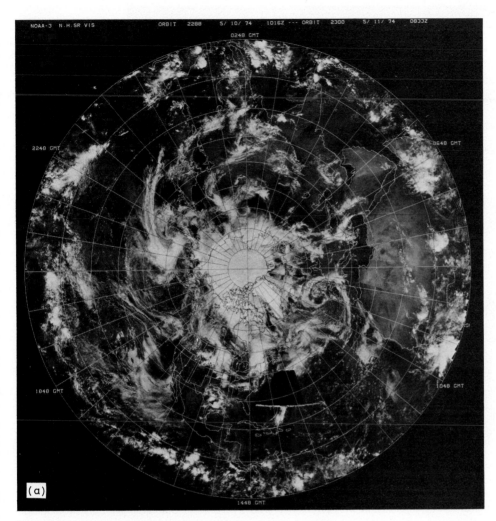

Plate 10.5 Computer-mapped polar stereographic Noaa images: (a) visible.

tropics. Very informative studies have been made of major weather features like the inter-tropical cloud band (ITCB) and the South Asian summer monsoon. One result is the new 'scale-interaction' approach to tropical meteorology. This has arisen from the realization that the basic building blocks of tropical disturbances (whether 'linear' or 'revolving') are relatively small, short-lived 'hot-towers' of tropical convection. The influences of these, the principal dynamic links between the lower and upper troposphere, range from the meso- to the sub-hemispheric scale. Geostationary satellites have a special part to play in the assessment of cumuliform convection, for such cloud phenomena are extremely dynamic: they may develop and dissipate within a matter of hours. Fascinating attempts are being made to measure mesoscale convective activity by classifying clouds according to the time evolution of their volumes. If such methods could be automated successfully, the charting of mesoscale convective activity would then be feasible. Techniques are also being developed to infer the nature and intensity of vertical mass circulations within and around cumuliform cloud clusters. Studies of the effects of larger scale circulations on cumulus convection off West Africa have revealed remarkable periodicities not only of the expected diurnal mode, but also in a time-frame of 4–5 days. It appears that large-scale processes must have an important effect on the space and time distributions of deep con-

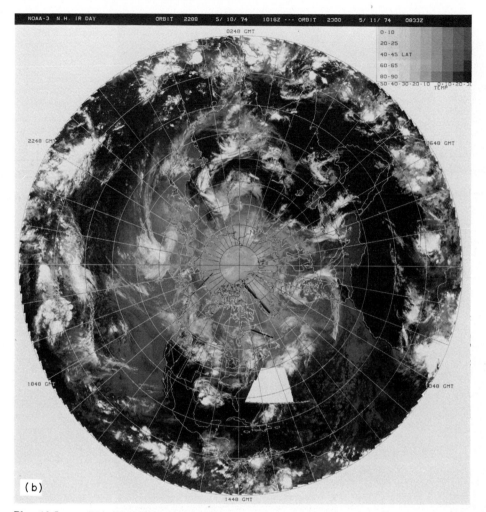

Plate 10.5 cont. (b) infrared. Each day one visible mosaic is prepared, for archival purposes, for either hemisphere, and two (one night and one day) infrared mosaics.

vection, with implications for both tropical and extratropical weather and climate, but the nature and relative significance of each process of this kind remains to be established.

Other areas in which weather satellites have much to offer the meteorologist include middle latitudes of the southern hemisphere, and the north and south polar regions. It has become apparent that southern hemispheric depressions differ in many respects of their structures and life-cycles from their northern hemispheric counterparts (see Fig. 10.7). Maps of cyclogenesis, vorticity distributions, and cyclolysis can be compiled much more satisfactorily from satellite than conventional data. Similarly the associated fronts, and their intrusions into polar latitudes (revealed well by the infrared imagery which can distinguish cloud from ice and snow, even during the polar night) can be studied best from weather satellite evidence. For these and many other research studies in synoptic meteorology the archive of computer-rectified, brightness-normalized polar-orbiting satellite imagery maintained at the Environmental Data Information Service (EDIS) at Camp Springs, Maryland, has proved to be a scientific pot of gold. Each day Noaa satellite imagery is mapped to polar-stereographic and Mercator map bases in the visible (one product in each format) and the infrared (two products in each format) (see Plate 10.5). Most geostationary satellite studies today are based on movie loops of visible, infrared or enhanced infrared imagery, with the emphasis on mesoscale weather phenomena. Such studies will probably dominate meteorological satellite research in the 1980s.

11 *Global climatology*

11.1 The nature of the problem

Near the end of the previous chapter, reference was made to studies of atmospheric variations which extend over months and years rather than hours or days. If our objects of interest vary over those longer time-scales manual extraction of key features of satellite imagery, automatic smoothing and reduction of the instantaneous data, or both, are usually necessary so that the more persistent features, patterns, or trends are not obscured by the statistical noise of short-term weather variations.

Studies of the atmosphere over periods upwards from about five days in length may be considered climatological rather than meteorological in nature and emphasis. As a consequence of the different time-scales involved the physical factors thought significant by climatologists may be different from those which are basic in meteorology. Although the advantages of satellite remote sensing systems for weather studies (see p. 151) are all advantages for studies of climates also, the advantages of such systems for studies which are specifically climatological are somewhat different in emphasis. The chief of these advantages are as follows:

(a) Weather satellite data are much more nearly complete on a global scale than conventional data.
(b) Satellite data for broad, even global-scale areas, are more homogeneous than those collected from a much larger number of surface observatories.
(c) Satellite data are often spatially continuous, in sharp contradistinction to data from the open network of surface (point-recording) stations.
(d) Satellite observations are complementary to conventional observations; each may throw extra light upon the other.

(e) Satellites can provide more frequent observations of some parameters in certain regions.
(f) The data from satellites are all made by objective means (unlike some conventional observations, e.g. visibility and cloud cover) and are immediately amenable to computer processing.

However, it should not be thought that satellite data are, as a result of these advantages, an immediate, universal panacea to the problems of the climatologist. There are several new problems posed to the would-be climatological user of such data by their intrinsic characteristics. The chief of these problems include:

(a) The vast quantities of new data points involved.
(b) The physical indications of satellite data; these are often different from those of conventional observations.
(c) The selection and development of appropriate techniques to analyse the new types of information.
(d) The resolution of the data. These are not always optimal for climatological uses.
(e) Degradations of components of the satellite – ground system complexes often complicate the evaluation and analysis of the data.
(f) It is often difficult to decide upon the best form for the presentation of results.

Notwithstanding all these problems, many climatological products of interest and value have begun to appear since the runs of satellite data have lengthened sufficiently to make them possible. These include inventories of parameters as significant as the net radiation balance at the top of the Earth's atmosphere (the primary driving force of the Earth's atmospheric circulation patterns) and the distribution of cloud cover (a big influence on the albedo of the Earth/atmosphere system and its component parts, and an indicator of horizontal transport patterns of latent heat). In pre-satellite days certain components of the radiation balance (short-wave (reflected) and long-wave (absorbed and re-radiated) energy losses to space) were established only by estimation, not measurement. The only comprehensive maps of global cloudiness compiled in pre-satellite days depended heavily on indirect evidence, and could not be time-specific. So, satellites are affording us more comprehensive and more dynamic views of global climatology than were possible before their day (see Fig. 11.1).

Satellites are also assisting the rise of applied macroclimatology. For example, active efforts are being made in several centres to develop computer models of the Earth's atmosphere, with a view to the improvement of methods of extended and long range weather forecasting. In such models the net radiation balance at the top of the atmosphere is one of the outputs. By comparing this with satellite-based net radiation patterns further insight is gained into the efficiency of the computer model, and those aspects of it which might be improved.

In the past, the development of climatology as a worthwhile and distinctive field of study had been singularly hampered by an inadequacy of data. Satellites are helping enormously to correct this chronic deficiency. Let us review in more detail the chief areas to benefit.

11.2 The Earth/atmosphere energy and radiation budgets

11.2.1 The global picture

A distinction is sometimes drawn between:

(a) An energy budget or balance, which is the equilibrium known to exist when all sources of heat gain and loss for a given region or column or the atmosphere are taken into account. This balance includes advective and evaporative terms as well as a radiation term; that is to say both horizontal movements of energy, and the part played by absorptive gases in the atmosphere, are considered as well as the vertical radiation interchange.

(b) A radiation budget or balance. This is the equilibrium which exists between the radiation received by the Earth and its atmosphere from the Sun, and that emitted and reflected by the Earth and its atmosphere in return.

One outstanding problem is that, since the advective processes involved in the heat balance include movements both in the atmosphere and the Earth's oceans, these are still hard to evaluate from satellite data alone. Estimates are usually used instead, based on rates of atmospheric and ocean water flux, established by *in situ* sensors. However, much progress has been made in reassessing the Earth/atmosphere radiation balance from satellite data sources. The net radiation balance depends on three quantities, namely:

(a) The solar content.
(b) The planetary albedo (the percentage ratio of that solar energy scattered and reflected by the atmosphere and Earth surface, to the total incident solar radiation). This depends largely upon the solar content, the inclination of the Earth to the Sun's rays, and the reflecting capabilities of the Earth surface and its cloud cover.
(c) Long-wave (re-radiated) energy losses to space.

Satellites are helping to evaluate (a) more accurately than before by measurements in the ultraviolet and visible wavebands. Infrared radiometers provide measurements of (b) and (c) through appropriate channels. Figs. 11.1(a), (b) and (c), illustrate mean annual geographical distributions of the parameters involved in the annual radiation budget. Table 11.1 integrates these parameters for global and hemispheric areas. The most significant conclusions from studies of these kinds include the following:

(a) Judged by one period of four to five years, the net radiation balance of the Earth (and either hemisphere separately) is in radiative balance. Thus there seems no requirement for an annual net energy exchange across the equator as assumed by many earlier workers.
(b) Both hemispheres are darker (mean albedo 30%) and warmer (average long wave radiation 0.340 cal cm^{-2} m^{-1}) than the widely accepted estimates by J. London (1957) in pre-satellite

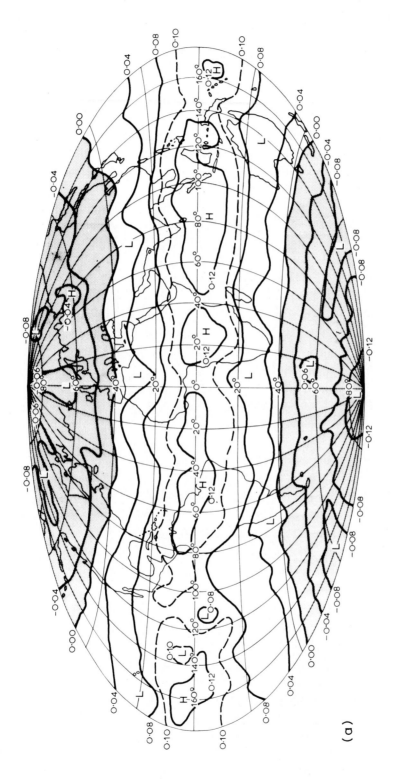

Fig. 11.1 Mean annual patterns of components of Earth atmosphere radiation budget ($RN_{EA} = I_0 - I_0A - H_L$), 1962–1965, from weather satellite evidence: (a) Planetary net radiation balance (RN_{EA}) at the top of the Earth's atmosphere.

(a)

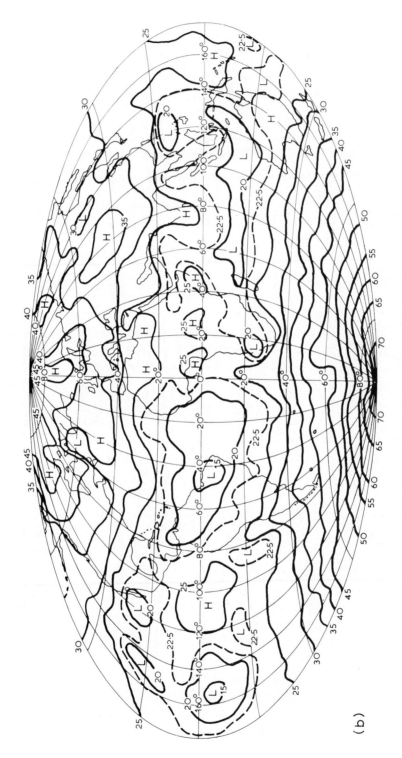

Fig. 11.1 cont. (b) Planetary albedo (I_0A) in percentages of incident radiation (I_0).

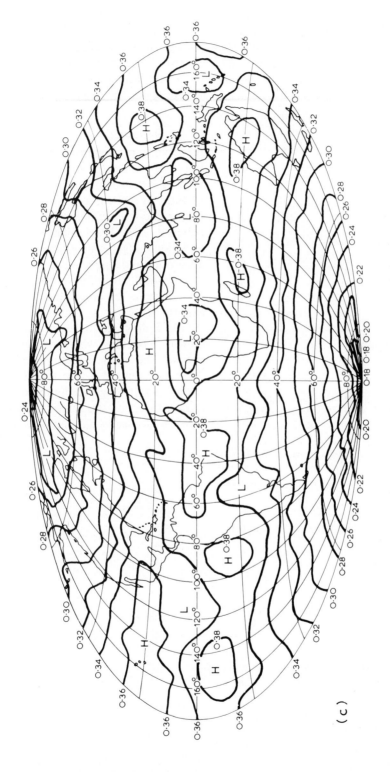

Fig. 11.1 cont. (c) Long wave radiation (H_L). (After Vonder Haar and Suomi, 1971.)

(c)

days (35% and 0.325 cal cm^{-2} min^{-1} respectively). If the Earth/atmosphere system is both warmer and darker than was previously thought, the system must accommodate – and probably transport – about 15% more energy in each hemisphere.

Table 11.1 Mean annual and seasonal radiation budget data for the Earth/atmosphere system, from first generation weather satellites. (From Vonder Haar and Suomi, 1971)

Northern hemisphere					
	DJF	MAM	JJA	SON	Annual
I_0	0.34	0.56	0.65	0.42	0.50
H_a	0.24	0.39	0.48	0.31	0.36
H_r	0.10	0.18	0.17	0.12	0.14
A	0.29	0.31	0.26	0.27	0.28
H_L	0.32	0.33	0.34	0.34	0.33
*RN_{EA}	−0.07	0.06	0.13	−0.03	0.02

Southern hemisphere					
	DJF	MAM	JJA	SON	Annual
I_0	0.69	0.43	0.32	0.58	0.50
H_a	0.46	0.30	0.25	0.41	0.35
H_r	0.22	0.13	0.07	0.17	0.15
A	0.32	0.30	0.22	0.29	0.29
H_L	0.33	0.32	0.32	0.34	0.33
*RN_{EA}	0.13	−0.02	−0.07	0.06	0.02

Global average					
	DJF	MAM	JJA	SON	Annual
I_0	0.51	0.50	0.49	0.50	0.50
H_a	0.34	0.35	0.37	0.36	0.35
H_r	0.16	0.15	0.12	0.14	0.15
A	0.31	0.31	0.25	0.28	0.29
H_L	0.32	0.33	0.33	0.34	0.33
*RN_{EA}	0.03	0.02	0.03	0.02	0.02

where I_0 = incident solar radiation (cal cm^{-2} min^{-1})
        H_a = absorbed solar radiation (cal cm^{-2} min^{-1})
        H_r = reflected solar radiation (cal cm^{-2} min^{-1})
        A = planetary albedo (per cent)
        H_L = emitted infrared radiation
           (cal cm^{-2} min^{-1})
RN_{EA} = net radiation budget of the Earth/
        atmosphere system (cal cm^{-2} min^{-1})
*Probable absolute error of ±0.01 cal cm^{-2} min^{-1}

(c) Each hemisphere has nearly the same planetary albedo and infrared loss to space on a mean annual basis. Since the surface features of the two hemispheres are quite different, clouds would seem to be the dominant influence on the energy exchange with space.

11.2.2 Regional details

Although some regional differences are apparent in Fig. 11.1, we need to view the world at a larger scale if more detailed pictures are required. Important findings in the tropics have been that the Earth is darker and warmer in low latitudes than previously believed. The lower albedoes now deployed in radiation budget evaluations are attended by higher absorption and long-wave radiation levels. These, in turn, indicate higher rates of energy exchange from the ocean to the atmosphere, and greater energy exports from low to higher latitudes. Thus quite simple findings can have profound implications for climate modelling where this involves latitudinal zones. The same is also true in an east-west direction. Fig. 11.2 portrays the variation of net radiation with longitude for three latitude circles. It is clear that there are some variations related to the contrast between land and sea, and others which

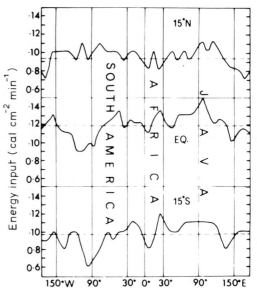

Fig 11.2 Longitudinal variations in net inputs of energy (RN_{EA}) (mean annual case), evaluated from five years of satellite data. (Source: Vonder Haar and Suomi, 1971.)

seem to be related to air mass and circulation differences.

The influence of even more localized complexities of geography on the net radiation balance is exampified by Figs 11.3(a) and (b). The radiation balance in middle and high latitudes of the northern hemisphere in a summer month is seen to assume a highly fragmented pattern. Some areas – even within the Arctic Circle – experience a net radiation gain, while the chief area of net radiation loss is over the Greenland ice-cap, neither over, nor very near, the pole. Generally the areas of net gain ('sources' of radiation energy for the general circulation) indicate little cloud cover, and the areas of net loss (the radiation 'sinks') indicate a high percentage of cloud, or, in the Arctic Ocean, permanently frozen surfaces.

In the same period winter held the south polar region strongly in its grip. Fig. 11.3(b) shows the whole area south of 40°S to be in deficit, though, curiously at first sight, the deepest radiation sink is not over Antarctica itself, but over the surrounding Southern Oceans. This is the zone in which most drastic environmental changes take place from one high season to the next. Most of Antarctica is always frozen, and radiation losses in winter are relatively small since there is little stored or advected energy available to be emitted as long-wave radiation (heat energy) to space. Meanwhile, the surrounding oceans freeze in winter, and much energy stored through the summer months is lost as long wave radiation, accompanying the phase change of water from liquid to solid. From the meteorological point of view, once frozen, the oceans act more like continents. In particular, their albedoes suddenly and dramatically increase, and the difference in received radiation and that lost by scattering and reflection tilts the net radiation budget to its winter level of considerable deficit.

11.2.3 Temporal details

Recently, data from geostationary satellites have been invoked for the first time so that short-term temporal detail may also be embroidered on our appreciation of the global radiation budget and its component parts. It is becoming clear that some unsatisfactory assumptions have had to be made when polar-orbiting satellite data have been used to evaluate such terms. Research in progress with Meteosat image data suggests that:

(a) Outgoing infrared flux and albedo show considerable diurnal variation.
(b) The variations differ according to the surface or type of cloud.
(c) Infrared flux tends to be fairly constant over sea surfaces and low clouds, but alters over land, especially over deserts, rising to a peak in early afternoon, and a trough in early morning.
(d) The albedoes of deserts, high and low clouds change little, but show a marked diurnal cycle over sea surfaces.

Such variations are significant, for they imply that biasing errors exist in all radiation budget parameters evaluated from polar-orbiting satellite data. In time, it is hoped that correction factors might be determined from geostationary satellite data so that polar-orbiting satellite radiation budgets might be made more accurate and realistic. Research into comparisons between budget parameters assessed by the two types of satellites is continuing.

11.3 Global atmospheric moisture distributions

11.3.1 Water vapour

Water vapour is of key significance as the raw material from which clouds and precipitation are formed. It is important, too, in the absorption of both short- and long-wave radiant energy in transit through the atmosphere, and in the transport of latent heat in both vertical and horizontal planes. Unfortunately, infrared radiation to space in the 5.7–6.9 μm water vapour absorption waveband can be difficult to analyse. First, this energy emanates mostly from water vapour in the absence of clouds, but partly from within the upper layers of clouds themselves when these are present. Consequently, both the cloud cover in each column and the temperatures of any cloud tops must be known if contaminating contributions from such sources are to be removed from water vapour waveband radiances, before these can be interpreted in terms of water vapour concentrations. Second, water vapour concentrations themselves (especially in

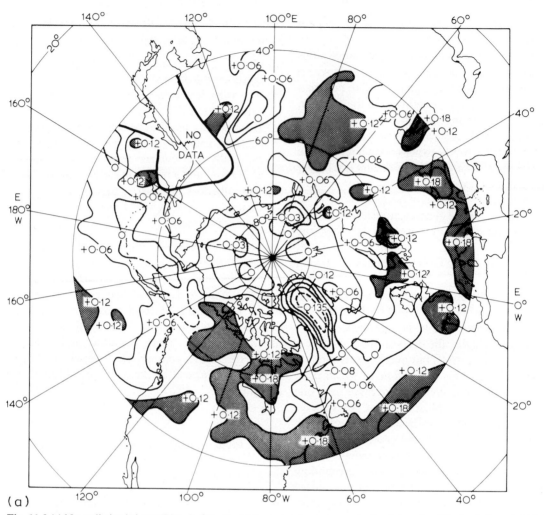

Fig. 11.3 (a) Net radiation balance (RN_{EA}) of the Earth/atmosphere system over the Arctic based on Nimbus 2 data, 1–15 July 1966 (cal cm^{-2} min^{-1}). Areas in surplus are stippled.

clear, i.e. cloud-free, columns of the atmosphere) are not easily assigned an altitude.

Data from MRIR and HRIR radiometers have been used in a rather restricted way for mapping global water vapour mass or relative humidity. Such studies have usually concentrated on the upper troposphere (above 500 mbar). Below 500 mbar the complications caused by clouds are considerable. Spectrometers like SIRS and IRIS have, in theory, the capability to profile the water vapour content of the atmosphere, but no climatological (time-averaged) results based on their data have been published yet. Meteosat 1 and 2, the European geostationary satellites have provided very valuable water-vapour absorption imagery for approximately

one-quarter of the global surface. These data are now being appraised both meteorologically and climatologically. The Tiros Operational Vertical Sounder (TOVS) on the new Tiros-N third generation operational polar orbiters provides a 3-level profile of atmospheric water vapour, and is already being used to improve routine maps of precipitable water.

11.3.2 Clouds

Several satellite-based methods have had the aim of producing mean cloud maps or other forms of time-averaged cloud displays. In the first decade of weather satellite operations the most popular

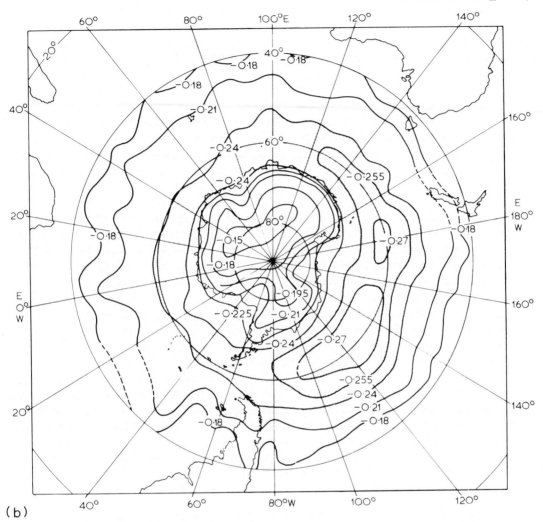

Fig. 11.3 cont. (b) As Fig. 11.3 (a) over the Antarctic. (Source: Barrett, 1974.)

techniques were based on nephanalyses (see p. 162), using their generalized indications of the percentage of sky which is cloud covered. Such techniques are based on sets of weighting factors through which the nephanalysis categories of cloud cover (commonly C, MCO, MOP, O and Clear) may be translated into acceptable cloud percentages. The techniques differ mostly in their sampling procedures. Some workers based their mean cloud maps on the nephanalysis indications at selected intersects of latitude and longitude, completing them by interpolating isonephs. Others have based their maps on estimates of the mean cloud cover in each grid square of given size. These maps have been presented either in chloropleth (area shaded) or isopleth form,

assuming in the latter case that the mean cloud cover in each square may be taken as representative of its centre point.

Any maps prepared from nephanalyses must be subject to errors arising from their own degrees of generalization, as well as those implicit in the nephanalyses themselves. The most obvious shortcomings of nephanalyses stem from the need for subjective judgements to be made by the compilers. It is widely accepted that nephanalysis-based mean cloud maps tend to over-estimate cloud cover at the upper end of the range, and to under-estimate it at the lower end. This is because the satellite imaging systems, and the nephanalysts, may eliminate small breaks in cloud fields where the proportion of cloud-covered sky is high, and fail to resolve and represent small cloud units where the sky is mostly clear. Despite these problems the patterns revealed by hand analyses of mean cloud cover have been generally acceptable, and of considerable interest where new climatological features (like the split ITCB, for example) have been revealed. On the credit side, such mean cloud maps are contaminated little by the 'background brightness' of surface phenomena which may be mistaken for clouds in automated ('objective') procedures.

An early objective technique for mean cloud mapping achieved some success by the late 1960s. This involved the production of 'multiple exposure average' pictures for a selected major region of the world for a chosen period of time. A photographic plate was exposed to each of a number of daily computer-rectified, full resolution picture products (4096^2 picture points) in turn. Multiple exposure averages, whilst being interesting climatological statements, suffer from two inherent character-istics. First, they do not differentiate between brightness due all or mainly to highly reflective Earth surfaces as distinct from that due to clouds. Second, they are qualitative, not quantitative, displays. More recently, efforts have been made to develop techniques which lack such deficiencies.

We mentioned in an earlier chapter that daily brightness values, derived from visible image signals received from meteorological satellites, have been processed to give daily cloud pictures of the sunlit Earth (p. 171). This 'mesoscale archive' has been used as the basis for perhaps the most comprehensive and successful approach to the mapping of relative cloud cover on a climatological time scale. The mesoscale matrix is comprised of 512^2 unit areas for each hemisphere compared with the 4096^2 points in the full resolution matrix. A grid square on the mesoscale matrix is one sixty-fourth the area of an American Numerical Weather Prediction – Global Weather Center grid square. Using a set of empirically-derived weights, the daily relative cloud cover may be estimated in oktas (eighths) for each mesoscale area (see Table 11.2). By saving the daily values by the month, for the entire period of the record, these values may be grouped in a 10-class frequency distribution including 0–8 oktas of cloud cover, plus one class (9 'oktas') for missing data. A range of mean ('relative') cloud cover maps has been prepared for 1967–1970, as illustrated by Plates 11.1(a)–(d).

Comparisons between picture brightness values (converted to 'relative cloud cover') and conventional observations have been made by staff at the USAF Environmental Technical Applications Center. The studies have revealed gross differences between point observations of total cloud amount made at the surface, and the satellite cloud cover estimates: the satellite and surface observations have

Table 11.2 Empirically derived weights for estimating the daily relative cloud cover. These were selected to achieve the best agreement in cloud amounts between manual (visual) estimated and automated estimates. (Source: Miller, 1971)

Original brightness range	Class	Contribution to total cloud amount %	Weights October–May	June–September
0, 1, 2	1	0	0	0
3, 4, 5	2	25	2	2.5★
6, 7, 8	3	88	7	7.5★
9, 10, 11	4	100	8	8
12, 13, 14	5	100	8	8

★The summer and winter weights were applied to the Northern and Southern hemispheres separately according to the hemisphere seasons. It was noted that the higher frequency of occurrence of small cumulus cloudiness in the summer hemisphere warranted the higher weighting during the summer months.

a similarity in pattern, but point-to-point values often differ. A general tendency is for the satellite data to suggest lower cloud amounts than concurrent surface estimates. These differences can be ascribed primarily to differences between the field of view of the surface observer and the angle of view from the satellite. To some extent the lower response threshold and resolving capability of the satellite sensor may also be important. Unfortunately, 'background brightness' is not extracted in the present archival programme, and strongly-reflective surfaces such as ice, sand and snow contaminate the 'relative cloud cover' maps. In future it may be possible to remove persistent brightness associated with the surface of the Earth, leaving the more mobile brightness caused by clouds. This has already been achieved in specialized studies, for example, ice margin 'mapping' as described on p. 198.

11.3.3 Rainfall

The possibility of mapping rainfall (rain-rate and distribution) from satellite data is now receiving very close attention. Rainfall is a key environmental parameter. Although it is not evidenced very directly by satellite observations its importance has led to the development of a range of techniques for the evaluation of rainfall from satellite data, especially satellite imagery. Several aspects of rainfall hydrometeorology are now amenable to improved analysis using data from satellites:

(a) Mapping the boundaries of areas likely to be affected by rain.
(b) Mapping rainfall totals accumulated through unit periods of time.
(c) Assessing extreme (intense) rainfall events.
(d) Assessing the climatology of rainfall distributions.
(e) Forecasting rainfall, especially in areas open to systems from relatively poorly-observed regions.

During the first decade of satellite meteorology the satellite was viewed generally as an *alternative* to the ground observing station for local environmental information. However, during the second decade, the two have been viewed more as *complements* to each other. Today, the general-purpose satellite rainfall monitoring methods best-suited to operational use are those which *integrate* evidences of rainfall from surface stations and satellite data, to give an analysis superior to those which could have been prepared from either type of information alone. The ground station (e.g. raingauge, streamgauge etc.) has the advantage of being able to provide quantitative data through time, but represents rainfall variations only at a single (point) location; the satellite has the advantage of being able to provide a really complete information, but only for separated points in time.

Most satellite data relate only indirectly to rainfall as measured on the ground. Therefore, the practice has grown to use satellite data to fill gaps in the conventional data network, evaluating those remote sensing data by reference to the units and values of the *in situ* observations. Methods of improved rainfall monitoring by satellite include the following:

(a) Cloud-indexing methods. Satellite cloud images are ascribed indices relating to cloud cover, and the probability and intensity of associated rain. Different methods have been used to calibrate the indices to give final rainfall estimates. In its most developed form the cloud-indexing approach has become a virtual extension of classical synoptic meteorology. Such methods are in quite widespread use for a variety of applications over land (see Fig. 11.4).
(b) Rainfall climatology methods. The basis of the most significant method is the relationship between climatologically-averaged rainfall and the long-term average contribution made to it by numbers of key types of synoptic weather systems (e.g. mobile mid-latitude lows in China). Climatology thus dictates the amount of rain deemed likely to be carried by a significant cloud system. Use of this method has been limited to support of general crop-information services in NOAA.
(c) Life-history methods. These are based on the twin premises that significant precipitation falls mainly from convective clouds, and that these clouds can be distinguished from others in satellite images. The estimation of rainfall then rests on suitable assessments of cloud performances, and its changes as the life-cycles of the clouds unfold.

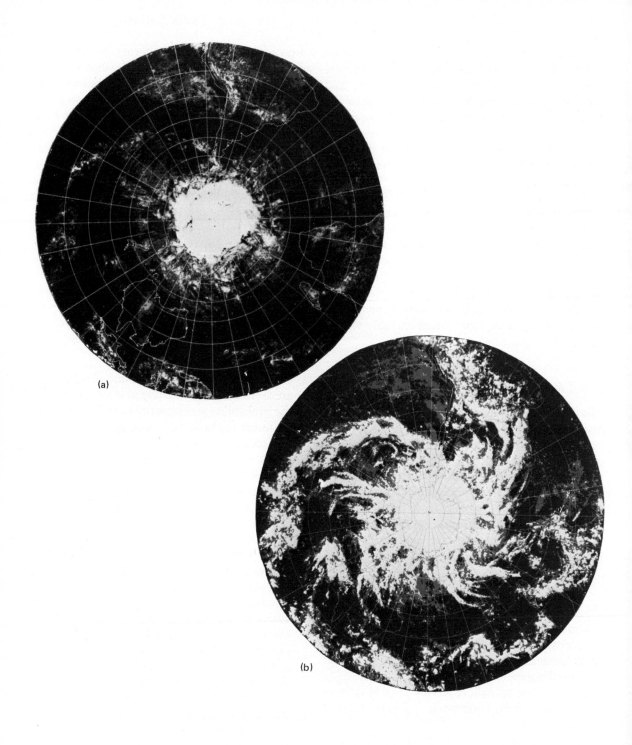

Plate 11.1 Brightness composite displays from Essa weather satellite photography. (Courtesy, ESSA.) (a) 5-day minimum brightness composite. (b) 5-day maximum brightness composite.

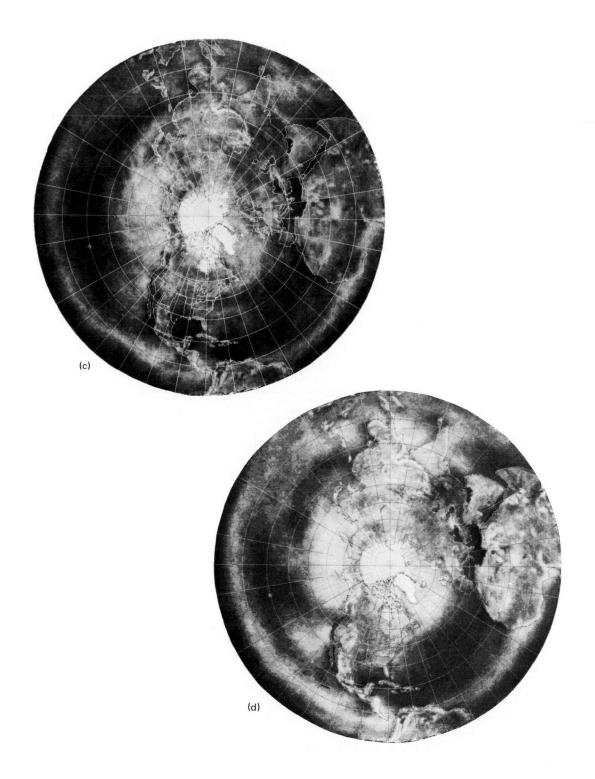

(c) 30-day average brightness composite. (d) 90-day average brightness composite.

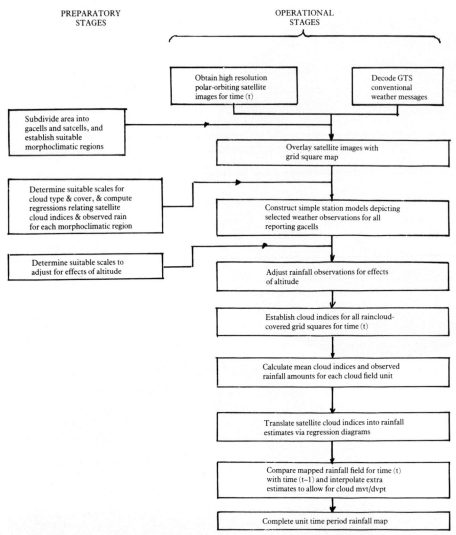

PREPARATORY
STAGES

OPERATIONAL
STAGES

Obtain high resolution
polar-orbiting satellite
images for time (t)

Decode GTS
conventional
weather messages

Subdivide area into
gacells and satcells, and
establish suitable
morphoclimatic regions

Overlay satellite images with
grid square map

Determine suitable scales for
cloud type & cover, & compute
regressions relating satellite
cloud indices & observed rain
for each morphoclimatic region

Construct simple station models depicting
selected weather observations for all
reporting gacells

Determine suitable scales to
adjust for effects of altitude

Adjust rainfall observations for effects
of altitude

Establish cloud indices for all raincloud-
covered grid squares for time (t)

Calculate mean cloud indices and observed
rainfall amounts for each cloud field unit

Translate satellite cloud indices into rainfall
estimates via regression diagrams

Compare mapped rainfall field for time (t)
with time (t–1) and interpolate extra
estimates to allow for cloud mvt/dvpt

Complete unit time period rainfall map

Fig. 11.4. Flow diagram for the Bristol Method of satellite-improved rainfall monitoring using a cloud-indexing approach.

(d) Bispectral methods. Here visible and infrared images are analysed objectively together to map the extent and distribution of precipitation. This is then scaled according to some form of ground truth (gauge and/or radar data). The central assumption is that heaviest precipitation falls from clouds which are both *bright* and *cold*.

(e) Cloud model methods. These, like bispectral methods, are essentially research techniques, being developed in man-machine modes in the search for more elegant formulations of relationships between clouds and rainfall. At present these methods are typified by a high degree of rigour but also by very complex algorithms.

(f) Passive microwave methods. Radiometers on recent Nimbus satellites have measured naturally-emitted radiation from the Earth and its atmosphere (see Plate 11.2, colour section). Some data (e.g. channels operating at 19.35 and 37.9 GHz) have been processed successfully to yield mesoscale rainfall intensity distributions over sea areas, but less successfully so over land where 'background' radiation is much stronger. Unfortunately it is over land that rainfall information is most urgently required for most

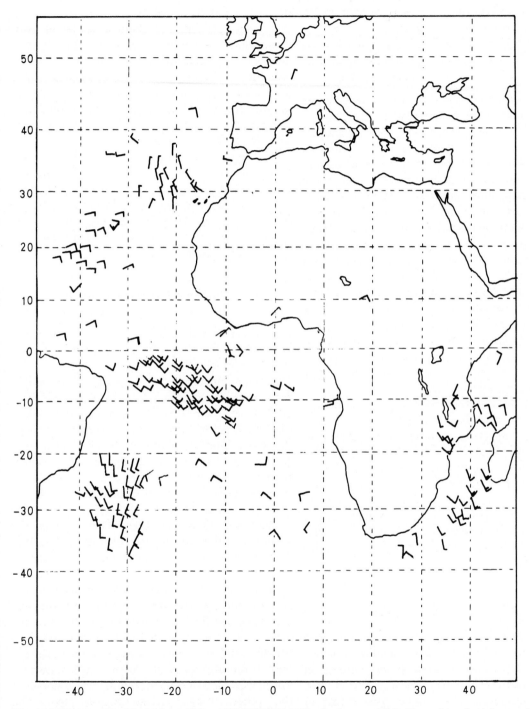

Fig. 11.5 'Meteosat Winds' for 2 July, 1979 at 1200 GMT (700–1000 mbar layer established by tracking cloud elements in successive images). Such maps are prepared routinely in the Meteorological Information Extraction Centre (MIEC) at ESOC from Meteosat data for the synoptic hours of 00 and 12 GMT, for high levels (700–400 mbar). (Copyright, ESA.)

hydrological purposes. Research with multi-band microwave data is continuing.

(g) Active microwave methods. These may become available in the future if tests of planned satellite-borne radar systems prove to be a success.

It is the *cloud-indexing* types which have, in one form or another, been most widely used, shown most flexibility, and yielded most results in support of continuous operational rainfall monitoring programmes. For example, they have been used in support of irrigation design in Indonesia, water resource evaluation and management in Oman, desert locust control in North-west Africa and general environmental assessment in tropical Africa and the Caribbean. Experience shows that rainfall maps based on both conventional and satellite data are more realistic in spatial detail than maps based on gauge data alone (see Chapter 18), and therefore reveal more accurately areas of average, above average and below average precipitation. Given suitable arrangements for the acquisition of input data, cloud-indexing methods can be tailor-made to suit local needs, and to provide information in near real-time if required.

Methods for rainfall monitoring by satellites are being very actively developed, refined and validated, especially through the American AgRISTARS (see p. 283) and Climate Programs. At present, different approaches are best for different regions. Over land, geostationary techniques ((c), (d) and (e) above) are applicable in low to middle latitudes, whilst polar orbiting techniques ((a) and (b) above) are necessitated in middle to high latitudes, and elsewhere if geostationary data are unavailable.

11.4 Wind flows and air circulations

There are several techniques for assessing the speed and direction of wind flow from satellite evidence, but these have been utilized almost entirely in meteorological, not climatological, contexts. For example, estimates of maximum wind speeds in hurricanes are made routinely by the US Weather Bureau (see p. 314). Similarly, the directions of wind flow can be established better in some areas and circumstances from satellite cloud images than conventional data, but the contributions of satellite-derived winds to climatological statistics have, so far, been too few and sporadic to be of great significance. Satellite observations for air flow analyses have their greatest potential in the tropics (see Fig. 11.5), where geosynchronous data can be processed to yield a variety of wind and circulation parameters, including wind speed and direction, relative and absolute vorticity, and horizontal divergence. Daily wind charts for tropical regions are now improved regularly through satellite inputs. Since mean monthly, seasonal and annual maps of isotachs and streamlines are prepared from such base data, satellite winds are already contributing implicitly to wind and circulation climatologies, especially in some of the more conventional data-sparse regions of the world.

11.5 The climatology of synoptic weather systems

Remembering that it is possible to identify many atmospheric weather systems in weather satellite images especially through their attendant cloud and radiation temperature fields, it is not surprising that many new climatological facts have emerged from studies of their time frequencies, and spatial distributions through extended periods. Perhaps the most interesting of such studies have been carried out for relatively inaccessible regions, though worthwhile results have also emerged for some better known land areas, and in the realms of jet stream climatology.

Particular light has been thrown upon the structure of the intertropical cloud band (ITCB), a prominent feature on visible and infrared images of the tropics. The position, range of forms, and secular shifts of the ITCB have been elucidated around the globe, and regional and seasonal contrasts described. Lateral to the ITCB, wave-like cloud features, associated with pertubations in pressure and wind fields, have been identified not only in the tropical North Atlantic (where easterly waves were first described) but in most if not all of the other tropical ocean regions too. The climatology of tropical revolving storms (especially hurricanes) has been much improved (see Chapter 18), as well as that of mid-latitude vortices and frontal bands especially in the southern hemisphere (Fig. 11.6). There is growing interest in satellite

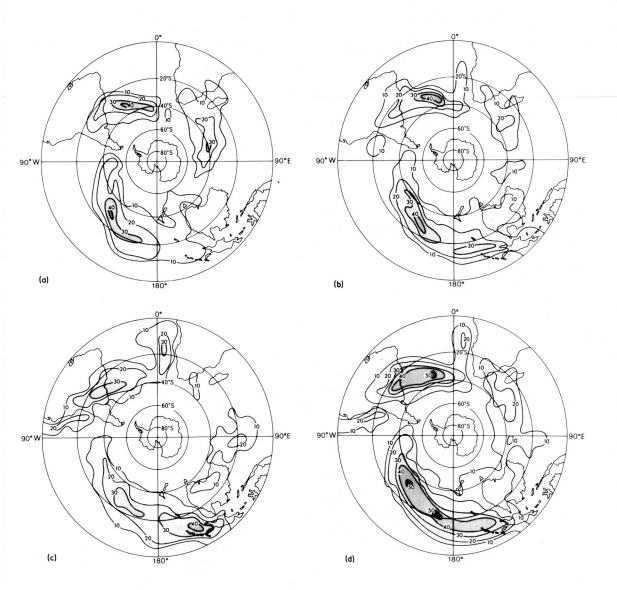

Fig. 11.6 Percentage frequency of 5-day averaged mosaics having axes of major cloud bands within a 5° latitude – 10° longitude square for (a) summer (December–March), (b) intermediate season (April, May, October, November), (c) winter (June–September), and (d) annual. Data based on November 1968–October 1971. (Source: Streten and Troup, 1973.)

evidence of atmospheric cycles in different parts of the world (Fig. 11.7), possible relations between them, e.g. in terms of the strengths and frequencies of synoptic weather systems ('atmospheric teleconnections'), and possible applications of such knowledge and understanding in climatic prediction.

11.6 The climatology of the middle and upper atmosphere

Many modern authorities believe that surface weather and climate is greatly influenced, or even largely controlled by, situations and events in the upper layers of the atmosphere. Thus *vertical*

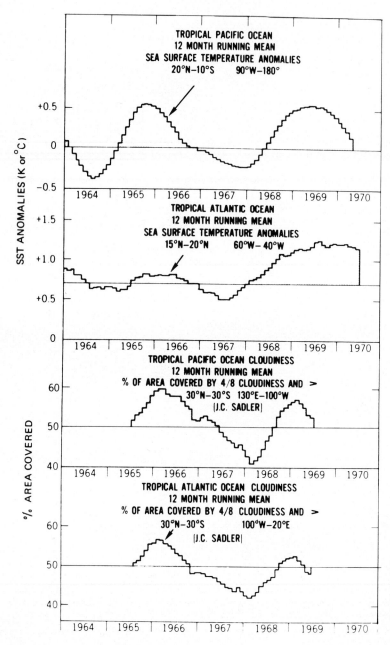

Fig. 11.7 A comparison of 12-month running means of sea surface temperature anomalies and satellite-derived cloudiness in per cent of area covered by $\geq 4/8$ cloudiness in specified sectors of the tropical Pacific. (Source: Allison *et al.*, 1972.)

interactions are being sought with increasing vigour. Since few radiosonde balloons penetrate far into the stratosphere, pre-satellite studies of the atmosphere above the tropopause depended heavily on expensive research rocket campaigns. From these, valuable vertical profile data were obtained, but understandably their spatial coverage was very poor. Today we have a much clearer picture of the

middle and upper atmosphere. Sensor systems like the CO_2-absorption waveband Selective Chopper Radiometers (SCR, see Chapter 3) of Nimbus satellites have provided data which have formed the bases for extended series of mean meridional profiles (Fig. 11.8(a) and (b)) and maps of selected layers high above the Earth's surface (Fig. 11.8(c) and (d)). Present high-altitude sounders (MSU and SSU)

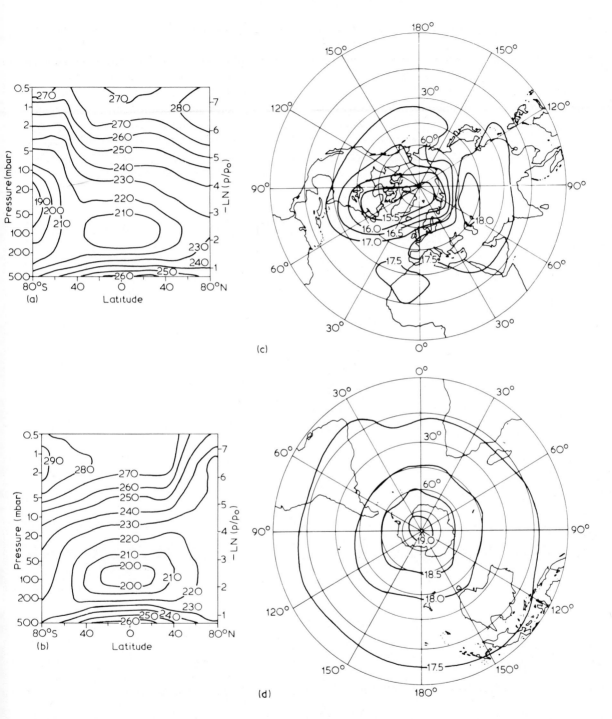

Fig. 11.8 Examples of meridional cross-sections of zonally averaged temperatures (Isopleths give temperature in K): (a) 16 July, 1970; (b) 21 January, 1971, obtained with Selective Chopper Radiometer (SCR) on Nimbus 4, and thickness of the 10–1 mbar layer at pressure surfaces from the same source: (c) Northern hemisphere; (d) Southern hemisphere, both on 4 July, 1970. (Source: Barnett *et al.*, 1973.)

described in Chapter 10 will help resolve some of the outstanding problems in upper atmosphere research. Prominent among these are questions concerning the directions and degrees of vertical interactions: does weather and climate develop from the top down, or the bottom up? The answers may lie in the middle atmosphere (the upper stratosphere and lower mesosphere). It is now clear that frictionally-driven waves propagate to such levels, but their relations with the mean flow has still to be understood in detail.

One thing is clear: at last, through the comprehensive eyes of a weather satellite system, we are able to study the whole atmosphere, either as a single, complex fluid continuum, or as a sum of many inter-related and more or less interdependent parts. The 1970s were particularly exciting years for the meteorologist; perhaps the 1980s will be a golden age of climatology from satellites.

12   *Water in the environment*

12.1 The importance of water

Water is the most ubiquitous yet most variable of the mineral resources of the world. Alone among the constituents of man's environment water may be found simultaneously in one area as a liquid, a gas, and a solid. Unlike most other Earth resources, water (on land and in the atmosphere) is continuously variable in its availability in one state or another. A further complication is that, geographically speaking, its patterns of concentration are always changing.

Apart from the purely scientific interest water arouses, man has a keen practical interest in its behaviour and distribution on account of its significance as a basic requirement of living organisms, and the consequent dependence of man himself and his economy upon it.

For these reasons – both academic and applied – the satisfactory monitoring of environmental water is one of the most difficult yet pressing problems confronting remote sensing as a distinctive discipline. Study of water in the environment is the science of Hydrology. This modern science is growing rapidly in importance because of ever-increasing demands for adequate and suitable supplies of water for human consumption, domestic use, agriculture and livestock husbandry, even for commerce and transportation. Hydrology has been defined by the WMO as *'the science which deals with the occurrence and distribution of waters of the Earth, including their physical and chemical properties, and their interaction with the environment'*.

For convenience, hydrology can be subdivided into:

(a) Hydrometeorology, concerned primarily with the exchange processes in hydrological cycles.

(b) Surface hydrology, which focuses on water at or near the surface of the land.

(c) Hydrogeology, concerned with water at or below the surface of rock and regolith.

(d) Oceanography, dealing with the nature and behaviour of the world's largest water bodies.

We will review the use of remote sensing in each of these fields in turn.

12.2 Hydrometeorology

The hydrological cycle of the Earth/atmosphere system can be represented in simplified form as shown in Fig. 12.1. This cycle is comprised of three types of components namely storages, transports, and exchange processes. By far the greatest reservoir of water is the oceanic girdle of the globe, although significant amounts of water are retained temporarily within the rocks of the Earth's crust, in ice caps and snow fields on the surface of the Earth, in soils and regoliths, in fresh water reservoirs like lakes and inland seas, and in the atmosphere. Water is transported by circulations in the atmosphere and oceans, and by movements over and near the surfaces of land areas chiefly under the influence of gravity. It is left to the exchange processes to effect transformations of water from one physical state (solid, liquid or gas) into another. These include precipitation, evaporation and evapotranspiration.

12.2.1 Precipitation monitoring

In any review of the hydrological cycle a convenient starting point is the water supplied to land and sea surfaces by the exchange process of precipitation. Since we have already reviewed the use of satellites for improved rainfall monitoring (see

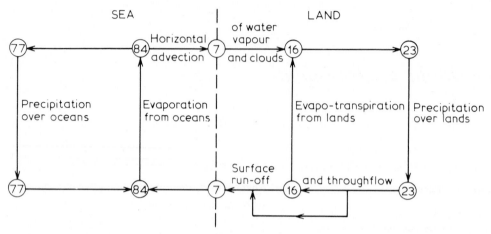

Fig. 12.1 A summary of the global hydrological cycle (assuming 100 water units).

Chapter 11), we can proceed to consider how surface-based remote sensing systems may be used for this purpose too, albeit on a more restricted scale. Microwave radar has been used for rainfall monitoring since the late 1940s. Today, operational use of rainfall radar is generally confined to parts of North America, Europe and the Far East, but it has a particularly vital part to play in meteorological and hydrological research. The most important applications are of the following kinds:

(a) Radar systems can provide real-time estimates of rainfall intensities over selected areas. For certain problems, such as river management and the control of soil erosion, the short-period rainfall intensity is an important factor.

(b) Radar systems can give estimates of the total rain which has fallen over, say, a basin or a catchment through a specified period of time. Such information is often vital to water management programmes generally, and predictions of the likely flows of streams and rivers in particular.

(c) 'Coherent' (Doppler) radar systems can give information on the movement and development of rain cells within clouds or cloud organizations. This is contributing significantly to our knowledge and understanding of the structure and behaviour of rainclouds and raincloud systems.

By established practice such rainfall data are conventionally obtained from networks of rain gauges, preferably of a continuous-recording type. However, it is common for the rainfall stations to be quite widely spaced. Especially when rain falls from convective clouds, e.g. thunderstorm clouds, the rain gauge data may be very misleading so far as average intensities of rainfall across wide areas are concerned. Dependent depth/area (volume) estimates of rainfall through periods of time may be quite inaccurate. Shortly after World War II it was recognized that radar was capable of observing the location and areal extent of rain storms, and that, in conjunction with some rain gauge data, more accurate maps of rainfall might be obtained from data from such a source.

If we know the technical specification and performance of a radar set, the radar equation basic to most rainfall studies may be written in simplified form as follows:

$$\bar{P}_r = \frac{cZ}{r^2} \tag{12.1}$$

where $\bar{P}_r$ is the mean received power (the 'signal') from the target, c is a constant (the speed of light), Z a reflectivity factor relating to the type of precipitation in the target area, (Plate 12.1), and r the range (or distance) between the radar and the reflecting shower of rain. The reflectivity factor is of particular significance, whether the output required is rainfall intensity or amount. Z is high where the raindrops are large and numerous, and low where they are small and few. Quantitative estimates of rainfall (R) may be derived from the

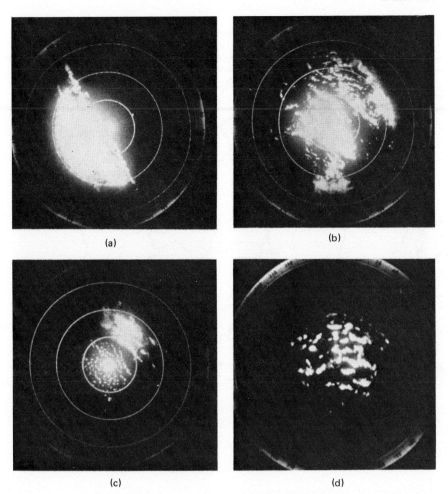

Plate 12.1 Typical maritime precipitation radar patterns interpreted in terms of the cloud characteristics with which they are likely to be associated. (PPI displays, 150 mile range). (a) Continuous stratiform cloud; (b) Ragged stratiform patches; (c) Small convective cells; (d) Large convective cells. (Source: Nagle and Serebreny, 1962.)

target echoes by evaluating the relationship $Z = aR^b$, where a and b are local constants established by empirical means. In middle latitudes, typical solutions for different types of rain are $200\,R^{1.71}$ for orographic rain, and $486\,R^{1.37}$ for thunderstorm rain. Such relations may be substituted for Z in Equation 12.1 so that, for operational usage, the radar equation (in simplified form) becomes:

$$\bar{P}_r = \text{constant} \times \frac{R^b}{r^2} \qquad (12.2)$$

The implication is that for different latitudinal zones and for different types of rain the radar signal is proportional to the rate of rainfall divided by the square of the range.

The principles have been employed by many workers on both sides of the North Atlantic. One recent project involved the Chester Dee in North Wales, being undertaken cooperatively by the UK Meteorological Office, the Water Resources Board, the Dee and Clwyd River Authority and Plessey Radar. The area covered measured approximately 1000 km². This was carefully instrumented to facilitate the necessary statistical comparisons between radar and conventional parameters (see Fig. 12.2 (a)).

The sequence of steps in the rainfall estimation process is set out in Fig. 12.2 (b). Sample results for a particular rainstorm are illustrated by Fig. 12.3. A good measure of agreement exists between

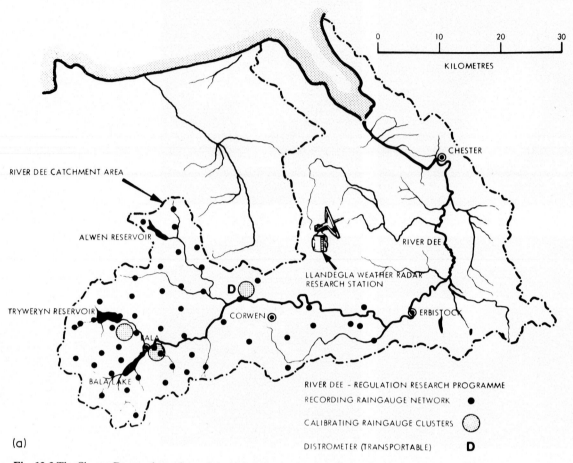

Fig. 12.2 The Chester Dee weather radar project. (a) Catchment instrumentation.

the two. Unfortunately in very mountainous, inaccessible terrain – the type of area from which radar estimates of rainfall might be most valuable – the narrow beam, low elevation type of radar currently used in work of this kind would be largely ineffective. If a higher elevation beam were to be used instead, the distance over which the rainfall estimates would be valid is relatively short, because ever higher regions of the atmosphere are sampled as range is increased. As a consequence, where there are large and important watersheds in mountainous regions, present systems and techniques can only cover a small area by each radar set. Perhaps new techniques – possibly involving a form of Doppler radar – can be devised to deal with this problem. Certainly for many hydrological purposes it is important to know as accurately as possible the scale, and variation through time, of this most important element in hydrometeorology.

12.2.2 Ice and snow monitoring

Our second important consideration in this section is frozen precipitation, especially through its surface manifestations of ice and snow. Two aspects of ice and snow distribution transcend the rest, namely the significance of frozen water within the hydrological cycle, and the influence of sea ice on maritime activities. A substantial portion of the world's fresh water resource is stored temporarily in the form of snow especially in high-mountain regions. Knowledge of the distribution of snow fields and their volumes in terms of water equivalents is required so that we can improve forecasts of stream flow and

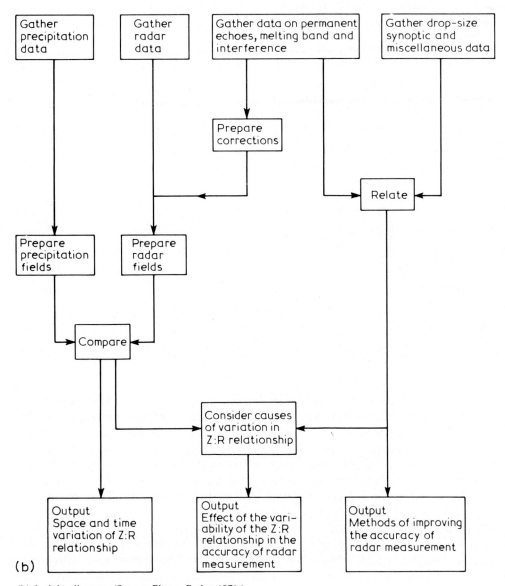

(b) Activity diagram. (Source: Plessey Radar, 1971.)

water storage. These in turn strongly affect power generation programmes, irrigation, water quality control, river management, manufacturing, recreation, and many other factors which contribute to a nation's economy. Sea ice imposes a seasonal control on shipping in many regions of the northern hemisphere. The opening and closing of ports is influenced thereby, and certain navigation routes can be used for only a part of each year, with or without the use of icebreakers.

Scandinavian and Alpine countries are prominent amongst those which have developed airborne techniques for estimating the volumes of snow accumulated on watersheds which serve hydro-electric stations at lower altitudes.

Several of these nations are now developing methods for the use of Landsat imagery to supplement or even replace surface or aircraft observations of ice and snow accumulations (Plate 12.2). It has been shown that such data can be

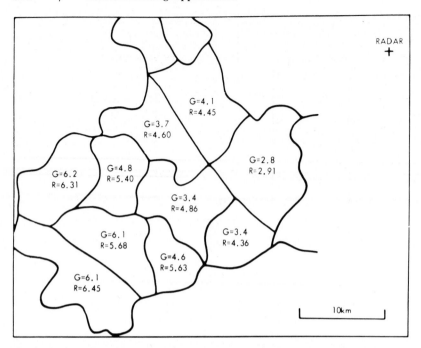

RADAR
+

G=4.1
R=4.45

G=3.7
R=4.60

G=2.8
R=2.91

G=6.2
R=6.31

G=4.8
R=5.40

G=3.4
R=4.86

G=6.1
R=5.68

G=3.4
R=4.36

G=4.6
R=5.63

G=6.1
R=6.45

10km

Fig. 12.3 The Chester Dee weather radar project. (c) Representative results: rainfall totals (mm) derived from radar (R), and raingauge (G), over sub-catchments for the period 1530–1630 GMT, 7 November, 1971. (Source: Plessey Radar, 1971.)

processed by computer to give snowcover area estimates with an accuracy of 93% or better down to the scale of individual pixels. Highly accurate positioning of the snowline is also possible – a most important consideration in run-off predictions and monitoring of the melting process in spring and summer. Perhaps the biggest remaining problem involves the separation of broken cloud from ice and snow. Research with the 13-channel Skylab-EREP-S-192 multispectral scanner has indicated that additional information from the 1.55–1.75 μm waveband makes clear discrimination of those features possible. This is covered by Channel 5 on the new Landsat Thematic Mapper (see Chapter 5).

Some broader-scale ice and snow surveys have also been based on Landsat imagery, for example improved maps of poorly-surveyed regions of Antarctica. However, for repetitive ice and snow mapping over large areas, weather satellites, especially of the Noaa family, are more widely used. The North American nations, Canada and the USA, have long operated an expensive programme of ice reconnaissance in the St. Lawrence region and northern coastal waters using aircraft to spot natural leads and passage through the ice. Satellite data were first used in support of this programme in the 'TIREC' (Tiros Ice Reconnaissance) Project in the early and mid-1960s. Today, Noaa imagery is analysed routinely by the US National Weather Satellite Service to provide a range of medium to small-scale ice and snow reports. These include:

(a) Northern hemispheric snow and ice charts: polar-stereographic maps of snowfield and ice-field boundaries and their relative reflectivities (three category classifications) on a scale of 1:50 m for use by the US Navy and various secondary users.

(b) Basin snow cover observations: per cent snow-cover messages sent via teletype, and maps sent via telecopier, for selected river basins in North America, primarily for use by the US Office of Hydrology, the Corps of Engineers, and the Soil Conservation Service.

(c) Great Lakes and Alaskan ice charts: detailed (1 km resolution) analyses of the boundaries and type or age of ice, revealing ice-fast and ice-free areas as well as ice concentrations and leads (navigational passageways). The primary users

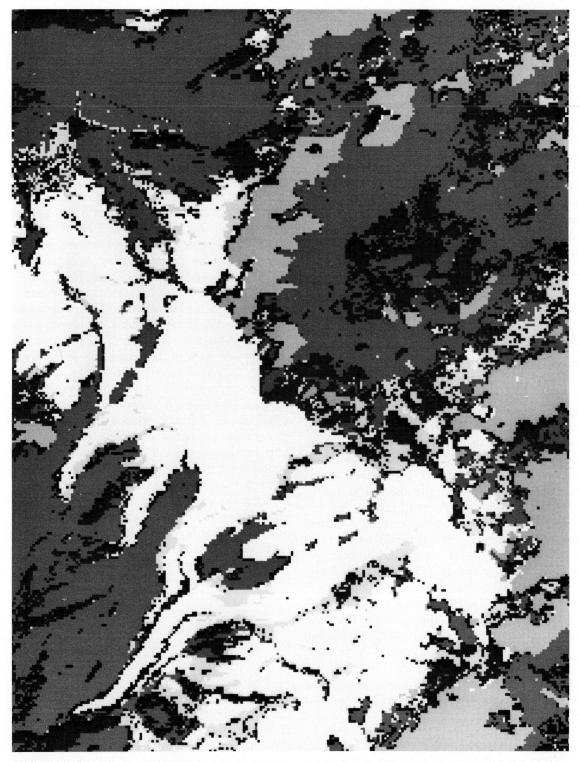

Plate 12.2 Digital mapping of snow and clouds in five main categories for test site 'Zermatt', Switzerland. White, snow in sun (and ice); Light grey, snow in shadow (and ice); Medium grey, clouds; Dark grey, snowfree area in sun; Black, snowfree area in shadow. (Source: Fraysse (ed.) 1980.)

include the US Navy and Coast Guard, the National Marine Fishery Service, and commercial marine transportation companies.

12.2.3 Evaporation and evapotranspiration estimation

Monitoring evaporation and evapotranspiration by remote sensing means has been described as '. . . the most challenging of all the applications of remote sensing to hydrometeorology' (A. Rango, 1980). These associated exchange processes play key roles in the hydrological cycle, and are of fundamental importance to agriculture through their effects on plant growth and performance. However, relatively little progress has been made towards the establishment of an acceptable monitoring method using satellites to provide the necessary data coverage in space and time.

The most promising approach to evaporation monitoring from satellites exploits the *thermal inertia* properties of soils as they are expressed through diurnal changes of surface temperatures, (see also p. 230 and p. 247). These changes are governed by the radiation budget (related to the external environment of the soil) and thermal inertia (related to the internal characteristics of the soil): the first can be modelled and evaluated with or without remote sensing inputs; the second can be assessed from infrared imagery when this is processed appropriately. We know that the thermal inertia of unsaturated soils is influenced greatly by soil porosity, and that any soil experiences a diurnal thermal inertia cycle which is closely related to its porosity. The greatest separation between the thermal inertia cycles of different soils is found about the local solar maximum and minimum, i.e. about 0200 and 1400 h local time. Much useful information relating to this problem has been obtained from the Heat Capacity Mapping Mission (HCMM), whose sun-synchronous orbit was planned to image at these times. The TELL-US model has been developed to yield the following parameters from twice-daily aircraft or satellite observations:

(a) Thermal inertia,
(b) Surface relative humidity,
(c) Daily estimated evaporation totals as illustrated by Fig. 12.4. Testing and development of this approach is continuing. Exploitation of the

more frequent radiation temperature measurements made by geostationary satellites like Meteosat should permit significant improvements in the accuracy of the results.

Turning next to evapotranspiration, a comprehensive estimation model designed to exploit satellite data has been developed from a resistance form of the energy balance equation, in which

$$LE = R_n - G - QC_p(T_c - T_a)/r_H$$

LE is latent heat flux (a measure of moisture in the air), R_n is net radiation, G is soil heat flux, Q is air density, C_p is the specific heat of air, T_c is plant canopy temperature, T_a is air temperature, and r_H is thermal diffusion resistance. Key variables which can be monitored by satellites include R_n, T_c and T_a. This is very demanding, for it requires the radiation intensities observed by satellites to be translated not only into radiation temperatures, but also corrected for temperatures of a thin layer of the atmosphere at its very base, as with the temperature in the upper foliage of a crop or stand of natural vegetation. Since atmospheric transmission and surface emissivities are quite variable, and canopy temperatures possess a high degree of spatial variation it is not surprising that results so obtained are thought to be too gross. It is recognized, too, that the evapotranspiration performance of a crop or vegetation area changes through the growing season. 'Leaf Area Indices' (relating the leaf area of a crop to the area of the ground on which it grows) evaluated from Landsat reflectance data are being invoked in attempts to tune such methods for these seasonal effects. However, to quote Rango again, '. . . only preliminary studies have been conducted, with inconclusive results. No widely acceptable or transferable technique has (yet) emerged.' So research continues, with the big prize of widespread application of the successful method at the end.

12.3 Surface hydrology

Some overlap is inevitable between this section and the subsequent chapter on 'Soils and Landforms'. However, the emphasis is different, for in the present context, the 'surface' is to be viewed only in respect of its significance for hydrology through run-off, infiltration, and/or standing surface water. We may review this large and rapidly-expanding

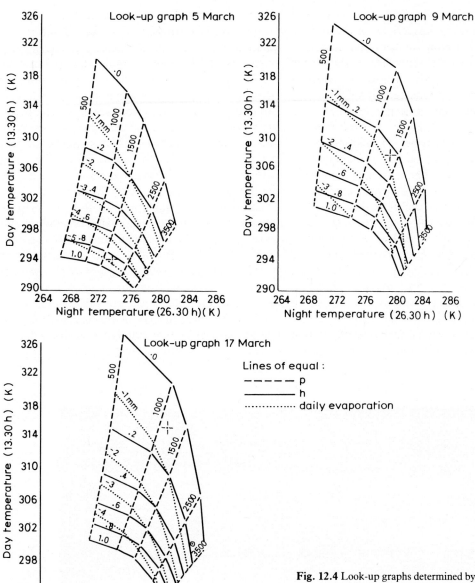

Fig. 12.4 Look-up graphs determined by the TELL-US Method representing 3 subsequent stages in drying of a test plot in the USA, March, 1971. The crosslets mark observed day and night temperature pairs. From the graphs, values of thermal inertia, surface relative humidity and daily evaporation may be read off. (Copyright, EEC.)

field under six sub-headings, leading us from matters of observation, through modelling, to conclude with possibilities of hydrological prediction.

12.3.1 Watershed parameters

The chief object of terrain analysis is to recognize and map 'recurring landscape features', especially through geomorphic, soil and land use character-

istics. Since terrain is, to a greater or lesser degree, dynamic, the advent of air survey and Landsat image analysis methods has contributed dramatically to the ability of applied hydrologists to take account of the change factor in their assessments of significant watershed parameters. For example, deforestation and urbanization both encourage increases in the rate and volume of basin run-off, even in the absence of any change in precipitation.

In areas which are heavily clouded, recourse must be made to microwave imagery in the search for more detailed and/or more frequent information on remote and inaccessible watershed regions than could be obtained readily at ground level. A benchmark project codenamed RAMP (Radar Mapping of Panama) was initiated in 1965 by the US Army Engineer Topographic Laboratories to demonstrate the performance of high-resolution radar imagery in lieu of optical photography in heavily clouded regions. The sensing system was a K-band radar (see Fig. 2.2). This has better than a 99% cloud penetration capability, although it does not penetrate heavy rain or vegetative cover. Once-over coverage of the eastern end of the Republic of Panama was achieved in approximately four hours of flying time. A number of interpretational overlaps were prepared from the final mosaic. These included surface drainage patterns, and types of surface drainage regions (Plate 12.3 and Figs 12.5 (a) and (b)) as well as general surface configuration (topographic provinces), vegetation and engineering geology.

It was possible to identify variations in the local and regional expression of surface drainage, for example differences in the density of streams and other drainage channels, and the depth of channel incisions. In turn such variations in the drainage patterns are diagnostic of specific terrain conditions, for example the nature and depth of soil cover, lithology, morphology, and the prevailing structural and tectonic influences. The patterns present themselves in an infinite variation of density and habit (or form). The *density* is determined mainly by the lithologic character of the rock traversed – itself a matter of considerable interest to hydrologists – involving its hardness, porosity, solubility and consolidation. The *form* relates mainly to structure, involving faults, fractures and the attitude of bedding. Thus the drainage pattern overlay was the first to be developed from the imagery, being basic to most of the others. Clearly side-looking radar (SLR) has a part to play in regional hydrological studies especially where cloud cover restricts the use of aircraft as platforms for conventional photography.

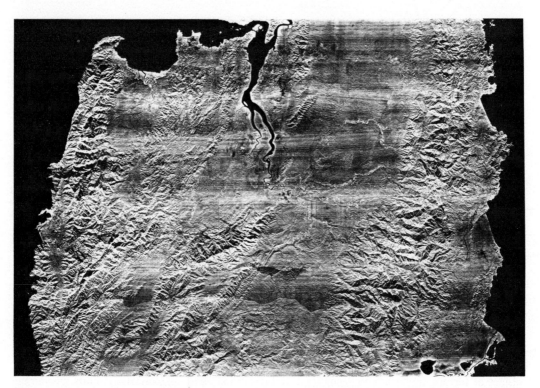

Plate 12.3 K-band radar mosaic of Darien (Panama) and north-west Colombia. (From Viksne, *et al.*, 1969.)

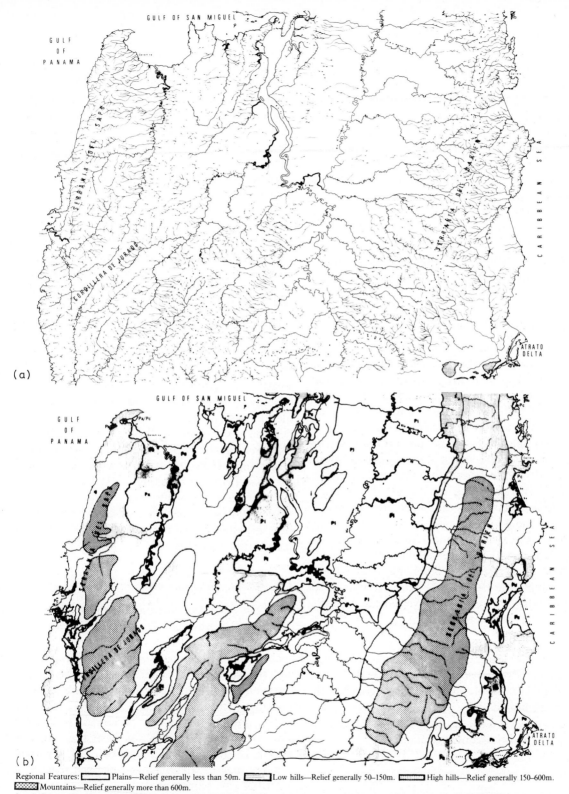

Regional Features: ▭ Plains—Relief generally less than 50m. ▭ Low hills—Relief generally 50–150m. ▭ High hills—Relief generally 150–600m. ▨ Mountains—Relief generally more than 600m.

Local Geomorphic Features: Pu—Upland plain, Pc—Coastal plain, Pi—Interior plain, Pa—Alluvial plain, ox—Oxbow lake, de—Delta, lg—Lagoon, be—Beach, ob—Offshore bar, nl—Natural levee, tf—Tidal flats, fp—Flood plain.

Fig. 12.5 Hydrologically-significant distributions mapped from a radar mosaic of the Darien region of eastern Panama, and northwest Columbia: (a) surface drainage network. (b) The surface configuration. (Source: Viksne *et al.*, 1969.)

At the scale at which the more remote satellite systems operate most efficiently, the complexity of the total hydrological problem, and the detail required by many operational monitoring programmes, comprise formidable hurdles to be overcome if remote sensing is to replace more direct 'contact' methods of assessing key variables. This is one of the relatively few fields for which further technological development is required before a useful potential might be realized. Remote sensing for everyday hydrological use has been likened to the computer; here we have very powerful new tools with the potential to yield large increases in efficiency in specifying and tackling environmental problems. This potential has yet to be fulfilled.

12.3.2 Drainage network features

Stream network analysis, involving the mapping, measurement, and classification of active drainage lines especially in terms of stream lengths, complexities of stream patterns, and the location of associated features including springs, ponds and lakes is an important aspect of modern hydrology. It has been shown that these can be assessed very satisfactorily from Landsat MSS imagery at scales of 1:250 000 or even better. Furthermore, changes in river channel characteristics can also be evaluated therefrom. Many rivers exhibit frequent changes of course and interactions between interlacing ('braided') channels. Manual analyses of Landsat scenes, displayed on simple colour additive viewers, have been undertaken profitably for many regions, for example the basin of the Kosi River in Bihar State, India, whose results revealed the following MSS band preferences:

(a) Band 5: flood plain features, e.g. abandoned channels, sandy islands, shoals etc., and contrasts between them and nearby vegetated areas.
(b) Band 7: active river courses (water is dark in the near infrared).
(c) Band 4, 5 and 7 false colour composites: alluvial areas, and classes of vegetation.

12.3.3 Flood plain mapping and monitoring of floods

Much of the world has still to be mapped on the ground at scales of 1:250 000 or better, affording much scope for Landsat imagery to be used to map important flood plain features either for the first time, or, over even wider areas, after each new major flood event. Unfortunately, because of its low imaging frequency, Landsat is less well-suited to the mapping of floods, although under optimum arrival time and sky conditions such satellites can and have been extremely valuable sources of vital flood statistics. Whilst similar restrictions also circumscribe the use of environmental satellite data for such purposes, these data carry the additional disadvantage of a relatively low resolution. However, Meteosat visible and infrared imagery have been used for the monitoring of change in large lakes and swamps, e.g. Lake Chad, and the Okavango Swamps in southern Africa. Passive microwave data from ESMR on Nimbus 5 have been analysed successfully in studies of the widespread floods in eastern and central Australia in 1974 (see Colour Plates 12.4 (a) and (b)); passive microwave systems have the advantage of a cloud-penetration ('all-weather') capability, and could be useful in synoptic-scale mapping of major flood events, to identify preferred aerial survey flight paths for more detailed assessments of flood extent and damage. Data requirements for flood monitoring in basins larger than 1000 km^2 were proposed at an international Planning Meeting in 1976. Table 12.1 summarizes the conclusions, with an added indication of those elements for which there is, even at present, either a significant or a reasonable satellite monitoring capability.

12.3.4 Wetlands monitoring

Here the object of attention is the hydrology of coastal and freshwater lake regions. Surface water area assessment is of key significance, involving such things as seasonal susceptibility to flooding, and the variation of water reserves in dams and reservoirs. As remarked earlier, surface water is prominent in Landsat Band 7 images, in which water bodies as small as one hectare can be located, and rivers only 70 m across have been directly traced. In coastal zones a wide range of important features may be identified and mapped reliably at 1:250 000 or better, including the marsh-water interface, the upper wetland boundary, and different plant communities within the marshy area. Successful estimates of water quality have been

Table 12.1 Data requirements for floods in basins larger than 1000 km². (After Ferguson *et al.*, in Salomonson and Bhavsar (eds.) 1980)

Hydrological element	Accuracy	Resolution (m)	Frequency	Present satellite capability
Flood plain area/boundary	±5% area	10	5 y*	Good
Flood extent	±5%	100	≤4 days**	Partial
Precipitation	±2 mm if <40 mm ±5% if >40 mm	5000	6 h	Partial
Saturated soil area	±5%	100	≤4 days**	Partial
Soil moisture profile	±10% field capacity	1000	1 day	Little or none
Snow covered area/snowline	±5% area	1000	1 day	Good
Snowpack water equivalent	±2 mm if <2 cm ±10% if >2 cm	1000	1 day	Partial
Snowpack free water content	±2 mm if <2 cm ±10% if >2 cm	1000	1 day	Little or none
Snow surface temperature	±1°C	1000	6 h	Partial

*And after each major flood event.

**Depends on degree of flooding: with major floods, daily monitoring would be desirable.

made through regressions of a 'quantitative brightness' parameter (the cube root of the product of standardized Landsat data in Bands 4, 5 and 6) against observed sediment levels in North American lakes.

12.3.5 Hydrologic basin models

There are many types of hydrologic basin models, including:

(a) Physical (laboratory) models, based on hydrologic processes.
(b) 'Representative Basin' models, based on empirical findings in a small number of well-instrumented basins considered to be characteristic of the broader areas in which they are set.
(c) Mathematical models, based on response, rather than causation. These include deterministic, parametric and probabilistic varieties.

In every case the aim is the same, namely to transform the *inputs* to a basin into *outputs* with the best possible representation of reality. Fig. 12.6 depicts some of the key variables which are involved, with the interlinkages between them.

As we have seen earlier in this book, satellite remote sensing can help to monitor precipitation inputs and, less well, outputs via streamflow and evapotranspiration. Levels of interception, soil moisture and groundwater storage are strongly influenced by topography and subsurface geology: aircraft and satellites can be used to map and assess the significance of these important influences on local basin cycles.

12.3.6 Mathematical modelling for flood hydrograph prediction

The ability to accurately forecast peaks on the stream hydrograph is, perhaps, the most conclusive test of applied hydrological knowledge and understanding. This ability is much sought after because of its significance for hazard advisories and warnings, as well as in river management and control. On the flood-flow scale, three processes are of particular importance, namely:

1. Infiltration into the soil surface, and through the soil profile.

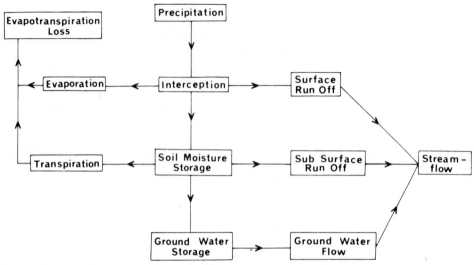

Fig. 12.6 A water-balance schematic for individual catchments. Certain of the components lend themselves more readily to assessment by remote sensing than others. (Source: Painter, 1974.)

2. Surface run-off.
3. Channel flow.

Five steps have been proposed for the development of a basin model to inter-relate these and thus predict flood peaks accurately. For each of these steps, preferred remote sensing approaches can be identified, as follows:

(a) Determine the physical extent of the basin (boundaries and area). Air or Landsat visible imagery should be used unless a high incidence of cloud necessitates radar imagery. A low angle of solar illumination is best, for the evidence of shadows is especially useful in watershed determination.

(b) Idealize the basin topography. Stereoscopic air photographs (or multifrequency radar images where slopes are shallow) are analysed for rectangular slope elements (areas and mean angles of slope).

(c) Idealize stream channel geometry. Near-infrared imagery (air-photo and Landsat MSS Band 7) may be used both for high water level determination and the analysis of stream patterns into rectangular line segments. Photogrammetric procedures must be followed for the analysis of stream cross-section at low water.

(d) Quantify the hydraulic roughness of slope elements and stream channels. Radar is the most direct source of information on slope roughness, because the wavelengths used are of the same scale as the surface variations. Visible image analysis (air-photo or Landsat) may be used to identify larger scale landscape or terrain units. Channel roughness may be assessed under clear water conditions by imagery in the blue/green region of the visible spectrum (approx. 0.5 μm); where water is turbid, much greater difficulties are encountered.

(e) Differentiate and delineate areas of impermeable surfaces and saturated soils; both these conditions result in surface flow of additional water. Impermeable surfaces may be identified and assessed in infrared images on account of their low thermal inertias. Many of these surfaces are of human origin, and have sharp, rectangular, outlines. Wet soils are dark in visible images, and radar at wavelengths in excess of 20 cm has some ability to integrate subsurface moisture values.

Our conclusion must be that remote sensing has a clear and varied applicability to surface hydrology, especially in poorly-instrumented regions, but no single approach can provide all the additional information required from remote sensing sources. Perhaps more than in most other fields of environmental enquiry a *multisystems* approach is strongly indicated.

12.4 Hydrogeology

Since hydrogeology is concerned predominantly with features at depth rather than at or very near the surface of the Earth this is the province of the geologist rather than the environmental scientist. Consequently our review of this economically important field will be very brief. Remote sensing texts relating more specifically to geology may be consulted for greater detail on this, and related, topics.

Groundwater is important to the hydrologist (as distinct from the hydrogeologist) because it is a store of water at least theoretically available for controlled surface exploitation. Now since groundwater is, by definition, subsurface, neither air nor satellite observations can be used to delineate aquifers directly. Rather, remote sensing may be of value in large-scale preliminary investigations of groundwater reserves. Important indicators of groundwater include anomalous soil moisture and morphologic (especially fracture) zones, vegetation types and densities, geological structure, and stream flow characteristics (especially the density of the drainage network). Under reasonable conditions all such phenomena can be elucidated by air-photo interpretation procedures and photogrammetry. Landsat image analyses are often profitable in large, poorly surveyed countries. Extensive groundwater surveys have been undertaken, for example, in India, based on Landsat MSS data as the primary data source. Ground surveys are planned for the areas of greatest promise, and test wells drilled in the most promising locations. Then subsurface information from the wells is correlated with the remote sensing indicators so that extrapolations can be made sideways into areas of apparently similar hydrogeology. Indeed, it may be remarked that this type of approach, which integrates remote sensing data with ground truth and then improves assessments of conditions between the ground control points using aircraft or satellite data alone is fast becoming the classic way of capitalizing on remotely-sensed data for environmental monitoring operations.

12.5 Oceanography

Oceanographic remote sensing can be subdivided into two broad categories related to the use or non-use of the dedicated oceanographic satellite, Seasat, which suffered such a disappointingly short life of 100 days, but which attracted great interest within the oceanographic community. For present purposes, therefore, it will be most convenient to review remote sensing applications in oceanography generally at first, before we summarize finally the contribution of Seasat to ocean science.

12.5.1 Water temperature and circulation patterns

Knowledge of sea-surface temperature distributions and changes over large areas is important for proper understanding and accurate evaluation of many oceanic and atmospheric processes and phenomena. These include the detection and monitoring of ocean currents, water fronts, upwelling zones and other thermal or motion systems, levels of air-sea energy exchanges and the initiation, development and dissipation of many atmospheric organizations at scales ranging from the local (1–10 km) through the meso, to the synoptic and even global scales.

Any unique interpretation of infrared radiance measurements from a satellite in terms of sea surface temperature requires a homogeneous target of known emissivity filling the radiometer field of view. Early studies of sea surface temperatures using Nimbus satellite infrared data employed synchronized cloud photographs to reveal the extent to which radiation temperatures of target areas were contaminated by clouds. These, of course, reduce the temperature levels. Aircraft-borne support systems have been used to assess the accuracy with which prominent features (e.g. the boundary of the Gulf Stream) were indicated by Nimbus isotherm maps. Fig. 12.7 shows a single scan line from a Nimbus HRIR analog record, along with an enlarged section of the analog record after digitization and the application of a numerical filter to remove oscillatory noise.

Later, temperature patterns were mapped, not only on individual days in various parts of the world, but also on a repetitive basis in selected areas in the search for secular patterns of temperature change. Fig. 12.8 illustrates an application of satellite thermal infrared data to the study of short-period temperature variations. Such studies reveal the dominant circulation patterns of water at different temperatures, and assist us in assessing the parts

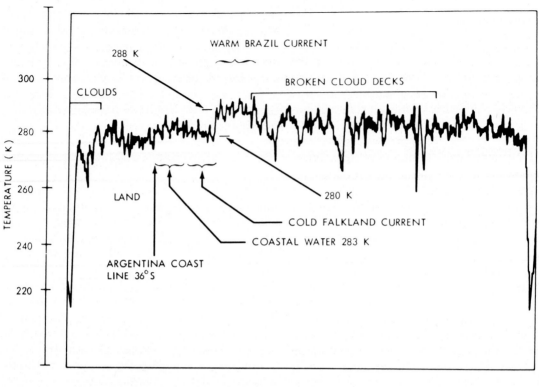

(a)   ← WEST   EAST →

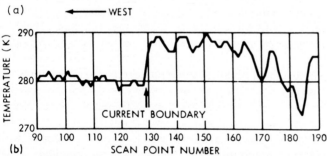

(b)

Fig. 12.7 Nimbus 2 observations of currents in the South Atlantic. (a) An analog record of an individual HRIR night-time scan-line taken on 17 August, 1966. (b) A section of the analog record in (a) after digitization and application of a numerical filter to remove oscillatory noise. (Source: Warnecke *et al.*, 1969.)

they play in related systems, for example the Earth/atmosphere energy budget. In this case (the Persian Gulf) the incoming water from the Gulf of Oman can be seen to compensate for the strong local excess of evaporation over precipitation. The inflow current is seen to spread along the Iranian coast until it turns cyclonically back along the Trucial coast. The water is warmed in this shallow region, and flows back eastward as warmer water to mingle with the currents along the Iranian coast. These circulations were previously unknown.

More recently, NESS have developed methods for computer processing of data from operational polar-orbiting weather satellites for routine sea-surface temperature mapping. The model used is known as GOSSTCOMP (Global Operational Sea Surface Temperature Computation). Surface temperatures are derived by a histogram technique applied to 1024 instrument measurements with partially overlapping fields of view in areas of roughly 100 km^2 around the retrieval point. Corrections for atmospheric attenuation are computed from VTPR data. The model develops Earth-located values of sea-surface temperature between 8–10 000 times daily. Computer products include photographic displays (see Plate 12.5) and gridded fields (maps showing each individual temperature estimate). The accuracy of the results is checked

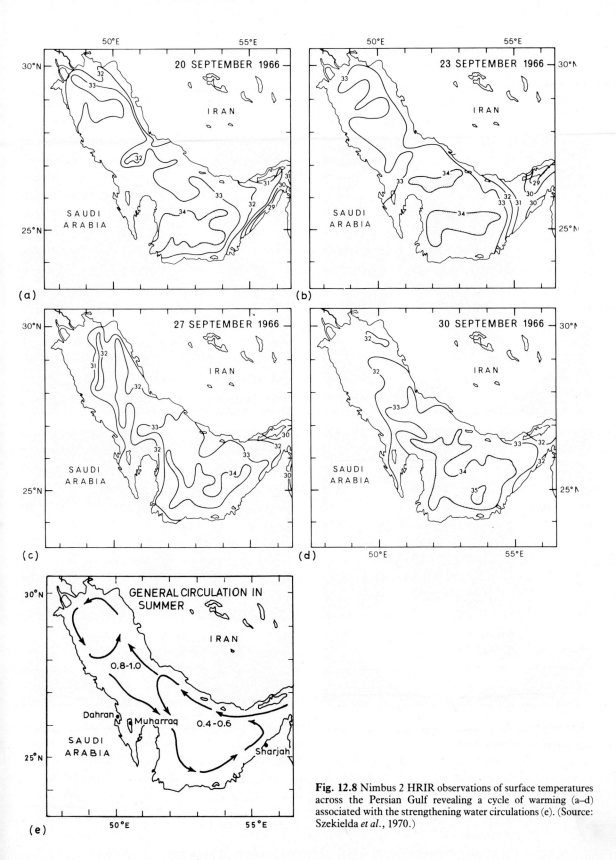

Fig. 12.8 Nimbus 2 HRIR observations of surface temperatures across the Persian Gulf revealing a cycle of warming (a–d) associated with the strengthening water circulations (e). (Source: Szekielda *et al.*, 1970.)

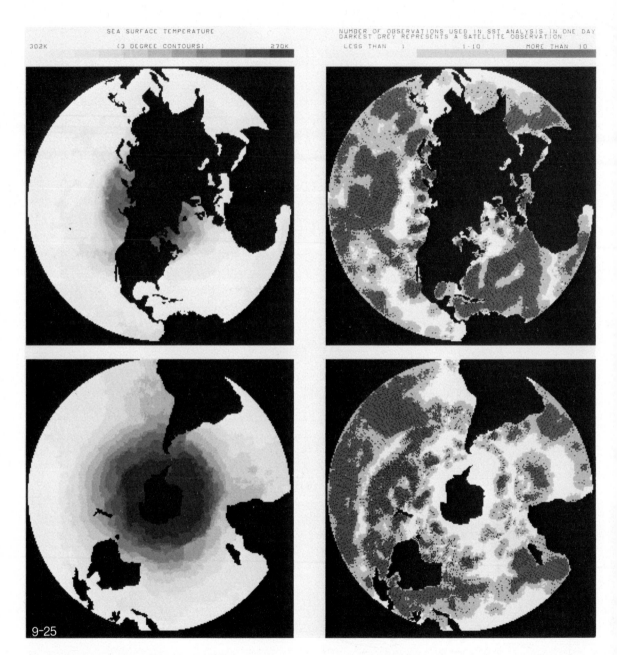

Plate 12.5 Example of GOSSTCOMP 'quick look' photographic display of sea surface temperatures and spatial distribution of observations. (Courtesy, NOAA.)

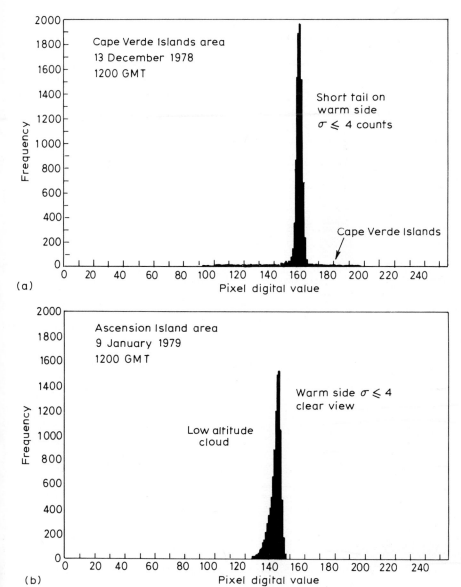

Fig. 12.9 Histograms for 100×100 pixel areas of Meteosat infrared window waveband data; (a) clear sea view, (b) clear sky plus low altitude cloud.

twice daily by comparison with ship-board observations. It is generally within ±1.5°C of 'sea-truth'. More detailed products from NESS include:

(a) High resolution (1 km) maps of surface temperatures of the Great Lakes.
(b) Bulletins on the location of the Gulf Stream Wall, where the current achieves its fastest flow.
(c) Experimental Gulf Stream analyses, to show current speeds and areas of cold and warm water eddies.

(d) Analyses of the US West Coast Thermal Front for the location of upwellings of cold water, where fish nutrients abound.

Routine methods for sea surface temperature generation have been developed also for geostationary satellite data. Of special importance is the interpretation of the initial histograms for clear-column radiances, which should be uncontaminated by the effects of clouds. Fig. 12.9 illustrates some common histogram forms.

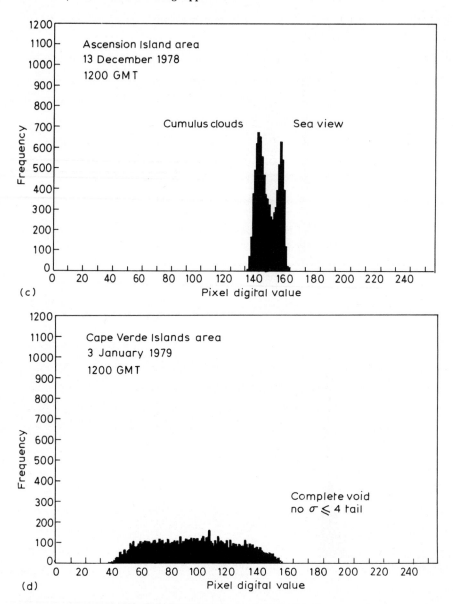

Fig. 12.9 cont. (c) clear sky plus partial cumuliform cover, (d) complete cloud cover. (Source: Williamson and Wilkinson, 1980.)

Where cloud cover is complete, or totally absent, the histograms are unimodal. In partly cloudy regions they tend to be bimodal or multimodal. The highest temperature mode corresponds to surface temperature, the lowest to extensive cloud contamination.

12.5.2 *Water quality and salinity assessments*

The turbidity of water is affected by a wide range of natural and anthropogenic factors, including suspended sediment content, biological activity (e.g. the abundance of plankton), pollution etc. Entire sub-disciplines have grown up to study water quality especially through reflectance properties of, and light propagation in, water bodies: *optical limnology* deals with freshwater lakes, and *optical oceanography* with oceans and seas. The theory of water optics has been exploited variously in the remote sensing of water quality. Of fundamental

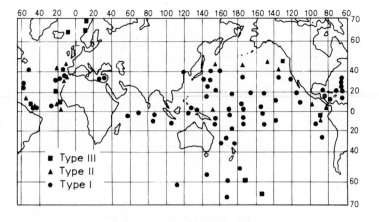

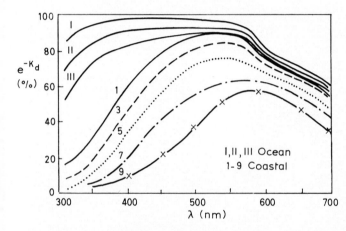

Fig. 12.10 Optical water types classified from aircraft MSS observations. (After Jerlov, 1976; from Fraysse (ed.), 1980.)

importance is the general rule that the spectral distribution of upwelling light from a water body contains quantitative information on the content of suspended and dissolved materials. We may consider two of the many possibilities which depend on this principle.

First, a 12-category classification of oceanic and coastal *water types* has been developed from aircraft multispectral imagery, as illustrated by Fig. 12.10. This reveals a trend of water clarity extending from the murky inshore waters of high to middle latitudes to the clearest areas of the tropical oceans. Coastal waters generally are much more turbid than open oceans not only because of higher concentrations of pollutants, but also because of large quantities of so-called 'Gelbstoffe' or 'yellow substance', which is the overall product of dissolved organic matter brought by rivers into these shallow water areas. Since the design of Landsat was optimized for land applications, not for oceans and seas, MSS data are of limited use for the mapping and monitoring of

water quality types, though Band 5 is useful for broad-scale optical wave type determination.

Second, the capacity for useful oceanic application of the Coastal Zone Color Scanner (CZCS) on Nimbus 7 may be illustrated through reference to monitoring of chlorophyll (phytoplankton) concentrations (Table 12.2). Preoperational tests suggested

Table 12.2 Wavebands investigated by the Nimbus 7 CZCS

Band number	Wavelength (nm)	Application
1	443	Multispectral detection of chlorophyll, suspended sediment and Gelbstoffe
2	520	
3	550	
4	670	
5	750	Comparison with Landsat MSS Band 6
6	11.5 (μm)	Ocean surface temperatures mapping

Plate 12.6 (a) Seasat SAR image of the Tay Estuary, Scotland, 19 August 1978, showing many coastal and land surface features.

that CZCS visible data should be suitable for estimating chlorophyll through known relationships between chlorophyll concentrations and CZCS band-differences, e.g. the difference between albedoes observed in Band 1 minus Band 2. Work on this topic is now proceeding with CZCS data, spurred by the knowledge that fishery advisories stand to benefit greatly from successful results.

So far as salinity monitoring is concerned, it has been shown that natural variations in the salt content of the oceans can be revealed by microwave methods, for, in the microwave region of the spectrum there are good relationships between the emissivity and absorptivity of water, and water temperature, dissolved gases and salinity. Fig. 12.11 illustrates comparative results for fresh and salt water. These could prove useful for studies of estuarine water circulations.

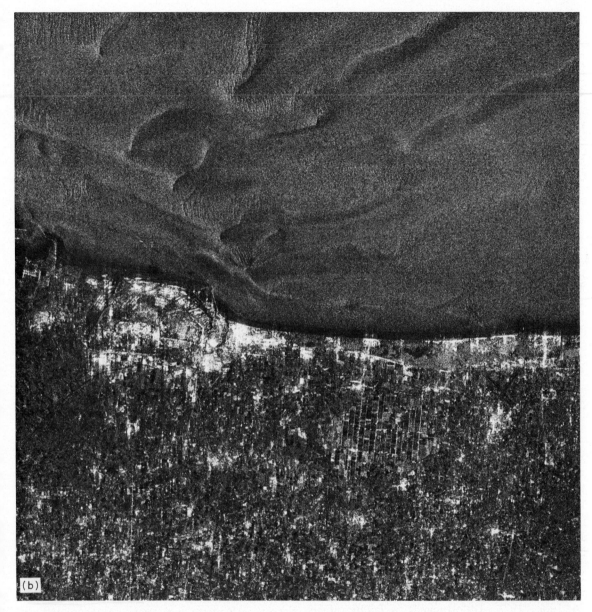

Fig. 12.6 cont. (b) Seasat SAR image of Dunkirk, France, 19 August 1978, showing sea surface features related to the topography of the sea bed. (Courtesy, Space Department, RAE, Farnborough; Crown Copyright Reserved.)

Finally, a word on pollution monitoring, a theme of growing international significance at the present time.

Contamination of water bodies by human activities lends itself more readily to investigation by remote sensing techniques since the effluents of commercial and industrial activity often contrast sharply with the waters which receive them. Aircraft have been used in many instances to identify and map the discharge patterns from sewage outfalls and industrial complexes. Oil spillages from ships at sea, and oil leaks from maritime wells can be located with particular ease from aircraft – and, potentially from satellites. There are four reasons why this is so:

1. The emissivity for petroleum products is significantly higher than for a calm sea surface.
2. Crude oil pollutants have increasing emissivities with increasing gravity.

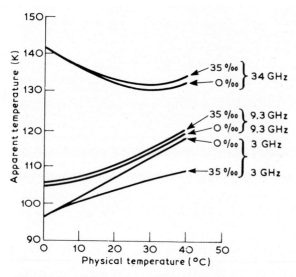

Fig. 12.11 Theoretically predicted curves of sea surface brightness temperature (vertical axis) versus physical temperature (horizontal axis) for different salinities (%) at selected wavelength of emitted microwave radiation.

3. The radiometric response of oil varies little with time of day or the age of the pollutant.
4. It is possible to design reliable microwave systems to map oil pollution.

Already, remote sensing data have been invoked in courts of law to obtain convictions of those responsible for significant pollution events. Further use of remote sensing in support of international law is sure to follow.

12.5.3 Seasat: the way ahead

The first purpose-built oceanographic satellite, Seasat 1 (see pp. 90 and 92), was launched into a near-polar circular orbit at an altitude of 800 km on 26 June, 1978. Unfortunately a failure of its electrical power subsystem occurred only 106 days later, on 9 October, and marked the end of its useful life. However, it had already demonstrated the great value of its instrument package, and had returned a rich supply of both land and sea surface data to Earth. Some researchers have remarked that the early demise of Seasat 1 was fortunate in that it has forced the principal investigators to examine a manageable quantity of data thoroughly, rather than an overwhelming flood of data very partially.

Seasat carried four microwave sensors, and an auxillary Visible-Infrared Radiometer (VIRR). The suite of microwave sensors, which gave the spacecraft its all-weather operating capability, included:

(a) A *radar altimeter* looking vertically down to measure wave height and the microtopography of the ocean surface.
(b) A *synthetic aperture radar* (SAR) which scanned a 100 km wide swath located some 250–350 km to the right of the sub-satellite track, primarily to measure wave length and direction, to provide data for the study of coastal processes, and to chart ice fields and leads.
(c) A *scatterometer* scanning out to 1000 km on either side of the spacecraft for wind speed and direction assessments.
(d) A passive *scanning multi-channel microwave radiometer* (SMMR) which recorded naturally-emitted microwave radiation from the target at five frequencies between 6.6–37 GHz, primarily for global measurement of sea-surface temperature.

It is generally agreed that these primary objectives have been well met. Many bonuses have also accrued, for example through the realization that much of the SAR imagery over land contains information of potential value for geological, hydrological, glaciological, and even land-use studies. Plate 12.6(a) illustrates the high resolution achieved by SAR, and the wealth of detail in the imagery over land and in coastal zones. An exciting and entirely unexpected finding has been that, under suitable conditions of sea-bed topography and tidal flow, seafloor features have visible expressions on the water surface. Plate 12.6(b) is a good example of this effect in the English Channel. Many other interesting phenomena may be evidenced also in Seasat SAR data, e.g. wave refraction and diffraction in coastal zones, current features, inlet plumes, and wetland characteristics.

Undoubtedly oceanography will benefit greatly from future satellites and instrument packages modelled on Seasat and its payload. Indeed, it is not too much to expect that oceanography, potentially, if not presently, the biggest subdivision of hydrology, will undergo revolutionary growth in scale and practical significance as ocean-monitoring satellites increase our knowledge of maritime conditions by a veritable 'quantum leap'.

13   Soils and landforms

13.1 The nature of soils

Soil is of major importance to mankind because it supports the plants and animals that provide man with food, shelter and clothing. It is also of importance as the foundation material for the buildings in which we live. Different concepts of soil have, therefore, been developed by agriculturalists and engineers.

The agriculturalist finds the whole of the soil profile significant to plant growth: the surface (A horizon) because it is the seat of biological activity and the principal source of nutrients; the subsoil (B horizon) because it affects drainage, soil moisture retention, aeration and root development of plants. Both A and B horizons are important in respect of their influence on soil temperatures. The underlying parent material (C horizon) is mainly significant because it may contain weatherable minerals yielding nutrients and its texture may affect permeability. Deeper parts of the regolith (D horizons) are generally of less importance to plant growth.

In contrast with the agricultural concepts of soil, the engineer normally defines soil as the unconsolidated sediments and deposits of solid particles derived from the disintegration of rock. Therefore in civil engineering the term soil includes all regolith material and a sharp demarcation between rock and soil is no longer made. In fact, the engineer is normally more concerned with what the agriculturalist would term the C and D horizons. From the engineering standpoint soil can be regarded as a three-phase system which is in dynamic equilibrium. The three phases are: soil (particulate organic and inorganic material), liquid (consisting of soil solutions containing various salts) and a gas phase containing soil air with changing amounts of oxygen, nitrogen and carbon dioxide.

The equilibrium between these phases changes continuously and the engineer studies the interactions of the phases and the physical properties of soils affecting soil engineering, e.g. soil stability.

13.2 The mapping of soils

The recognition of soils as organized natural bodies is a development of nineteenth-century science and is generally credited to the Russian scientist Dokuchaiev who first explained the relationships between soil profiles and the environment. It was not until the 1920s, however, that soil units were eventually mapped on the basis of the soil profile using A, B, C horizon notations. With further experience of soil mapping it was recognized that soils were very variable in their properties and several kinds of soil might occur within a mapping unit. Even though a major part of a mapped unit might accord with the definition of a particular soil type, not all of the area would conform. Thus it became usual to recognize that a percentage (up to 15%) of soils within a mapping unit would show variation from the described attributes of the unit. In general, the user wishes within-class variance to be as small as possible, or below a certain threshold value. A number of studies of soil variability have been made. These include variances and intra-class correlations for various topsoil properties in Oxford, England, which showed mechanical properties were acceptable in terms of within-class variance, whereas chemical properties were not.

Studies of soil units by remote sensing methods must, therefore, take into account both different concepts of soil bodies held by agriculturalists and engineers and the variability of topsoil characteristics, particularly in respect of chemical properties.

The soil categories used by agriculturalists that are significant to remote sensing studies include those of great soil groups, sub-groups, soil series and soil phases. The great soil group brings together soils having similar horizons arranged in similar sequence e.g. Typorthod (Podzol). Thus a broad zone of Podzolic soils extends across northern Canada and Russia over a variety of rock types. This unit is, therefore, suitable for small scale (1:1 000 000 or less) or regional maps. At medium scales (c. 1:50 000–1:100 000) the soil series category is more commonly used. A soil series consists of a group of soils of similar profiles derived from a particular parent material. The soil series is often subdivided into soil phases. The basis of subdivision into phases may be any characteristic or combination of characteristics potentially significant to man's use or management of soils. The most common differentiating attributes for soil phases are slope, degree of erosion, depth of soil, stoniness, salinity, physiographic position and contrasting layers in the substratum.

Soil classifications for engineering purposes are primarily based on the sieve size of the soil material. Thus the extended Casagrande Soil Classification as used by engineers has as its major divisions, coarse-grained soils and fine-grained soils, subdivided into gravelly, sandy and fine-grained soils. Finer subdivisions in the classification use criteria such as uniformity of deposit, grading of deposit, plasticity, shrinkage or swelling properties, drainage characteristics and dry density.

13.3 Remote sensing of soils and landforms by photographic systems

13.3.1 Conventional air photography

It has been shown above that engineers and agriculturists adopt different definitions and classifications of soils; as a result, remote sensing techniques are required to serve several purposes based on different definitions of soil and different basic concepts. The first major postwar study of the recognition of soil conditions, made at Purdue University, USA, was aimed mainly at identification of parent materials and was chiefly of value to those interested in soils engineering. It was based on two assumptions. First, that once photographic characteristics for soils at one site had been determined the identification of similar sites elsewhere on the photographs could be used to identify similar soils. Second, that by study of the aerial photograph such features as landform, surface colour (or tone), erosion, slope, vegetation, land use, micro-relief and drainage patterns could be used to deduce the general character of the soil. Further development of this latter approach by the International Training Centre for Aerial Survey in Holland led to a technique of pedological analysis. In this, each element of the landscape was mapped out separately (e.g. slope, vegetation, land use) and a map was prepared from a composite of the boundaries. The boundaries which were coincident with more than one element of the landscape were given additional weight in the final interpretation of soil boundaries.

The widespread recognition of the close association of soils and landforms promoted interest in landscape mapping as a means of reconnaissance soil mapping. Research was based on the idea that knowing detailed characteristics for one terrain unit, one can apply them to another analogous terrain unit elsewhere. The problem facing terrain analysis was (and continues to be) one of classification and definition of the units to be recognized on the photograph. It was necessary to have some means of identifying the terrain unit and storing data relevant to the unit so defined so that other workers could recognize it elsewhere. Thus a system of hierarchical classification was proposed as shown in Table 13.1.

This 'land system' approach has been widely applied, for example, in Australia by Land Research and Regional Survey teams, South Africa by engineers and Africa by agriculturalists. It is best applied to those areas which display distinctive and well-defined facets of landscape. Where units are ill-defined and where cultivation has largely destroyed the natural vegetation cover it is less easily applied.

The development of computing techniques has led to investigations of the parametric description of landforms on an entirely numerical footing. Computer analysis of landform now seems to be within our grasp and future developments of the

Table 13.1 Terrain unit classification

Unit	Description
Land element	The simplest part of the landscape; uniform in lithology, form, soil and vegetation
Land facet	Consists of one or more land elements grouped for practical purposes
Land system	Recurrent pattern of genetically linked land facets
Land region	Consists of one or more occurrences of land system local forms which are generally contiguous
Land division	An assemblage of surface forms expressive of a major continental structure (morphotectonic)
Land zone	The world extent of a major climatic type

land system approach are likely to be in the form of automated cartography using some form of supervised classification based on knowledge of soil-landscape relationships.

The accuracy of soil mapping from the air photograph has proved to be very variable. Comparisons of soil maps made from ground observations with photo-interpretation maps have revealed that increasing complexity of geological deposits or increasing importance of properties not closely associated with landforms greatly reduces the accuracy. Also there is increasing awareness that the timing of air photography is critical. Photographs taken at inappropriate times may provide an unsatisfactory yield of information. Under European conditions photographs to record soil-tone patterns and soil changes should be taken in winter and spring, the best months being March and April (Plates 13.1(a) and (b)). Crop patterns which reflect soil boundaries are best recorded in July and early August. These requirements must, however, be fitted in to the period when conditions are suitable for aerial survey and so the planning of flights and placing of contracts should be carried out well in advance.

Suitable statistical methods for the testing of the goodness of soil boundaries drawn by photo-interpretation have become necessary following the increasing use of air photography. Multiple correlated attributes of the soil profile can be reduced to a single variate expressing a large proportion of available information by principal component analysis. This variate, the first principal component, can be plotted against distance on a linear transect and approximate positions of maximum (positive) and minimum (negative) slope found by inspection. These positions are then pinpointed using Student's t statistic. They represent the points where the soil boundaries, the maximum rates of change of soil with respect to distance, cross the transect. Beckett (1974) has already reviewed the statistical assessment of resource surveys by remote sensors, particularly in respect of the precision of information and the costs of obtaining data on soil conditons at different scales.

The use of colour photography as an aid in soil mapping only became important after 1960 following advances in technology by film manufacturers. Studies of the use of colour film for engineering soil and landslide studies showed that it gave a saving in interpretation time of fifty per cent in the USA. Research by British and Russian scientists indicates that colour aerial photography yields more data on soil conditions than panchromatic photography. Work has shown, however, that soil colours as recorded in the field and on aerial photography do not correspond exactly in their Munsell colours, indeed overlap of hue, value and chroma is quite frequent. The influence of different soil hues on the optical densities of aerial colour and aerial infrared colour has subsequently been studied by densitometric methods. Statistical tests applied to density data obtained from a sample of twelve soils have shown that the soils can be separated into two groups on the basis of significant differences in densities. One group consisted of those soils with low chroma (soils grey or neutral in colour) which were best distinguished by infrared colour. The

Plate 13.1 Soils on a former estuarine marsh and adjacent upland, near Martin, Lincolnshire: (a) under crops, July 1971. (b) Bare soil, April 1971. (Note the clear portrayal of the patterns of creeks which formerly traversed the marsh. The tonal variations in the photograph along the creek lines are primarily due to variations in soil texture and moisture content. (Source: Cambridge University Collection: Copyright Reserved.)

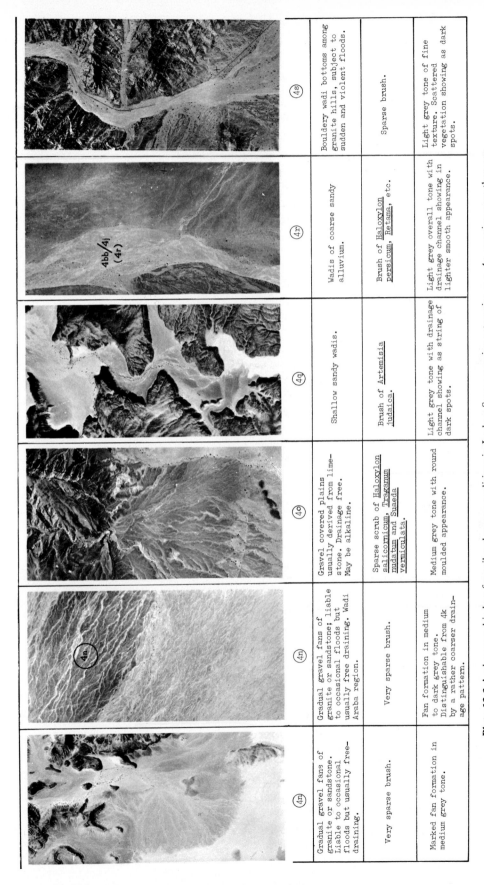

	4m	4n	4o	4q	4r	4s
Soil	Gradual gravel fans of granite or sandstone. Liable to occasional floods but usually free-draining.	Gradual gravel fans of granite or sandstone; liable to occasional floods but usually free draining. Wadi Araba region.	Gravel covered plains usually derived from limestone. Drainage free. May be alkaline.	Shallow sandy wadis.	Wadis of coarse sandy alluvium.	Bouldery wadi bottoms among granite hills, subject to sudden and violent floods.
Vegetation	Very sparse brush.	Very sparse brush.	Sparse scrub of Haloxylon salicornicum, Traganum nudatum and Suaeda vermiculata.	Brush of Artemisia judaica.	Brush of Haloxylon persicum, Retama, etc.	Sparse brush.
Tone	Marked fan formation in medium grey tone.	Fan formation in medium to dark grey tone. Distinguishable from 4k by a rather coarser drainage pattern.	Medium grey tone with round moulded appearance.	Light grey tone with drainage channel showing as string of dark spots.	Light grey overall tone with drainage channel showing in lighter smooth appearance.	Light grey tone of fine texture. Scattered vegetation showing as dark spots.

Plate 13.2 A photographic key for soil and range conditions in Jordan. Some mapping categories, e.g. 4n, require more than one photograph to illustrate the appearance of a particular soil and range condition. (Courtesy, Aerofilms, Hunting Surveys Ltd., Boreham Wood, Herts.)

second group included soils with high chroma (intense soil colour) which were best distinguished with colour film.

Infrared colour transparencies require suitable techniques for viewing and annotation and suitable light tables and filter frames have been developed as more and more infrared colour has been used. Techniques have also been developed for the production of panchromatic internegatives and positive prints from infrared colour film. There remains, however, the fact that transparencies are less convenient for use in manual photo-interpretation and this may partly explain the somewhat slow adoption of infrared in user agencies despite its considerable information yield in respect of soil conditions and crop vigour. A summary of the advantages and disadvantages of different types of film used in aerial surveys is given in Table 4.2.

In order to communicate the photographic characteristics of soil and terrain units to other users photographic keys of various kinds have been devised. Many are available for use, including keys for vegetation, landforms, forest sites, and soil conditions. An example of a photographic key for soil and range conditions in Jordan is shown in Plate 13.2. The continuing interest in site selection is reflected by recent publications dealing with terrain evaluation, some of which are listed in the further reading recommended for this chapter.

13.3.2 Multispectral photography

Since soils are composed of mineral and organic constituents together with varying amounts of soil moisture it can be assumed that different soils will show different reflectance characteristics as these constituents vary in amount and kind. There was, therefore, early work by soil scientists on the reflectance characteristics of different soil types, particularly in the infrared part of the spectrum. These studies made in the 1950s were not related to remote sensing techniques but with the increasing use of multispectral aerial cameras and multispectral scanners there has been renewed interest in spectral signatures of surface objects, including soils. Recent laboratory measurements within the wavelengths of 0.32–1.0 μm have been made on soils in wet (almost saturated) and dry (oven dried at 43°C) states.

From some 160 sets of curves three general shapes emerged (Fig. 13.1). Type 1 curve was characteristic of chernozem-like (Mollisol) soils, Type 2 of pedalfer soils (Spodosol), Type 3 of lateritic soils (Oxisol). An important part of the analysis of these results showed that reflectance measurements made at only five wavelengths would suffice to predict with sufficient accuracy the other 30 wavelengths measured. It was recommended that the five selected wavelengths should be 0.4, 0.5, 0.64, 0.74 and 0.92 μm. Much more extensive studies of spectral signatures have now been made in the field as well as in the laboratory under NASA contracts. These consist of airborne and ground measurements

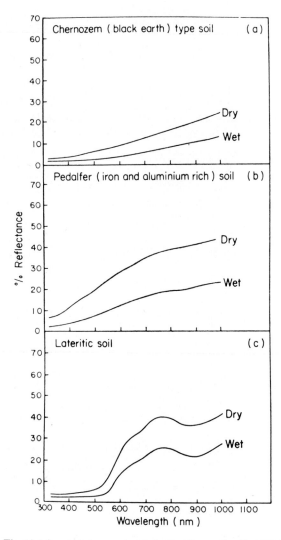

Fig. 13.1 Spectral characteristics of soils. (Source: Condit, 1970.)

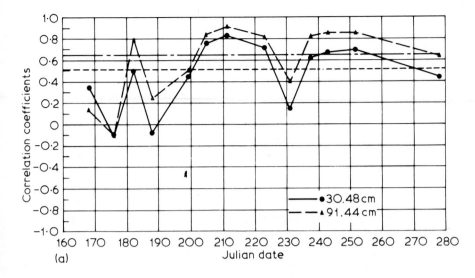

(a)

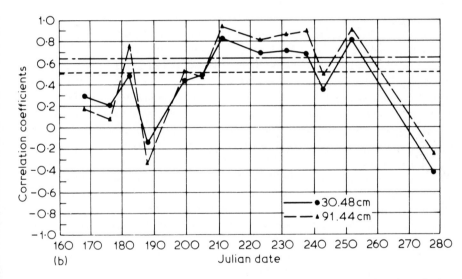

(b)

Fig. 13.2 Mission correlation coefficients plotted against time for the available soil moisture in the sorghum versus film density (a) colour infrared film, and (b) green filtered black and white film. (Source: Werner *et al.*, 1973.)

over variable parts of the spectrum within the range 0.3–45 μm. Similar studies have been carried out by Russian scientists. For example, densitometric studies of spectral reflectance of sands in the Karakum Desert have shown that isopleths drawn by trend surface analysis of the spectral data can be used to distinguish three types of sand with differing mineralogical characteristics.

Spectral reflectance data has also been examined for the purpose of evaluating soil moisture and water use by plants by airborne remote sensing methods.

Relationships were investigated between remote sensing imagery and soil moisture in grain sorghum. Three spectral ranges were used, 0.47–0.61 μm (Green), 0.59–0.70 μm (Red) and 0.68–0.90 μm (near Infrared). It was found that soil moisture differences could best be detected in the red portion of the spectrum. The near infrared region of the spectrum gave little information concerning the soil moisture status. The green spectral band gave good results during the time of active plant growth. Examples of correlation coefficients plotted against

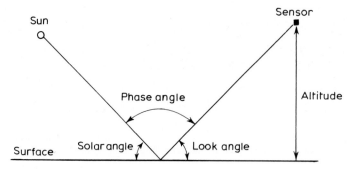

Fig. 13.3 Geometrical relationships and terms for the use of polarized photography. (Source: Curran, 1981.)

time for the available soil moisture in the sorghum versus film density for red filtered (8403 film) and infrared (2424 film) are shown in Fig. 13.2.

13.3.3 Polarized photographic data

Although multispectral sensing for soil moisture assessment has received much attention, little work has been focused on the use of polarized visible light. When the unpolarized light waves are reflected from a soil surface the degree of polarization at a high phase angle (see Fig. 13.3) is much affected by surface soil moisture. Where free water stands at the surface, light will not be scattered but reflected specularly in one vibrational plane. For rough surfaced soils below saturation the reflectance is diffuse and polarization low, but it will increase as saturation is approached. The intensity of polarized visible light reflected from the surface is related to a very thin surface layer of less than 1 cm. Laboratory studies have shown the relationships between soil moisture content, per cent soil moisture and phase angle (Fig. 13.4).

Early studies revealed that polarization of reflected sunlight provides the basis for a sensitive technique for remote sensing of soil surface conditions. In particular the polarization percentages have been shown to be related to both water content and soil type (Fig. 13.5). The technique seems particularly useful when soils are at maximum moisture retention capacity, or near to this state. However, it should be noted that in some circumstances the degree of polarization is more dependent on soil type than on soil moisture and variations in soil aggregation are particularly important.

Some success has been obtained by analysing data obtained from a light aircraft using a camera equipped with a polarizing filter. Photographs were taken with the focal axis set to face the sun and angled towards the ground surface. Two exposures were made of each scene, both through a polarizing filter onto a panchromatic film. The camera operator can obtain the record of maximum polarization by turning the lines of the filter until they are first parallel with the ground surface and then at right

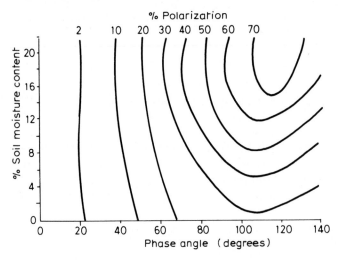

Fig. 13.4 The degree of polarization (PVL) recorded using a photometer–polarimeter in the laboratory for a range of soil moisture contents and phase angles. (Source: Steg and Frost, 1971.)

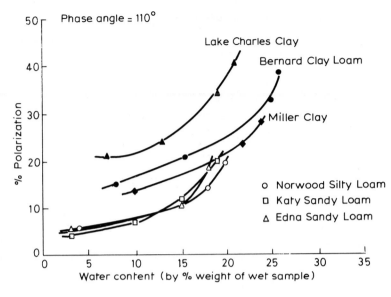

Fig. 13.5 Polarization related to water content and soil type. (Source: Stockhoff and Frost, 1972.)

angles to the surface. The densities of the imaged areas can then be measured quantitatively to show per cent polarization (Fig. 13.6) by using a densitometer, and standardization of densities for each scene are made by reference to the density of a marker such as a road or vehicle roof within the scene (Fig. 13.6). Such a technique is, however, limited to very low altitudes (50–100 m) due to atmospheric effects and to variability in soil area sampled. Also such factors as soil particle size, soil albedo and the range of moisture values will affect the correlation of polarized light to moisture content. In brief, the technique is best suited for obtaining an overview of moisture content in small sites where surface roughness is constant and data can be obtained in sunlit conditions.

13.4 Remote sensing of soils and landforms by non-photographic systems

13.4.1 Visible and near-infrared scanner data

Data from all three Landsat satellites have been used extensively in terrain studies. In particular it has been useful in Third World countries where the topographic and thematic mapping base is often poor. The unique, near orthographic perspective of 34 000 km² of the Earth's surface provided by each Landsat frame gives good opportunities for soil and land systems to be interpreted. The interpretation of satellite imagery for terrain analysis is essentially an extension of the conventional technique of air-photo interpretation discussed in Chapter 8. It involves

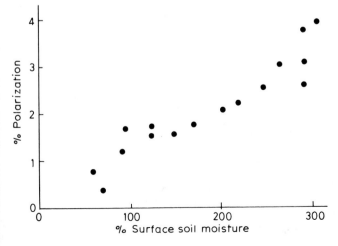

Fig. 13.6 Per cent polarization (PVL) recorded for a range of surface soil moisture contents. The peat soil is recorded from a light aircraft using a camera with polarizing filter. (Source: Curran, 1978.)

evaluation of a number of image characteristics such as tone, or colour, texture, size, shape, contrast and shadow. The scale differences between aerial photographs and satellite images pose an initial problem for the interpreter of terrain features, but this can be quickly overcome.

Land system and soil boundaries are normally interpreted from 1:500 000 and 1:250 000 false colour composite prints derived from MSS bands 4, 5 and 7. Clear plastic overlays placed over the image are used as the drawing medium.

In a study of land systems in Western Sudan, Hunting Technical Services Ltd compared the 'standard' land system map prepared by field and air-photo interpretation methods with a map prepared from Landsat 1 imagery. As can be seen in

Table 13.2 Western Sudan: Land system area percentage error.

Land system	Standard map area (km²)	Landsat map area (km²)	Difference (km²)	Error (%)
Qoz	9877	10 190	+313	+3
Baggara	5513	5 228	−285	−5
Basement	6381	6 261	−120	−2
Bahr	70	72	+2	+3

Table 13.3 Western Sudan: soil agricultural potential groupings percentage error. (After Parry, 1978)

Soil units – agricultural potential	Ground survey soil units (km²)	Landsat interpreted soil units (km²)	Interpreted soil units (% error)
Sandy soils cultivated	8 965	9 364	+4
and suitable for	462	648	+40
cultivating using	240	271	+13
traditional hand	405	375	−7
tillage methods			
	10 072	10 658	+6
Alluvial soils	2 239	1 413	−37
suitable for	1 289	1 756	+36
mechanized	1 078	1 258	+17
agriculture	44	74	−8
	5 450	5 201	−5
Alluvial soils in	2 310	2 416	+5
major wadis suitable	70	72	+3
for irrigation			
	2 380	2 488	+5
Basement soils	1 355	1 630	+20
unsuitable for	2 584	1 774	−31
cultivation			
	3 939	3 404	−14

Table 13.2 the analysis showed that no system interpreted from satellite imagery has an areal difference compared with its equivalent on the 'standard' map of more than ±5%. This level of accuracy was considered adequate for a reconnaissance survey. When soil mapping units were compared for this same area of the Western Sudan it was found that there was a greater discrepancy between the 'standard' map and satellite interpretation. The former contained 122 mapping units compared with 102 units derived from satellite interpretation. Nevertheless, the overall accuracy improved if the soils were grouped according to their agricultural potential (Table 13.3).

It must be recognized that soil mapping from satellite data is, at present, severely limited by the resolution of the sensors employed. Therefore the mapping units which can be determined are those of major soil associations. Furthermore, the successful application of satellite data has been mostly in semi-arid or sub-humid areas of the Third World where terrain units often show close relationships to the soil associations. For example, North West India has been well covered by Landsat imagery recording seasonal changes in soil state. In one study of the irrigated areas in Haryana District west of Delhi it was possible to identify nine terrain types with different soils and drainage characteristics (Fig. 13.7). The soil groups present ranged from Psammustents on dune fields to Camborthids and Calciorthids in depressions. The interpretation was carried out by examination of a 1:250 000 false colour composite with the aid of a monoscopic magnifier. The boundaries were drawn on a transparent overlay. Digital printouts or densitometric plots can, of course, be used as an aid to the mapping of soil associations. For example, Landsat 1 imagery of Obion County, West Tennessee, was scanned by a high-speed digital scanning microdensitometer using a 25 μm raster. Subsequently the computer printouts were able to separate major soil associations such as the Memphis occurring at the break between loess soils and the delta soils of the Mississippi floodplain (Fig. 13.8). It was noted that water and soil moisture are most effectively imaged in the near infrared, therefore, Band 7 is the most useful one in such studies.

Separation of small soil associations in intensive row crop agricultural areas was found to be much more difficult at the scale of Landsat imagery. In this study the reflectance was thought to be at a minimum at low moisture contents (about 2 bars tension) whereas maximum reflectance was obtained at moisture levels slightly below field capacity (1/3 bar tension).

Studies of Landsat imagery of South Dakota have also shown that soil associations can be identified from film colour composites of bands 4, 5 and 7 when viewed over a light table with magnification. Specific comparisons between boundaries shown on the Landsat composites and soil association maps are shown in Fig. 13.9. It was found that soil association boundaries were mainly distinguished through the identification of boundaries around landscape features. Vegetation differences were also found to assist the placing of boundaries around the soil associations.

The work outlined above was mainly concerned with soil identification. Other studies have been made, however, to determine whether general terrain units such as water, exposed rock, grasslands, coniferous forest and lowland marsh areas can be identified. An example of such a study is that carried out in respect of terrain classification maps of Yellowstone National park. Training sets of the five categories just described were defined by ground observations. Samples were selected for each of the terrain types and the signature of each type was extracted from the digital record of the Landsat imagery. At first individual (i.e. from one band only) signatures were obtained. Later, signatures from several bands were combined. The optimum channels for classification were then determined by calculating the average probability of misclassification (i.e. an estimate of the percentage of points which will be incorrectly classified in a terrain recognition map). The results were as shown in Table 13.4.

It will be seen that using bands 5, 6 and 7 was only slightly less accurate than using all four bands. Thus economy of computer effort would indicate that three bands only might be used.

13.4.2 Thermal infrared imagery

Although the scanners in the visible and near infrared wavelengths are producing good results there is considerable research into the applications of

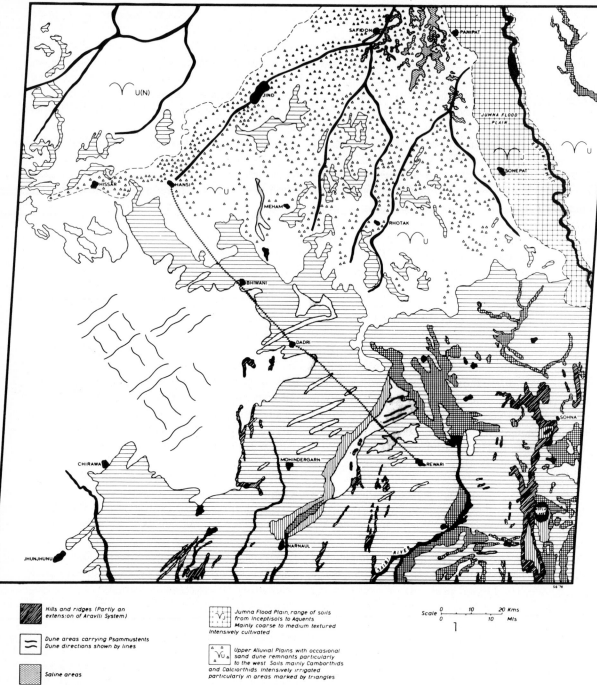

Fig. 13.7 Irrigated areas in Haryana district west of Delhi. (Source: Curtis, 1976.)

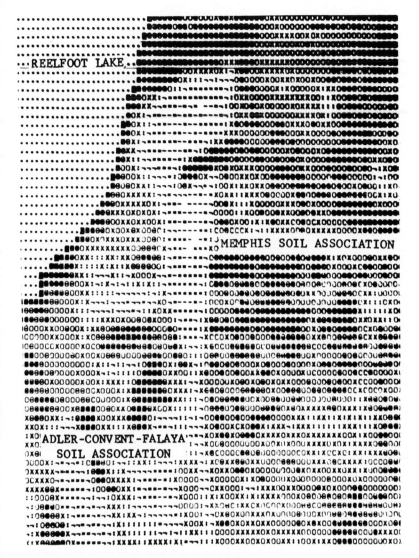

Fig. 13.8 Landsat image and computer print-out of soil association map. (Source: Parks and Bodenheimer, 1973.)

thermal infrared and microwave scanners for soil studies. The use of thermal infrared for soil temperature studies has been investigated by NASA investigators using aircraft-borne sensors.

Detailed soil surveys were related to the thermal infrared imagery in a manner which suggested that surface soil temperatures can be indicative of sub-surface soil conditions. Such results are to be expected in view of the effect of soil texture, composition, porosity and moisture content on soil temperature regimes in different soils. Thermal infrared scanning is particularly sensitive to moisture content at the very surface of soils. This is particularly the case when surface winds evaporate moisture from the surface layers. In these circumstances the cooling of the surface in exposed areas contrasts with the higher surface temperature in areas protected from the wind. An example of these conditions is shown on Plate 13.3 which shows part of an area of Somerset under conditions of 11°C air temperature and 4.5–7.9 m s^{-1} wind strength. This is an area of reclaimed estuarine clay which is under pasture and requires extensive field drainage. The open drainage ditches are readily distinguished by dark tones reflecting the lower temperatures of the water surfaces in relation to the land. The lines of buried field drains can also be detected as a result of the wetter soil conditions along these sub-surface

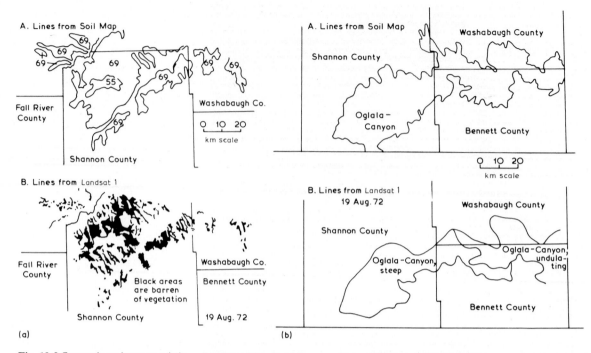

Fig. 13.9 Comparisons between existing soil maps and Landsat 1 imagery. (Source: Weston and Myers, 1973.)

channels. The white toned areas in the lee of field boundaries reflect the shelter effects of the boundaries and the lower evaporation and hence higher surface temperatures existing in these sheltered places.

Table 13.4 Average probabilities of misclassification for five category recognition (Yellowstone Park)

MSS Bands	Average probability of misclassification
5	0.086
5, 7	0.031
5, 7, 6	0.028
5, 7, 6, 4	0.027

The principal space investigation of the relationships of soil temperatures to soil conditions has been through the Heat Capacity Mapping Mission as developed from the project for Applications Explorer Mission-A (AEM-A). This system provided for a circular, sun-synchronous orbit at 600 km altitude. The major sensing system was the Heat Capacity Mapping Radiometer (HCMR) which in-

cluded two bands (0.5–1.1 and 10.5–12.5 μm) and provided a resolution of 500 m with a swath width of 700 km. The HCMM satellite was launched on 26 April 1978. One of the principal investigations of the mission was that made by European scientists in the Tellus project. The specific objective was to map diurnal soil temperature variations so that, by correlation with ground truth observations, thermal inertia models could be developed for the prediction of variations in soil moisture content of the top soil layers. The principle adopted was that soils of high water content would show lower diurnal variations in surface temperature than would drier soils. Two digital models were used in the Tellus project sponsored by the Joint Research Centre of the European community. One model, the 'Tergra' model, is for use in grassland areas and uses one measurement of the soil surface temperature. The second, the 'TELL-US' model, developed by Rosema, estimates daily evaporation and thermal inertia by measurement of both day- and night-time temperatures (see also Chapter 12). Estimated and calculated values for evaporation and soil moisture for a site in Buckinghamshire, England, are shown in Table 13.5.

Plate 13.3 Infrared linescan imagery of estuarine lowland near Huntspill River north of Bridgwater. Note dark tones of water and clear representation of field drains by dark lines of lower temperature. (Courtesy, Royal Signals and Radar Establishment, Malvern; Crown Copyright.)

13.4.3 Microwave radar data

Knowledge of the spatial and temporal distribution of soil water is of economic and scientific significance to many agricultural, hydrological and meteorological applications. In view of the effects of soil moisture on the dielectric properties of surface materials there is growing interest in microwave sensors for the detection of moisture states at the ground surface. Also the dynamic nature of water conditions in the landscape makes the microwave sensors attractive because they can be used on a repetitive basis due to their relative immunity from atmospheric effects.

In summary, it can be suggested that the user or environmental scientist is mainly concerned with four aspects of soil moisture:

1. The amount of moisture stored at a given time. This can best be expressed on a volumetric basis or as a percentage of field capacity.
2. The availability of soil moisture to the higher plants. The available water is determined by the water tensions occurring at the time of observation.
3. The movement of water into and within the soil. This movement is due to positive and negative pressures within the soil.
4. The periodicity of waterlogging in a given soil. The Soil Survey of Great Britain classifies soil into drainage categories on the basis of the relationships of soil profile morphology to periods of waterlogging (Table 13.6).

Table 13.5 Measured and estimated soil moisture and daily evaporation for Grendon Underwood, Bucks, England, 13 September 1977. (Source: Gurney, 1979)

Model	Evaporation (mm)		Soil Moisture (%)	
	Estimated	Measured	Estimated	Measured
Bare Soil (Tellus)	0.85	0.62	29.6	30.1
Grassland (Tergra)	0.53	0.62	49.6	47.7

In order to use radar effectively for soil moisture determination, it is not only necessary for the backscatter coefficient, $\sigma°$, to be dependent on soil moisture, but also that its dynamic range in response to soil moisture should be much larger than its dynamic range in response to other surface features such as roughness and vegetation. Much of the initial work to determine the optimum choice of frequency, polarization and angle of incidence of the radar sensor was carried out by the University of Kansas. A Microwave Active Spectrometer was used from a Cherry-Picker platform to acquire a wide range of data beginning in 1972. The multi-year data, containing 190 data sets of bare fields and 165 data sets of vegetation covered fields was used to generate an empirical model for backscatter coefficient as a function of the volumetric moisture content normalized to percentage field capacity at ⅓ bar (Fig. 13.10).

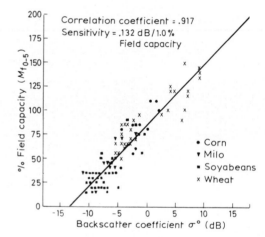

Fig. 13.10 Per cent field capacity in the 0–5 cm soil layer as a function of backscatter coefficient at 4.25 GHz, HH polarization, and 10° incidence angle for corn, milo, soybean and wheat data sets combined. (Source: Ulaby, 1980.)

Table 13.6 Soil moisture regime classes – wetness classes – duration of wet states. (Source: Curtis, 1977, after Soil Survey of England and Wales)

Class	
I	The soil profile is not wet within 70 cm depth for more than 30 days★ in most years†.
II	The soil profile is wet within 70 cm depth for 30–90 days in most years.
III	The soil profile is wet within 70 cm depth for 90–180 days in most years.
IV	The soil profile is wet within 70 cm depth for more than 180 days but not wet within 40 cm depth for more than 180 days in most years.
V	The soil profile is wet within 40 cm depth for more than 180 days and is usually wet within 70 cm for more than 335 days in most years.
VI	The soil profile is wet within 40 cm depth for more than 335 days in most years.

★The number of days specified is not necessarily a continuous period.
†'In most years' is defined as more than 10 out of 20 years.

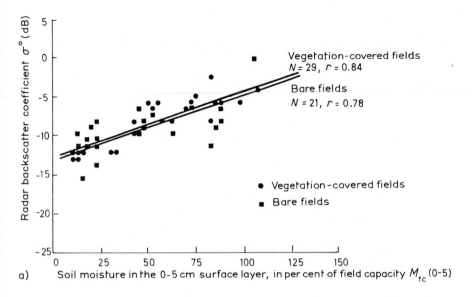

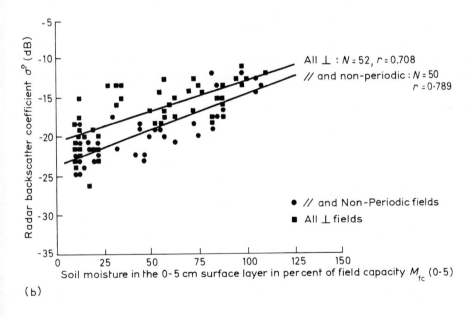

Fig. 13.11 (a) Comparison of radar soil moisture responses to bare and vegetation-covered fields. There is a minimum dependence of the response to vegetation at radar parameters of 4.75 GHz, HH polarization, and 10° incidence angle. (b) Aircraft radar cross-polarization response to soil moisture for two categories of agricultural fields. Radar parameters are 4.75 GHz, HV polarization, and 15° incidence angle. (Source: Bradley and Ulaby, 1980.)

At present only a limited number of airborne and spaceborne investigations of radar response to soil moisture are available. However, campaigns for the evaluation of aircraft radar response have been carried out in the USA by the University of Kansas and in Europe by means of the SAR 580 missions sponsored by the Joint Research Centre of the EEC and the European Space Agency.

The analysis of aircraft radar response to soil moisture made by the University of Kansas used a site south-east of Colby, Kansas, which afforded flat terrain with relatively uniform soils. It was instrumented with 39 rain gauges and three meteorological stations and formed part of a site in the US Department of the Interior High Plains Experiment (HIPLEX). Three airborne radar sensors were used at operating frequencies of 1.6, 4.75 and 13.3 GHz from a C-130 aircraft operated by NASA/JCS. The radars collected data at incidence angles of 5–60° and dual polarization facilities were available, though both polarizations could not be obtained simultaneously. The study showed that the aircraft response to soil moisture is optimum at C-band frequencies and at angles of incidence of 10–20°. Like polarization (HH) radar response is unaffected by vegetation but has been found to be dependent on the character of row-tillage patterns (Fig. 13.11(a)). Cross polarized (HV) data also was found to be unaffected by vegetation cover but was also independent of tillage patterns. (Fig. 13.11(b).)

Interest in microwave methods also springs from the fact that theoretical considerations suggest that some penetration of the soil may be achieved. Information in respect of subsurface moisture or soil layering can, in principle, be obtained by a time domain measurement (Fig. 13.12). For a given range of soil relative dielectric constants and surface roughnesses, approximately one half of the energy of the incident wave will be reflected from the surface. Some portion of the wave energy which penetrates the surface layer will be reflected back to the observing platform from the first subsurface discontinuity or moisture accumulation gradient. The reflected energy from this interface is reduced with respect to the transmitted energy by the transmission loss through the surface layer and the three reflections the wave must undergo. In order to detect the discontinuity it would require extremely good range resolution to detect the presence of the weak second reflection. This places a requirement on the radar system that the pulse duration be very short, and makes the engineering tasks involved

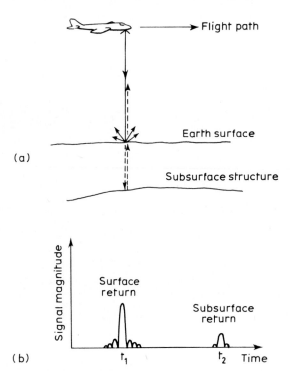

Fig. 13.12 Time domain measurement for subsurface features. (Source: Barringer Research Ltd, 1975.)

very severe. At the present time there is no evidence that systems are available which can achieve this purpose.

It seems likely that interest in remote sensing systems for soil and terrain studies will be strongly sustained as world population continues to grow. Two main themes can be expected to develop. First, the extended use of terrain maps for planning purposes where developments of agriculture, irrigation or forestry are envisaged and second, the rapid development of sensing systems providing temperature and moisture estimations which will be employed in predictive models of crop yields, crop disease, flood conditions, irrigation need and biomass production (see Chapter 16).

14 Rocks and mineral resources

14.1 The use of air photo-interpretation in geology

The use of remote sensing techniques in geology is long established through the use of aerial photography in photo-geologic studies. Sources of further information on the methods used in photo-geology are cited in the bibliography at the end of the book.

Aerial photographs have provided a great deal of data for geological studies in all parts of the world. They can be interpreted to give information on structure and lithology of rocks. In the study of structural geology such features as bedding, dip, foliation, folding, faulting, and jointing can be observed. Aerial photographs provide evidence of bedding through the occurrence of ridges in the stereomodel and differences in tonal response where beds differ in their mineral constituents. Sometimes certain beds can be recognized by a constant lithological interface which is so distinctive that they can be regarded as marker beds or marker horizons. The dip slopes of rocks can often be recognized more reliably on a stereomodel than on the ground because of the synoptic view of an area of dipping sediments obtained from the air. Accurate measurements of dip can be made by photogrammetric measurements on stereopairs. In regional mapping, however, geologists normally assess the dip by eye and place the dip into categories (e.g. $<10°$, $10-25°$, $25-45°$, $>45°$, $<90°$). (Plate 14.1 and Fig. 14.1.)

Any discussion of foliation in rocks is made more difficult by the fact that the term foliation is used in different senses in Britain and America. British

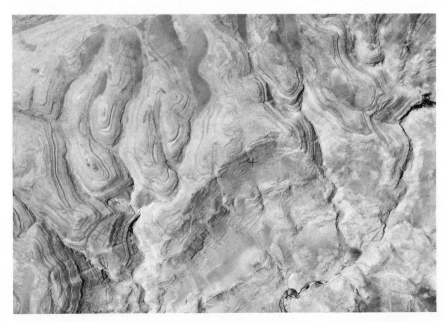

Plate 14.1 An area of gently dipping sediments occurring in Jordan. Note how beds of differing lithology are characterized by different tones on the photograph. Field systems can be seen on the gentler slopes. (Courtesy, Hunting Surveys Ltd, Boreham Wood, Herts.)

geologists distinguish between schistosity and foliation so that for them segregation of minerals into
thin layers or folia is foliation whereas parallel
orientation of such minerals is referred to as schistosity. On the other hand American geologists give
the term foliation wider meaning so that lithological
layering, preferred dimensional orientation of
mineral grains and surfaces of physical discontinuity
and fissility resulting from localized slip may be
included in the term. In this chapter the American
usage of the term will be adopted. Lineaments
resulting from foliation are generally parallel to one

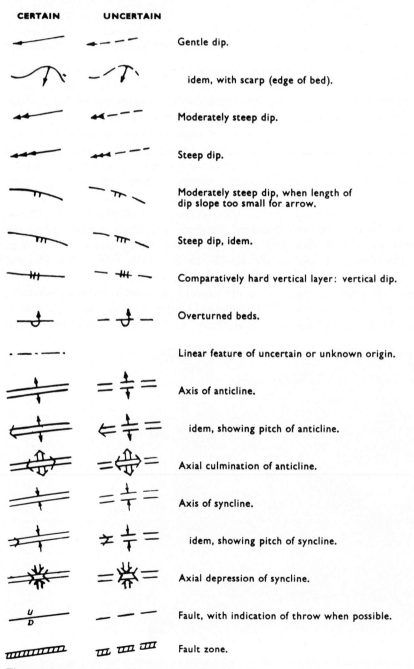

CERTAIN	UNCERTAIN	
		Gentle dip.
		idem, with scarp (edge of bed).
		Moderately steep dip.
		Steep dip.
		Moderately steep dip, when length of dip slope too small for arrow.
		Steep dip, idem.
		Comparatively hard vertical layer: vertical dip.
		Overturned beds.
		Linear feature of uncertain or unknown origin.
		Axis of anticline.
		idem, showing pitch of anticline.
		Axial culmination of anticline.
		Axis of syncline.
		idem, showing pitch of syncline.
		Axial depression of syncline.
		Fault, with indication of throw when possible.
		Fault zone.

Fig. 14.1 Selected symbols for photo-interpretation in geology. (Source: Shell Petroleum Company Ltd.)

another but are short and do not normally consist of long continuous ridges or valleys (Plate 14.2).

Folding in rocks is often apparent on air photographs where it would not be noticed in the field by a ground surveyor. The geologist can normally see a representation of the whole fold in the stereomodel so that the axis and plunge of a fold structure can often be seen very readily (Plate 14.3).

Faulting in rocks occurs where there has been a fracture along which rocks have slipped relative to one another. Generally faults provide fairly straight features and since the fracture is a zone of weathering and weakness the surface manifestation is often one of negative relief. Where mineralization has

taken place along a fault line, however, there may be positive rather than negative surface features. In most instances the most reliable evidence of faulting is displacement of bedding along negative linear surface features (Plate 14.4).

Joints form patterns in rocks which are very similar in photographic appearance to faults, i.e. they often provide fairly straight negative features. The distinction between joints and faults can sometimes be made by careful observation to see if relative movement has taken place in the beds. If relative movement can be seen then the feature can be classified as a fault, and conversely if no movement can be detected it is better to record the feature

Plate 14.2 Oblique aerial photograph of the Fraser River area, Canada (56° 45′ N. 63° 35′ W). Note the lineaments in the surface which are made easily visible by the snow collecting in the lineament depressions. Both short and long lineament features are displayed. (Courtesy, Royal Canadian Air Force.)

as a joint. Jointing often plays a part in determining the patterns of river networks (Plate 14.5). It can also be associated with topographic forms characteristic of granite areas and in such cases the boundaries of granite intrusions can often be plotted with some accuracy.

Whereas structural geology can often be interpreted with considerable certainty from aerial photographs, lithological interpretation is less easy. If the researcher is familiar with a particular field area and has used air-photos extensively it may be possible to recognize rock type with considerable accuracy. However, where an attempt is made to recognize rock types in unfamiliar areas from photo features alone, the task becomes more difficult and uncertain. Various strands of evidence must then be used – in other words the principle of convergence of evidence must be applied. It has been suggested that the following stages may be included in interpretation of lithology and structure:

(a) The recognition of the climatic environment, e.g. temperate, tropical rain-forest, savannah, desert, etc.

(b) The recognition of the erosional environment, e.g. very active, active, inactive.

(c) The recognition and annotation of the bedding traces of sediments or metasediments (altered sediments).

(d) The recognition and delineation of areas of outcrop that do not indicate bedding e.g. intrusions in horizontally bedded rocks.

(e) The recognition and delineation of areas of superficial cover that do not indicate bedding.

(f) The restudy of the bedding traces around the noses of folds to determine if possible the approximate position of the axes of folds.

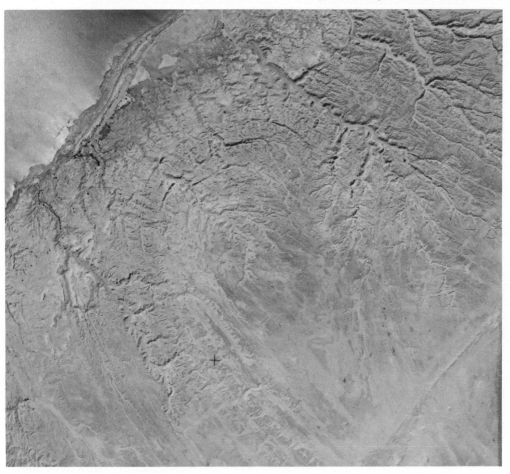

Plate 14.3 A plunging fold structure near the Dead Sea, Jordan. (Courtesy, Hunting Surveys Ltd, Boreham Wood, Herts.)

(g) The study of lineaments transverse to the bedding traces to determine whether they represent faults, dykes, joints, or combinations of these.

It is important to use a recognized set of symbols for annotation of prints in geological studies so that work can be interpreted by co-workers. In Figs 14.1 and 14.2 the symbols used for photo-geological work by the Royal Dutch/Shell Group are illustrated. An example of a restricted working legend used by the Overseas Division of the Institute of Geological Sciences, London, is shown in Fig. 14.3. This set of symbols was used in the interpretation of the geological structure of an area of folded sediments in Australia (Plate 14.6) for which the final photointerpretation is also shown in Fig. 14.3.

It has long been recognized that certain plants are characteristic of soils which are found over rocks containing particular minerals. These plants are termed indicator plants and they have been used as guides by mineral prospectors seeking the occurrence of ore bodies. There is now a considerable literature dealing with the use of geobotanical techniques for mineral exploration and since vegetation boundaries are often well displayed on air photographs attempts have been made to trace indicator plants on the photographs and thereby outline potential mining areas. These methods have been successfully applied in Africa and Australia and the development of multispectral photography together with infrared colour photography has made such geobotanical studies of growing interest.

The literature dealing with the use of air photography in the study of rocks and mineral resources is now voluminous and most major exploration companies rely heavily on photogeological work. There are also separate photogeological departments in various governmental institutes e.g. Institute of Geological Sciences, Britain.

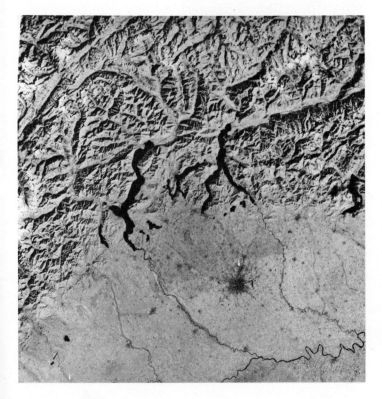

Plate 14.4 Faulting in Alpine Structures fringing the North Italian plain as seen in Landsat 1 imagery obtained in Band 7, 7 October, 1972. (From Boriani, Marino and Sacchi, 1974.)

Plate 14.5 Faulted terrain in an arid environment in Jordan. Note the sharp changes of direction in the wadi system resulting from the effects of faulting and jointing. (Courtesy, Hunting Surveys Ltd, Boreham Wood, Herts.)

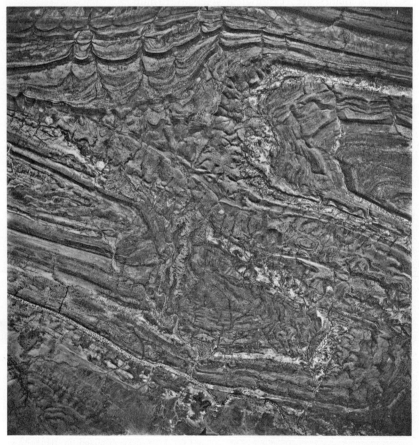

Plate 14.6 Folded sediments in Australia. Note the pitching anticline on the left of the figure and the dipping sediments on the right. The photo-geological map of this area is shown in Fig. 14.3. (Courtesy, Dr E. A. Stephens, Institute of Geological Sciences, London.)

Plate 14.7 Infrared line scan imagery of a mountainous area of North Wales. Note the ability of the sensor to detect different rock strata. (Source: Laird, 1977 courtesy Royal Signals and Radar Establishment, Malvern; Crown Copyright reserved.)

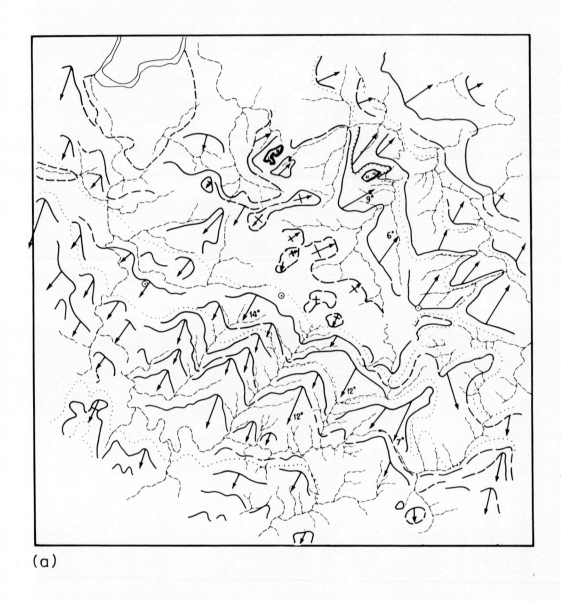

(a)

Fig. 14.2 (a) Photogeological interpretation of Tropical Rain Forest areas in New Guinea. Scale of original 1:40 000. (Source: Shell Petroleum Company Ltd.)

It is of interest to note that the early prospectors recognized that ore bodies were marked by indicator plants and differences of soil temperature. With the use of sophisticated geobotanical and bio-geochemical techniques there is now increased interest in indicator plants.

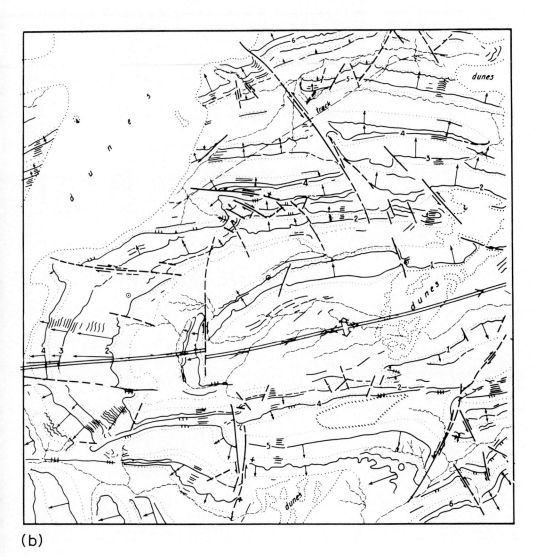

(b)

Fig. 14.2 cont. (b) Photogeological interpretation of North African Desert region, Algeria. Scale of original 1:33 000. (Source: Shell Petroleum Company Ltd.)

Aerial photography is particularly well-suited for large-scale coverage of a target area at a resolution suitable for geobotanical studies. In the next section we shall consider the role of the temperature sensors.

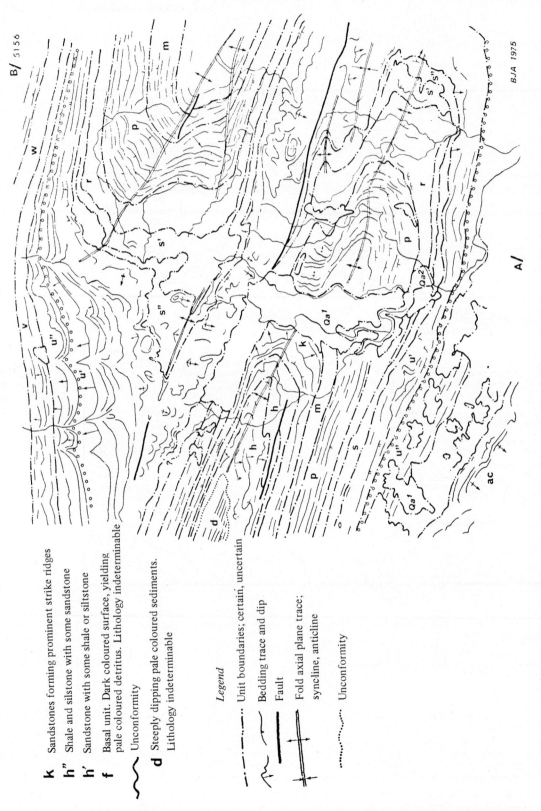

Fig. 14.3 Photogeologic map prepared from a vertical photograph of folded sediments in Australia. (Source: Institute of Geological Sciences, London.)

k   Sandstones forming prominent strike ridges

h″ Shale and silstone with some sandstone

h′ Sandstone with some shale or siltstone

f   Basal unit. Dark coloured surface, yielding pale coloured detritus. Lithology indeterminable

    Unconformity

d   Steeply dipping pale coloured sediments. Lithology indeterminable

Legend

—··— Unit boundaries; certain, uncertain

      Bedding trace and dip

      Fault

      Fold axial plane trace; syncline, anticline

······ Unconformity

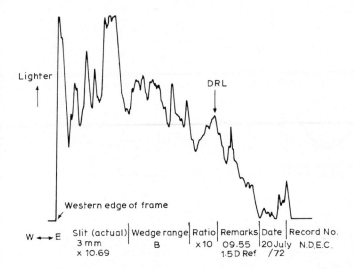

Fig. 14.4 Microdensitometer scan line of the infrared line scan imagery of the Dugald River Lode area. (Source: Custance, in Cole *et al.*, 1974.)

14.2 The use of non-photographic sensors in geology

14.2.1 Thermal infrared scanning

The development of non-photographic sensors has provided remote sensing data in the infrared (thermal) and microwave regions of the spectrum which are potentially very important for geological studies. Infrared scanning has been used for studies of Italian volcanoes and volcanic deposits. These have been overflown with a two-channel Daedalus thermal scanner. The ratio of two thermal channels when plotted can be considered as a relative emissivity map of volcanic materials. It has also been found that such a ratio method can be used for mapping the texture of volcanic materials.

In the Soviet Union airborne infrared images have been used to study sand desert features in the Repetek region. Landscape features such as barchan ridges and sand hills have been found to show different thermal patterns according to differences in slope steepness, density and composition of the vegetation cover, solar elevation and wind speed and direction. The temperature graduations were found to have a large range from 55°C on the sunlit slopes of barchans to 16°C in the shade of desert bushes.

Another application of thermal scanning arises from the fact that mineral prospectors have long recognized that there are differences in soil temperature over ore bodies, therefore thermal infrared line scan techniques are potentially useful for mineral exploration. An example of this is shown in Fig. 14.4 where the microdensitometer scan across the Dugald River Lode in Australia indicates that it is marked by a fairly clearly defined zone of high emission in line scan imagery.

The thermal response of a material to temperature change is termed thermal inertia (see also Chapter 12). For most materials thermal inertia increases linearly with increasing density. During a diurnal solar cycle (Fig. 14.5) some surface materials with lower thermal inertia, such as shale, loose pumice or gravel, may reach relatively high temperatures during the day but cool to relatively low temperatures at night. In contrast, rocks or deposits with higher thermal inertia such as sandstone, basalt or granite may be cooler in the daytime but warmer at night. This response can be used to differentiate between different strata using data such as that portrayed in Plate 14.7 from Snowdonia, Wales showing part of a sequence of contorted beds of Palaeozoic slates and sandstones.

14.2.2 Microwave radar

Perhaps the most extensively used non-photographic system for geologic studies is the microwave radar system. Side-looking radar systems are particularly valuable in equatorial and maritime regions where persistent cloud renders photography difficult to acquire. Furthermore the radar system illuminates the ground from the side so that

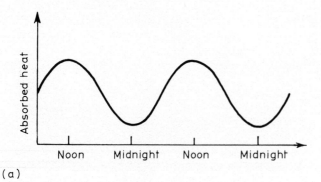

(a)

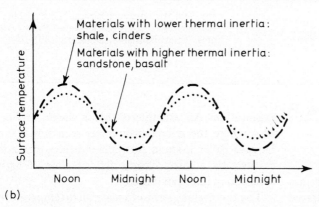

(b)

Fig. 14.5 Effect of differences in thermal inertia on surface temperature during diurnal solar cycles. (a) Solar heating cycle. (b) Variations in surface temperature. (Source: Sabins, 1978.)

shadowing effects are produced similar to those of low-sun photography (Fig. 15.8). These are well shown in the radar imagery of the Malvern Hills illustrated in Plate 4.6.

Shape, pattern, tone and texture are used in the interpretation of radar imagery but the side illumination of radar systems places an emphasis on patterns, i.e. lineaments. Tones may vary greatly according to look angle and so must be used with caution. Textures in the imagery mainly reflect the physical form, i.e. roughness of the surface.

Lineaments are often well displayed in radar images (Plate 12.3) and regional fault and fold patterns can be conspicuously displayed and readily mapped. The orientation of the radar system can, however, give rise to bias in the representation of linears. Also where areas are in the radar shadow (see Plate 4.6) ground information is lost.

14.3 Space observations for the study of rocks and mineral resources

The advent of the satellite platforms has provided

the geologist with a new and wider perspective on geological structures. Thus, for example, photographs of New Mexico obtained from the Apollo-Saturn 6 spacecraft launched in April 1968 were examined with interest by geologists. They found a major north-east-trending lineament which was not shown on the most recent geologic map available for the New Mexico area (Fig. 14.6(a)). However, the major impact of the satellite platform has come with Landsat satellites with their MSS data. Much geological work has been achieved with this data and some selected examples will be discussed below. The general achievements of the Landsat programme have resulted in keen interest on the part of geologists. The most significant results can be listed as follows:

(a) The broadest use of Landsat imagery in geologic mapping lies in the construction of regional structural (or tectonic) maps. Major geologic features, contact between distinctly different, thick rock units, and land form types can be effectively mapped from Landsat images with details comparable to and sometimes superior to

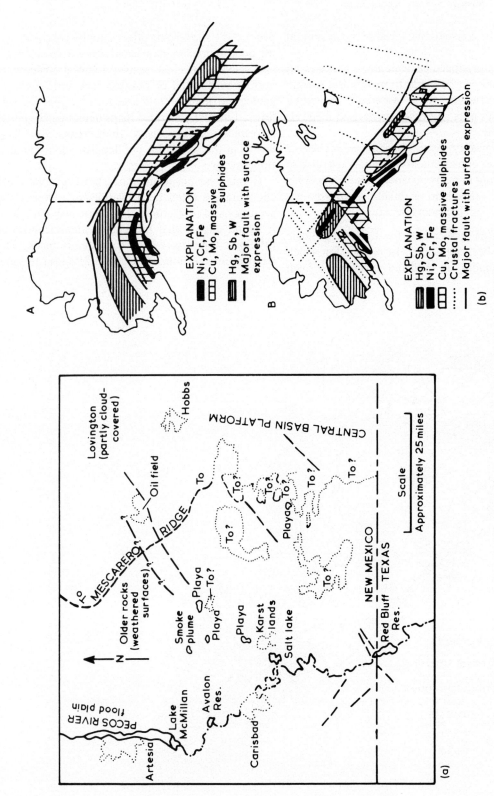

Fig. 14.6 (a) An example of linear features in the New Mexico area mapped from an Apollo photograph obtained in 1968. (Source: Carter and Meyer, 1969.) (b) Geologic conditions in Alaska interpreted from Nimbus 4 data. (A) Conventional concept of the lithologic belts. (B) Alternative linears seen on the Nimbus image. (Source: Lathram *et al.*, 1973.)

mapping by conventional aerial photo or ground methods.

(b) Computer produced geologic maps at scales up to 1:24000 can be made from Landsat images with surprisingly good accuracy, providing training set data and other supervised methods are applied.

(c) Winter imagery, with foliage-free scenes or snow-cover enhancements, provides useful data for mapping in many instances. Such imagery is especially valuable in seasonally vegetated areas and areas with considerable relief.

A number of applications for Landsat geologic studies have been identified. In regions for which only poor quality or outdated maps have been produced Landsat data offers a method of constructing good general maps at small scales. Existing maps can also be checked against Landsat imagery to correct mislocated or omitted rock unit contacts, geologic structures (fold axes for example) and lava flows. Landsat generated maps depicting structural information (particularly lineaments) also have utility in the search for ore deposits, oil accumulations and ground water zones.

It must be borne in mind, however, that field checking has often been insufficient to verify the accuracy and correctness of Landsat-map data. Also it would appear that no reliable identifications of lithologic types have been consistently made. The band ratio techniques (e.g. MSS 7/5) lead to some

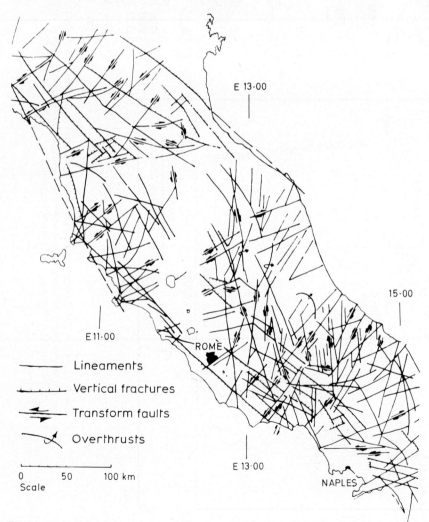

Fig. 14.7 Main fault systems in central Italy, as interpreted from Landsat 1 data. (Source: Bodechtel *et al.*, 1974.)

enhancement of visual images that improves the separation of rock types. However, the ranges of ratio values for many common rocks and minerals overlap. Unique spectral signatures have not always been found to occur within the MSS sensing wavelengths. In fact the ratio interval for haematite and serpentine – strikingly different minerals – is almost identical.

It is also found that individual stratigraphic units are sometimes thinner than the linear resolution capability of Landsat 1 (70–100 m.) Thus remote sensing 'units' are often groupings of stratigraphic units and they sometimes have limited applicability to standard geologic mapping procedures. With these advantages and limitations of Landsat imagery

in mind one can examine some selected examples of geologic interpretations in order to illustrate some of the techniques employed.

An interesting application of the mapping of linears and faults is provided by workers comparing Nimbus and Landsat imagery. Prior to launch of Landsat 1 a cloud free image of Alaska was obtained by the Nimbus 4 Image Dissector Camera System. This image showed a set of northwest and northeast trending linears that suggested previously unrecognized geologic structures deep in the Earth's crust.

Linears and faults on Landsat 1 images have corroborated and added detail to the initial geological trends noted on the Nimbus imagery. These linears have been compared to the relation of known

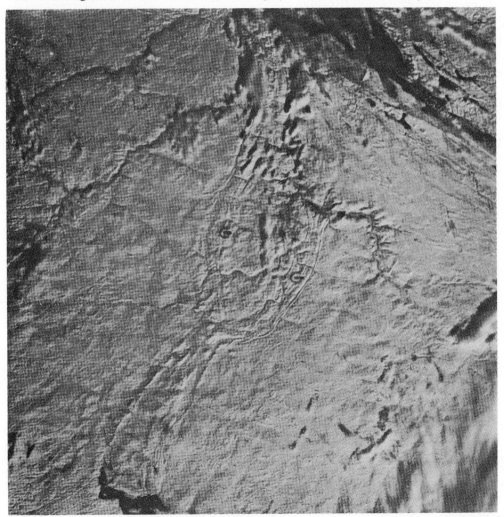

Plate 14.8 Landsat 1 image; early winter, 9 October 1972; sun elevation 25°; no geometric correction; MSS, Band 6, of New England, USA. Linears of quartzite marked G. (Source: Gregory, 1973.)

mineral deposits and fundamental fractures in the Canadian Cordillera. As a result it has been possible to provide an alternative map to guide mineral resource application in Alaska and Western Canada. (Fig. 14.6(b)). The great detail obtainable from Landsat images is also borne out by the detailed mapping of main fault systems revealed by Landsat 1 data for the Italian peninsula as shown in Fig. 14.7.

Many researchers have noted that snow cover does not obliterate linears, indeed major structural features may be accentuated (Plate 14.8). Observations in New England, USA indicate that a heavy blanket of snow (22.5 cm; 9 in) accentuates major structural features whereas a light dusting of snow (2.5 cm; 1 in) accentuates more subtle topographic expressions. Comparisons of snow free and snow covered Landsat 1 images for the same area have shown that snow covered imagery allows more rapid fracture analysis and provides additional fracture detail.

Some researchers have used the methods of spectral ratioing of reflected radiances of selected pairs of Landsat multispectral channels and production of analog ratio images for the mapping of large exposures of iron compounds in Wyoming. The data were combined with laboratory data as training sets wherever possible.

Enhancement techniques were employed in which the digital data of one channel were divided by the data from a second channel. For example where channel 7 data were divided by channel 5 data the resultant ratio digital graymap is given the notation R_{75}. When the analog ratio images were printed in colour it was found that green areas represented vegetation, violet areas primarily rock and vegetation, blue areas rock outcrop and red represented iron rich outcrops. Darkest red was found to occur only in an iron mine area and along pond edges, where muds and tailings were present. Other images were also created by analog ratio images. For example the analog R_{74} showed iron mines as unique dark areas in the scene. These digital ratio and analog ratio maps can be used to focus the attention

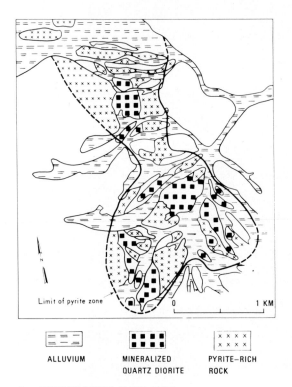

A. GEOLOGY MAPPED IN THE FIELD.

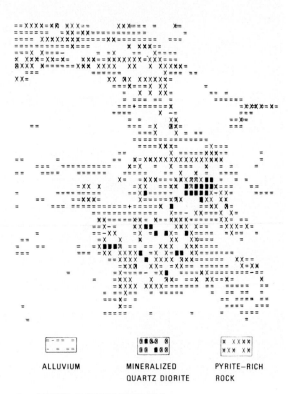

B. DIGITAL CLASSIFICATION MAP.

Fig. 14.8 Training area for digital classification of Landsat CCT data. Saindak copper deposit, Pakistan. (Source: Sabins, 1978.)

Table 14.1 Digital classification table for Saindak training area. (Source: Sabins, 1978)

Rock type	Classification reliability	Computer symbol
Mineralized quartz diorite	High	0
	Low	X̄
Mineralized pyritic rock	High	X̶
	Low	X
Dry wash alluvium	High	=
	Low .	–
Boulder fan	High	+
Eolian sand	High	·
	Low	,
Dark rock outcrops, desert-varnished lag gravels, and black sand		1
		#
		H

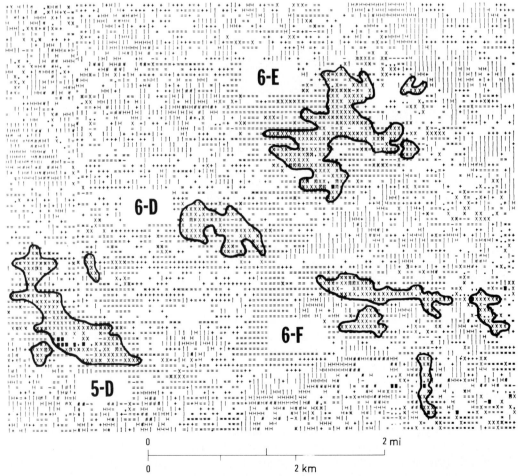

Fig. 14.9 Copper prospects identified in western Pakistan. The categories developed at Saindak training area were used in this classification. Field work indicates area 6-D is the most promising prospect. (Source: Sabins, 1978.)

Plate 14.9 Po valley, Landsat 1, 28 February 1973, MSS Band 6 (0.7–0.8 μm) (a) normal display from magnetic tape. (b) Enhanced display from magnetic tape. (Source: Bodechtel and Haydn, 1977.)

of the geologist or scientist on areas in the scene where chemical properties are different from those of adjoining or surrounding localities.

Landsat data have been widely used in mineral exploration. For example, a training set for digital classification of rock types with potential for copper deposits was derived from a Landsat image of Saindak copper deposit, Pakistan as shown in Fig. 14.8. This training set was given computer symbols (see Table 14.1) and then used to make a digital classification map of Western Pakistan (Fig. 14.9). In this manner large areas can be examined at a lower cost than would be incurred for conventional field reconnaissance. Subsequent field work can then be selective in its approach and the most promising areas for copper extraction can then be determined (Fig. 14.9).

Substantial improvements in the quality of an image can be achieved by a variety of computer techniques (see Chapter 9). These include principal component analyses and contrast enhancement algorithms. For example, work carried out at the Zentralstelle fur Geo-photogrammetrie und Fernerkurdung, Munich, has demonstrated the value of image enhancement. An image of the Po valley, Italy, obtained in Band 6 from Landsat is shown in Plate 14.9(a) and the enhanced version of the same area obtained by transformation of data in Plate 14.9(b). The enhanced image emphasises the relationships between recent fracture systems and the distribution of soil moisture in the Po valley lowlands.

Automatic pattern recognition by computer analysis of the LARSYS type described on p. 273 can, of course, be applied to geological materials. However it is necessary to make certain assumptions if such mapping is to be undertaken. First, it must be assumed that subsurface materials will manifest themselves as spectrally separable classes at the Earth's surface. Since subsurface rocks are normally veneered by soil and vegetation it must also be assumed that spectrally separable surface features are correlated with subsurface variations. Secondly, the assumption that lithologic types are naturally segregated into a limited number of discrete compositional and textural categories which can be recognized in some classification system must be made. This assumption is clearly wrong, as it is for many soils. However it does provide a basis for grouping similar lithologies into discrete classes. Where intergrades occur e.g. sandy-shale the classifier may assign to either sandstone or shale in a random fashion.

Where these automatic recognition techniques have been applied to Landsat images of southwest Colorado a correlation of 89.9% has been obtained with existing geologic maps. It is apparent, however, that extensive ground truth observations and aircraft underflight data are needed before definitive statements can be made concerning the reliability and accuracy of machine-made maps. Nevertheless the vast amount of information now coming from satellite sensors makes it necessary to think in terms of automatic mapping procedures for the future.

15 Ecology, conservation and resource management

15.1 Introduction

There is now a growing awareness that man's use of nature's processes and resources must be adjusted to the limitations and requirements which nature sets for us. Thus 20th century man has become more interested in the ecological changes which human settlements and industry have brought about in the local and the world environment. In particular he is now concerned with the conservation of species in danger of extinction and the maintenance of ecological balance in environments which have been altered to suit his economic needs. The important and fundamental fact remains that all animal life, including man, ultimately depends on the plant life which alone is able to synthesize elements into the form of food.

Wherever conditions are sufficiently favourable the climax vegetation cover consists of forest. In some areas, notably in savanna regions, forest clearance has led to an extension of grassland where forest formerly occurred. A summary of data for the Earth's cover of natural vegetation is shown in Table 15.1 from which it will be seen that approximately 42% of the total land area is potentially forest land, 24% potentially grassland and 34% essentially desert.

Timber has been cut for a variety of purposes, mainly for fuel, construction material and pulp. Many values are associated with forests besides the lumber and pulp that can be obtained from them. These include the maintenance of watersheds, oxygen production, their function as reservoirs for a variety of plant species, fish and wildlife, and the recreational and aesthetic pleasure they can provide.

The grasslands carry the vast numbers of livestock on which much of the world depends for protein-foods. Most of the critical problems in the management of grazing lands (range resource

Table 15.1 Natural vegetation cover. (Source: Roberts and Colwell, 1968)

	Area in square miles	Per cent of total land area
Forests		
Tropical rain forest	3 800 000	7.5
Temperate rain forest	550 000	0.9
Deciduous forest	6 500 000	12.0
Coniferous forest	7 600 000	15.0
Monsoon (dry) forest	2 000 000	3.8
Thorn forest	340 000	0.6
Broad sclerophyll forest	1 180 000	2.1
Total forest	21 970 000	42.0*
Grasslands		
High grass savanna	2 800 000	5.3
Tall grass savanna	3 900 000	7.5
Tall grass	1 580 000	3.1
Short grass	1 200 000	2.4
Desert grass savanna	2 300 000	4.3
Mountain grass	790 000	1.4
Total grassland	12 570 000	24.0*
Deserts		
Desert shrub and grass	10 600 000	21.0
Salt desert	30 000	—
Hot and dry deserts	2 400 000	4.7
Cold desert (tundra)	4 400 000	8.3
Total desert	17 430 000	34.0

*Some areas of potential grazing land occur within areas of forest and desert. A more realistic figure for total land area of potential grazing is 46%.

management) and husbandry of grazing animals occur in remote areas where least is known about native grazing land. In most cases what is needed is better information on the following:

(a) Acreages of useable grazing land by ecologically appropriate classes.
(b) The ecological characteristics of each kind of range (phytosociology, plant sucession, present range condition, autecology of the important species).
(c) The special management problems associated with different grassland ranges.
(d) Indices of potential productivity of each kind of range.

Information is also needed on the numbers and kinds of animals that make use of grassland resources. These include not only the farm animals but also wild herbivores.

Ecological studies require detailed biotic and abiotic information, some of which can only be obtained by ground survey, such as chemical data for the environment and organisms concerned. However, remote sensing has provided valuable support for ground sampling methods. It has been particularly useful for mapping plant associations and estimating the populations of larger animals occupying remote regions.

In the case of plant ecology it may be noted that remote sensors are especially affected by variations in the growth forms of plants. Thus the broad vegetation categories of trees (Phanerophytes), shrubs (Chamaephytes) and perennial forms such as the grasses (Hemicryptophytes) can normally be recognized readily. Also, the habitat of the community being studied can be viewed by examination of the images of the surface of the area. In this way the inter-relationships of topography, local climate, soil, surface water, rock formations, human artefacts and vegetation cover can be assessed.

Seasonal changes in plant forms usually produce marked differences in the images obtained by remote sensing. For example, winter and summer images of deciduous forest vary greatly from each other. The term *phenology* has been applied to that branch of science which studies periodic growth stages in the plant and animal world depending upon the climate of any locality. Phenological studies can be greatly assisted by remote sensing observations. Conversely the interpretation of remote sensing data can be aided considerably by applying knowledge of local phenological changes to the study of images obtained at different times of the year.

The reflectance of light from a vegetated ground surface is determined by several factors which may include leaf geometry, morphology, plant physiology, plant chemistry, soil type, solar angle and climatic conditions. Plant structure, soil background and the surface condition of the plants'

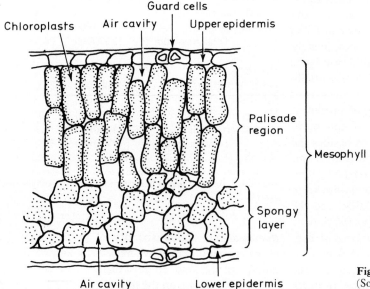

Fig. 15.1 A generalized diagram of leaf structure. (Source: Curtis, 1978.)

reproductive and vegetative parts are particularly important in determining spectral reflectance, and will now be considered.

15.2 Spectral reflectance from vegetated surfaces

The characteristics of light reflectance and transmittance can be explained mainly on the basis of critical reflection of visible light at the cell wall–air interface of both the palisade and spongy layers of the mesophyll (Fig. 15.1). It has been shown that reflectance increases with an increase in the number of intercellular air spaces. This is because diffused light passes more often from highly refractive hydrated cell walls to lowly refractive intercellular air spaces.

The leaf structure is important in that the upper and lower epidermal layers together with the palisade cells and spongy mesophyll each play a part. Spongy mesophyll is important because it scatters near infrared light. On the other hand, the

lower sides of leaves reflect more light than the upper side and this is thought to be due to the absence of palisade cells on the lower side. Often the leaf becomes more spongy with age with the result that the mature leaf displays less reflectance in the visible bands (about – 5%) and more in the infrared (about + 15%).

Frequently the different parts of the spectrum are affected differently by variations in plant composition and structure. The 0.5–0.75 μm band is characterized by absorption by pigments consisting mainly of chlorophylls a and b, carotenes and xanthophylls (Fig. 15.2). The 0.75–1.35 μm band is a region of high reflectance and low absorption which is greatly affected by the internal leaf structure. The 1.35–2.5 μm band is influenced somewhat by internal structure but is more particularly affected by water concentration in the tissue. In general the spectral *transmittance* curves for mature and healthy leaves are similar to their spectral

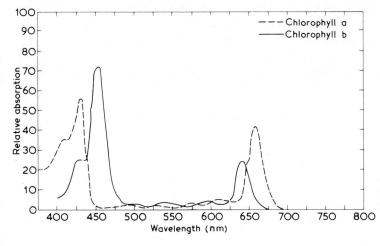

Fig. 15.2 Curves showing relative absorption of chlorophyll a and chlorophyll b as a function of wavelength. The combined absorption would be a summation of the two curves at each wavelength. (Source: Colwell, 1969.)

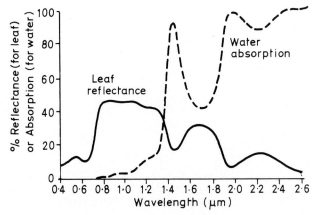

Fig. 15.3 A generalized spectral signature for leaf reflectance. Note the characteristic infrared 'plateau' at 0.75–1.2 μm and the regions of water absorption. (Source: Curtis, 1978.)

reflectance curves for the 0.5–2.5 μm bands but are slightly lower in magnitude (Fig. 15.3).

Leaf senescence occurs as the leaves end their functional life. Most herbaceous annual plants also have a progressive senescence from the older to younger leaves. During leaf senescence, starch, chlorophyll, protein and nucleic acid components are degraded. Thus the familiar autumn colours are caused partly by loss of green chlorophyll and the build up of yellow and orange carotene and red anthocyanin pigments. Usually light reflectance increases markedly in the 0.55 μm (green) band when chlorophyll degradation takes place. Changes in leaf water content can usually be correlated with near-infrared reflectance but, in general, dehydration leads to increased reflectance over the whole range of wavelengths. Leaf senescence, however, leads to decreased infrared reflectance and the infrared plateau at about 0.75 μm is usually reduced considerably.

The reduction in infrared reflectance which characterizes senescence also occurs where disease strikes and the leaves lack their functional condition (Fig. 15.4). It is for this reason that infrared sensors have proved particularly valuable in studies of plant disease. It must be recognized, however, that the resolution capability of the sensor is important. Where there is a non-uniform distribution of dead and dying vegetation along with patches of more healthy vegetation, the classification becomes difficult if the pixel contains a mixture of elements.

Catalogues of the spectral properties of leaves have been derived mostly from laboratory determinations (see Fig. 15.5) but the field spectral response may be rather different. As a result, analytical models of the natural scene have been developed and tested against real conditions. In many early studies the plant canopy was assumed to be an ideal diffusing medium having a uniform distribution of light scattering and absorbing components. More recent studies have envisaged a layered canopy, the biological components being represented by three, flat, 'Lambertian' plane sections. These planes are mutually orthogonal projections, one horizontal and two vertical.

15.3 Phenological studies

Models of the scattering and absorption by plant canopies require estimates to be made of the leaf area affecting the response. Such estimates have been made by measuring the cumulative one-sided leaf area per unit ground area projected from the canopy top to a plane at a given distance above ground level. Such measurements are termed the Leaf Area Index (LAI) and can be measured for horizontal and

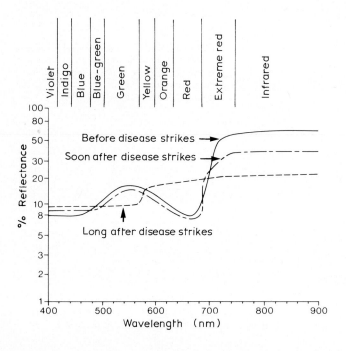

Fig. 15.4 Reflectance characteristics of healthy and diseased leaves. (Source: Colwell, 1969.)

Sycamore, live

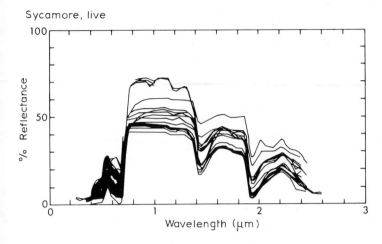

Fig. 15.5 An example of laboratory determinations of spectral reflectance curves. (Source: Leeman *et al.*, 1971.)

vertical projections. In practice such LAI measurements can be correlated with plant stage and vegetation height (see Fig. 6.3). Thus modelling of vegetation reflectance is closely concerned with the phenology of the plant types and the amount of biomass characteristic of different stages of growth. These relationships have been exploited in studies of the productivity of different ecosystems by remote sensing. For example, remote sensing data for the shortgrass prairie was used as an input into biosystem models for the IBP (International Biological Programme) Grassland Biome programme in the USA.

The synoptic view provided by satellites can be used effectively in studies of the development of growth stages in vegetation over wide areas. For example, NASA investigators have described a phenology satellite experiment using Landsat data. In their study two phenological sequences were observed:

1. The Green Wave. A succession of images were used to record the geographical progression with time of foliage development (greening stage) as spring conditions extended over wide areas (Fig. 15.6).
2. The Brown Wave. A record of the geographical progression with time of vegetation senescence (maturation of crops, leaf colouration and leaf abscission). This progression plays the analogous role in the autumn to the Green Wave.

Results from the Phenology Satellite Experiment show that it is possible to develop phenoclimatic models showing the progression of vegetational change during a particular year. For countries with highly developed agricultural or grazing economies such information is very valuable. Hence the LACIE programme (see p. 278) has been used for determinations of crop status, crop yield and for management planning. Outside the cropped areas the ecological conditions of grassland ranges, particularly the steppe, prairie and savanna regions, are of great significance. The most important use of Landsat imagery for grassland ranges is that of monitoring changes in the forage conditions and development of grazing areas as the season progresses (i.e. Green/Brown Wave effects). It is possible to determine areas and dates when plant growth ceases due to drought or soil moisture depletion.

15.4 Conservation in National Parks

We can now turn to consider some examples of the application of remote sensing to conservation management. Interest in conservation has grown steadily as the effects of economic exploitation of animal, plant and land resources become increasingly apparent throughout the world. In some measure this interest has been linked with growth in leisure time in the developed nations and the often expressed need for urban man to find places which are relatively remote and retain elements of wildness.

The National Parks in various countries provide areas of refuge for wildlife and recreation for mankind. They differ widely in their characteristics. Some are virtually untouched by human beings and

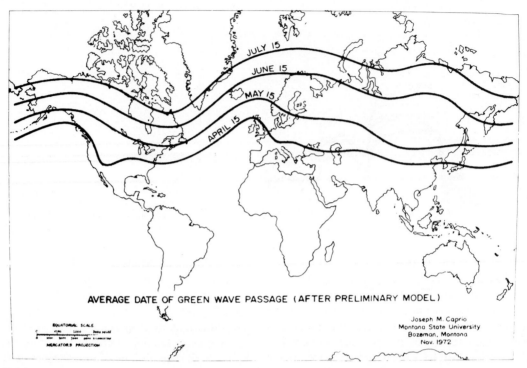

Fig. 15.6 Average date of the Green Wave (greening of grass) passage in the northern hemisphere. (Source: Dethier *et al.*, 1973.)

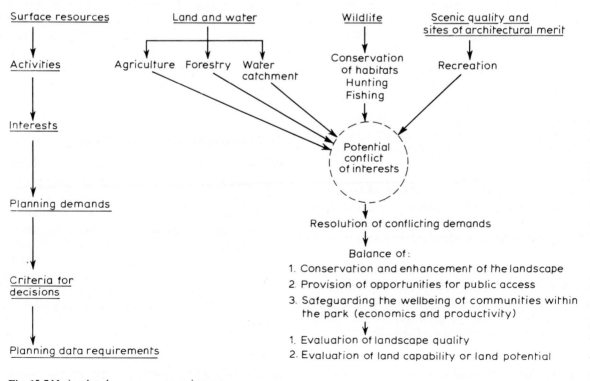

Fig. 15.7 National park management requirements.

stringent precautions are taken to maintain them as wild areas. Others, such as the National Parks in England are areas of outstanding landscape and wildlife interest which also support local farming populations. These Parks might aptly be termed *protected landscapes*.

In protected landscapes the preservation of natural beauty and wildlife requires careful management, and sometimes control, of several competing uses of the land and wildlife resources (Fig. 15.7). For good results it is necessary to start with an inventory of existing wildlife and then to monitor changes in the natural environment over periods of time.

The important role of remote sensing can be illustrated by the inventory of moorland made by Exmoor National Park when concern had been expressed at the loss of moorland habitats as a result of ploughing of the moorland and its conversion to agricultural use. In order to determine the scale of the problem and to provide an inventory of existing moorland a Moorland Map has been constructed by interpretation of infrared colour photographs and back-up field work. The vertical infrared colour photography of 1:10 000 scale was obtained using a Wild RC-8 camera flying at approximately 1800–2000 m. The photography was examined on light tables using both stereoscopic and monoscopic viewers.

Moorland could easily be separated from permanent and temporary pasture by colour and texture differences (Plate 15.1, see colour section). Of some 70 000 hectares of land within the National Park boundary, approximately 19 500 hectares were classified as grass and heath moor. In order to obtain a comprehensive record of the surface conditions an uncontrolled black and white photo-mosaic was constructed from black and white internegatives derived from the original infrared colour photography. The mosaic was printed at a scale of 1:25 000 approximately and was subsequently marked up to show boundaries of the areas of moor and heath vegetation. It provides a record of the vegetation in great detail which enables farmers and landowners to identify small scale features within fields and blocks of land which they manage. Both the Moorland Map and the photo-mosaic offer a valuable basis for discussions between farmers and Exmoor National Park Authority as to the best

options for conservation and sheep grazing policies. The map is now widely used by field officers of the National Park, Agricultural Advisory Service, farmers and their agents as an acceptable statement of the field conditions existing at the end of 1979.

Although the preparation of the moorland map was based on air photography, examination of Landsat 1 and 2 images of Exmoor using the IDP 3000 digital processor (Plate 15.2) at the UK Remote Sensing Centre, Farnborough, shows that the principal boundaries of the moorland vegetation can be defined (Plate 15.3, see colour section), but ground resolution of Landsat in this case is insufficient to allow for detailed mapping of complex patterns of heath and grass vegetation. However, the principal blocks of heath (*Calluna* and *Erica*) can be readily separated from the grass (*Molinia* and *Nardus*) moorland. The potential use of Landsat data in such work is important since conditions for air photography in upland areas of Britain and Europe are rarely good. Indeed in the UK only about eight days in any year are perfect. As a result there are many abortive aircraft missions and a proportion of occasions when the aircraft is not positioned to take advantage of clear weather in a particular area.

15.5 Resource management: the example of forestry

Remote sensing studies by aerial photography have been used for several decades in the field of forestry. Air photo-interpretation has been used for classification of forest stands and types, survey of mortality and depletion, planning of reforestation, inventory of timber and other forest products and assessment of property taxes.

15.5.1 *Tree stand inventories and forest mapping*

As well as using photographic tone, texture and colour to identify different tree stands foresters can make precise measurements of tree height, crown diameter or stand density. Tree height is closely correlated with tree volume and stand volume and can be measured on photographs in a number of ways. Measurement of shadows, measurement of parallax, and measurement of relief displacement on single large-scale vertical or oblique photographs are

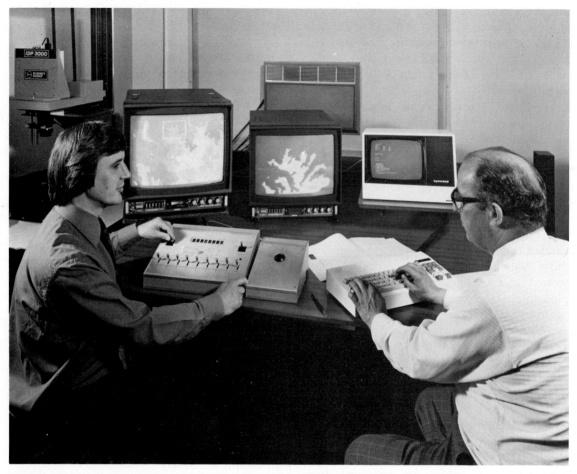

Plate 15.2 Plessey IDP Image Processor in use at UK Remote Sensing Centre, Farnborough. (Courtesy, Space Department, RAE, Farnborough; Crown Copyright Reserved.)

some of the methods used. Tree shadow measurements are normally made with a micrometer scale consisting of a finely graduated series of short lines one of which is matched with each shadow visible on the air photograph. Measurement of crown diameter can be made by either micrometer or dot type crown wedges. The micrometer wedge used normally consists of two converging lines calibrated to read intervening distance to the nearest thousandth of an inch. The dot type wedge usually consists of a series of dots differing in diameter by 0.0025 in. It is laid alongside the image and moved until the dot which just matches the size of the crown is identified. Alternatively the dot images can be moved over the crown until the appropriate size which covers the crown is found. The accuracy of crown measurement is largely dependent on the

scale of the photographs. The error may be about 1 m with either kind of measuring device on photographs of 1:12 000 scale.

Transparent dot templets are probably the most widely used area measuring instruments in forest inventories made from aerial photographs. The density of dots in a templet varies from 1–65 in^{-2} depending on the intensity of the survey and the scale of the maps on photographs used. The ratio of dots in a given class to dots in the entire tract gives the proportion of the tract occupied by that class.

These elementary devices for the measurement of size and area can now be replaced (at a cost) by automated scanning equipment. If very large areas are to be studied such equipment is essential for data to be available within a reasonable period of time.

Once average height, crown diameter and crown

cover are known it is possible to estimate the stand volume of timber from look-up tables. Stand volume tables have been compiled for a variety of species and species groups occurring in specific regions. They usually give average volumes for stands, often in 10-foot height classes, 10% crown-closure classes, and 5-foot crown width classes. Statistical analysis has shown that little volume variation is associated with variation in crown width in some areas. This variable is, therefore, often omitted and volume is classified by stand height and crown closure alone. The accuracy of the estimate will depend on the accuracy of the measurements made from the image and on the quality of the correlations of tree height, diameter and crown cover with volume of timber. In ground surveys the wood volume is estimated from individual tree measurements, usually diameter at breast height (dbh) and the total height of each tree. Breast height is standardized at 140 cm (4.5 feet).

Forest inventory techniques using air photographs normally fall into three basic classes: combined air and ground surveys, aerial surveys using stand volume tables, and aerial surveys using tree volume tables.

Qualitative and quantitative forest estimates often include a combination of descriptive classes of cover type as well as size and density classes. The number of classes and categories may be quite varied according to scale, quality and type of photographs, purpose of the survey, skill of the interpreter, and allowance cost per unit area.

Cover type is usually based on species and species groups that provide relevant information for a particular inventory. The classification could be a simple one of proportion of conifers (softwoods) and broad-leaved deciduous (hardwood) species. For example the classes might be as shown in Table 15.2. Alternatively, detailed species analysis may be used or the concept of utilization may be introduced into cover-type classification. For example in a study of a commercial forest in the northern latitudes the classification might include sawtimber, poles, saplings or reproduction. Species identification is a complex problem of interpretation and various photographic keys have been produced over the years. Table 15.3 is an example of a key for the identification of the northern conifers of New England and the Maritime Provinces of Canada. The key was devised for use with panchromatic photography at scales from about 1:6000 to 1:16000. It would require considerable revision for use with colour photography or other scales of images.

When large-scale photography is used there may be problems of image movement during the period of exposure. Certain 70 mm cameras have been developed for low flying aircraft to provide faster shutter speeds than are normal with air survey cameras.

Scalar distortions can be important in this work. For example, if the flying height is 400 m (1312 feet) above mean ground height the scale on a hill only 20 m (66 ft) high will be 5% greater.

Accurate estimation of scale can be made if a radar altimeter is used to determine the height above ground at the photographic centre, unaffected by forest canopy. Alternatively stereoscopic pairs may be used photogrammetrically to obtain true scale. The use of two Hasselblad (70 mm) cameras with synchronized shutters mounted 4.57 m (15 feet) apart on a boom slung beneath a helicopter has been reported. The preferred flying height was 100 to 200 m above the ground.

At present, the majority of forest inventory is carried out using black and white photography but the usefulness of colour is generally accepted. Infrared colour photography is less commonly used in forestry studies at present but there is evidence to show that it offers many advantages: infrared colour film is always exposed with a yellow filter (Wratten 12 or similar) and is far superior at high altitude where natural colour is often affected by haze. A wildland vegetation and terrain survey in California compared infrared colour at scale of 1:120000 with a survey of the same area using black and white at scale 1:16000. The results were comparable but the infrared colour was more efficient because the work was carried out in half the time. This was partly due to less handling of photography – 3 infrared colour as against 78 at the larger scale in black and white.

Table 15.2 Classification of forests

Symbol	Composition
S	More than 70% by volume of softwood
M	Mixed – 30% to 70% softwood, by volume
H	Less than 30% softwood by volume

Table 15.3 Air photograph interpretation key to northern conifers. (Source: Hilborn, 1978)

1(a) Crowns proportionally large, more or less irregular especially in mature specimens, spreading. Branches often protruding, tops not narrowly conical, usually rounded or flattened   2

1(b) Crowns smaller, regular, conical to narrowly conical or almost columnar. Tops conical, more pointed, well defined   6

2(a) Light-toned, irregular crowns, prominent branches or ragged and open crowns   3

2(b) Darker toned more irregular shaped crowns   4

3(a) Prominent branches, 'star' shaped in plan. Solitary or in small groups in mixed stands. Pure stands often on sand plains or well-drained sites   *white pine*

3(b) Ragged, open crowns. Branching sparse, not prominent. In mixed stands solitary or scattered, often taller than other species. Sites well drained   *overmature white spruce*

4(a) On sand plains or dry sites. Broad, rounded, rough, open crowns   *red pine*

4(b) On moist or ill-drained sites, rarely in pure stands   5

5(a) In swamps. Crowns irregular, open or ragged, if conical, blunt-topped. Some specimens may lean precariously (see also 10)   *white cedar*

5(b) On moist sites. Dense, broadly conical to irregular crowns. Crowns appear fuzzy. Very dark shadows   *eastern hemlock*

6(a) Mostly in uniform, pure stands on sand plains or excessively dry sites. Crowns small, irregular, pointed. Tones medium   *jack pine*

6(b) In pure or mixed stands, rarely on excessively dry sites. Crowns conical to columnar   7

7(a) On moist to boggy sites. Crowns narrowly conical or columnar   8

7(b) On moist to well-drained sites. Crowns conical, rounded, or pointed-topped   9

8(a) Crowns deciduous, light-toned especially in autumn, open. In mixed stands often taller than other species   *tamarack*

8(b) Crowns darker, more dense, narrowly conical to columnar. Stands may be irregular in height   *swamp type black spruce*

9(a) Crowns very symmetrical, smooth. Height irregular. Tops pointed, rounded, or blunt   10

9(b) Crowns narrowly conical, not so symmetrical, not noticeably rounded or pointed   11

10(a) Crowns symmetrical, round-topped, dense, light-toned. Mostly in pure stands and clumps on upland, limestone sites   *white cedar*

10(b) Crowns symmetrical, pointed in front lighting (blunt in back lighting). Dark-toned shadows   *balsam fir*

11(a) Crowns narrow conical to columnar, rougher branching than balsam fir. In stands heights may be irregular   *black or white spruce*

11(b) Crowns more broadly conical, larger. Branching tends to be more prominent   *red spruce*

Note: Red spruce has the broadest, most luxuriant crown, while black spruce has the narrowest. Black spruce will occupy drier and poorer sites. Red and black spruce hybridize making distinction of the eastern spruces very difficult.

In the infrared wavebands there is strong reflectance from deciduous trees but lower reflectance from coniferous trees. Thus both infrared black and white and infrared colour can be used to make broad distinctions between deciduous and coniferous trees (see Plate 15.1 for infrared colour, and Plate 4.2 for infrared black and white images of trees).

Radar imagery has been used in mapping forest areas of Nigeria where the weather conditions point to radar as the most practical sensor in many regions. The whole country was imaged onto radar at a scale of 1:250000 within a five month period. Strips of imagery were then mosaiced together to form sheets which coincided with an existing 1:250000 map sheet series of the country. The project showed that radar could be used successfully for forestry. While the more arid northern areas required more field work and presented greater difficulties, the southern high forest areas gave a good correlation between interpreted units and ground observations. The survey will enable Nigeria's Department of Forestry to make an inventory of different forest types, primary stratification units, and existing reserves. It will also be able to assess potential for plantation development and firewood areas in heavily cultivated zones.

It will be appreciated that the radar beam is directed at an oblique angle away from the aircraft and in areas of surface roughness this produces a radar shadow (Fig. 15.8(a)). In order to counter this, two simultaneous beams are sent out on each side of the aircraft. The eventual flight lines are then arranged so that each area is covered by two 'looks' of imagery, both north and south (Fig. 15.8(b)).

Due to the oblique angle of observation, a further complication is that a uniform vegetation unit may show differences in the signal return according to the range of the image. Care must be taken, therefore, to define and cross reference the signatures for particular areas (Fig. 15.8(c)).

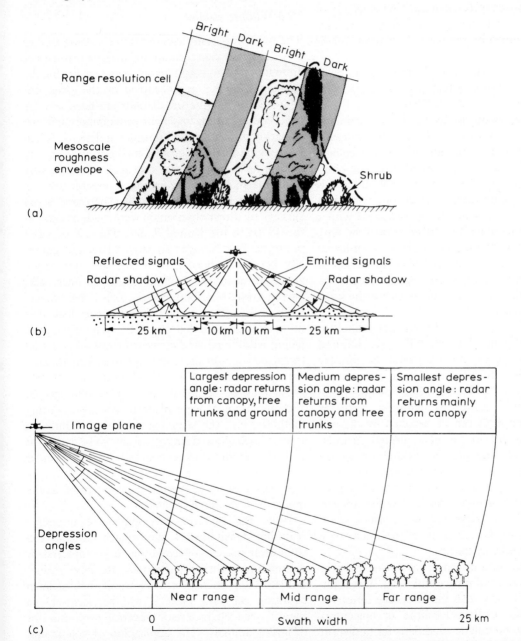

Fig. 15.8 Radar shadows and radar signatures in forest areas. (a) Formation of image texture. (If the amplitude of the canopy roughness on the scale of resolution is large, then there will be very different contributions to the scattering from adjacent resolution cells, giving a corresponding variability in the image brightness.) (b) Image acquisition by SLAR, using two simultaneous beams, showing areas of radar shadow. (c) SLAR signature differences in a uniform vegetation unit on level terrain. (Source: Parry and Trevett, 1979.)

15.5.2 Disease and fire hazards

One of the first indications of disease in a plant is discolouration of the leaves. Similarly, where plants are suffering stress due to attack by pests the changes in leaf structure and colour often provide a guide to the worst affected areas. The detection of disease and insect damage in forest areas can, therefore, be aided by remote sensing using colour photography.

The change in reflectance in the near infrared wavelengths as disease affects a plant can be very considerable (see Fig. 15.4). Therefore the use of infrared colour photography for the detection of forest disease is growing. Healthy broad-leaved trees image in red or magenta in infrared colour, whereas dead or dying trees usually image in blue–green. Red autumn leaves appear yellow and yellow autumn foliage appears white. It will be evident that distinct shades of colour can be associated with a particular species at a given season of the year. Any photography for the purpose of disease monitoring should, therefore, be carried out at the period of maximum growth and where regular checks are required the data should be obtained for the same phenological stage.

The role of non-photographic imagery is likely to increase in the next decade. Thermal infrared imagery has already proved useful in forestry because of the moisture and local temperature effects which can be observed. For example, small smouldering fires started by lightning strikes or abandoned campfires can be detected by thermal infrared detectors on regular aircraft patrols; conventional photography is affected by smoke plumes, but these detectors can identify 'hot spots' when a major fire is under way. Infrared imagery from aircraft or helicopters can help in the efficient deployment of fire crews and on occasions guide them away from dangerous positions where they might be overtaken by the fire.

In Canada the infrared sensor has proved useful in fighting infestations of spruce budworm. This is the most destructive insect in the fir-spruce forests of eastern Canada. Certain localities or epicentres provide optimum climatic conditions for budworm development but climatic observations from special towers erected in the canopy are expensive and liable to be only partly representative of the overall pattern of climate. As an alternative, thermal sensing of the canopy has been carried out by an airborne sensor providing a series of strip records of temperature variation along a metre transect. In this way the potential epicentres were identified.

15.6 Wildlife studies

The most important uses of remote sensing studies in wildlife management are for making censuses of animal populations and mapping and evaluating the vegetation in the area occupied by the game. Inventories of wildlife populations are necessary for the planning of management programmes and formulation of fishing and hunting regulations. When surveys are made by ground methods animals may be counted twice or perhaps not at all. Aerial surveys minimize such errors by making it possible to cover an area quickly and completely. Therefore, aerial surveys by direct observation were made as early as the 1930s in the United States. The use of aerial photographs developed somewhat later but gradually became important. The animals counted by air-photo methods include antelope, deer, elk, barren-ground caribou, moose, muskoxen, waterfowl, seals, and in certain circumstances fish. It is interesting to note that one of the main difficulties facing wildlife researchers was the task of counting thousands of animals. This was particularly the case with photographs of waterfowl where some 14 000 geese may be recorded on a single photograph. Various attempts were made to develop counting methods, e.g. the scanning device of Kalmbach, 1949, which allowed successive slits to be examined and the animals were counted by a binocular microscope. Another method consisted of placing transparent overlays over the print and pricking the image of each animal counted.

It is clear that the development of automatic scanning densitometers allows such tedious (and somewhat inaccurate) counting to be superseded. Automatic counting techniques are now readily available for such work. However before such counting can take place it will be necessary to identify the characteristic spectral responses for different animals. The application of colour and black and white photography to the inventory of livestock has been studied by workers in the USA. A comparison between image counts (IC) and ground

Table 15.4 Comparison of image counts (IC) with ground enumeration (GE) of livestock numbers for area sampling units – cultivated stratum. (Source: Roberts and Colwell, 1968)

Sampling unit number	Livestock species							
	Cattle		Sheep		Other		All species	
	IC	GE	IC	GE	IC	GE	IC	GE
1	45	47	0	0	0	0	45	47
2	0	0	0	0	0	1	0	1
3	0	0	0	0	0	0	0	0
4	22	25	184	180	0	3	206	208
5	135	182	0	1	8	1	143	184
6	202	255	0	0	15	0	217	255
7	0	0	0	0	0	0	0	0
8	0	0	0	0	0	0	0	0
9	0	0	0	0	0	0	0	0
10	2	0	0	0	0	4	2	4
11	0	0	0	0	0	0	0	0
12	58	61	316	608	6*	0	378	669
13	0	0	0	0	0	0	0	0
14	0	0	1373	1000	0	0	1373	1000
15	0	0	0	0	0	0	0	0
16	41	43	0	0	5*	1	46	44
Totals	505	613	1871	1789	34	10	2410	2412

*Cattle incorrectly identified as horses.

enumeration (GE) for an area of about 1000 square miles in Sacramento Valley, California is shown in Table 15.4.

In the great National Parks of East Africa such as the Serengeti in Tanzania the success or failure of wild life conservation is measured in terms of the population figures. Millions of animals are involved and regular censuses of the more common species can only be achieved by aerial survey. The terrain is often semi-arid, travel is slow over land and the wild animals are elusive. Aerial photography has provided the ability to monitor the sizes of the populations of the commoner species to new limits of accuracy.

Two methods have been reported. Total counts, using oblique overlapping photographs, have been made for very gregarious animals such as wildebeeste and buffalo. Other animals showing a more scattered distribution have been counted by sampling techniques. In this method the number of animals is determined for random sample strips of photography of known surface area. The results of a census of wildebeeste in the Serengeti region of Tanzania have been reported as 322000 ($\pm$7%), 381875 ($\pm$6.5%), 334425 ($\pm$7.5%), and 350000 ($\pm$8.5%) for the years 1963, 1965, 1966 and 1967. The size of these populations underline the difficulties ecologists would face in making reasonable estimates by means other than remote sensing.

Other counts reported using aerial survey include those of crocodile and hippopotamus and estimates of the physical condition of large herbivores such as the buffalo.

Infrared scanning techniques are of interest in livestock studies in that they offer a potential means for counting and studying the distribution of animals at night. The body temperatures of animals normally contrast with the surface ground temperature (Plate 4.5). It is clear that thermal scanning used alongside aerial photographic methods could also be a means of estimating mortality rates. This idea has also been proposed for counting dead seals after culling operations.

Radar studies of migrating swarms of locusts began when they were accidentally seen on radar over the Persian Gulf. It is well known that long distance migrations are made by swarms of sexually immature locusts. Apart from these swarms, however, locusts can also exist in the solitary phase (see p. 317) and the solitaries fly by night and rest by day. From field studies using radar by night in the Niger Republic of Africa it has been shown that solitary locusts are able to orient themselves downwind in contrast to the essentially random orientation within a swarm.

Although extensive survey by radar over large areas is not economic at present locust research workers have suggested that radars placed at strategic points – for example on frequently used migration tracks or regions where control operations are taking place – could enable population estimates to be made. In this way protective measures could be planned (see also Section 18.6.2).

We may conclude that the application of remote sensing methods can do much to monitor the conditions within a wildlife habitat. It is possibly in this area that most assistance will be forthcoming for those concerned with wildlife management. The problem facing many conservators is that measurements of changes in an ecosystem are time consuming and often occur at a few points. Also, the sites where change takes place may be remote and expensive to reach. It is for these reasons that most conservators now include in their budgets some provision for air photography or other imagery. The savings in staff time, together with the value of a permanent record of transient conditions, make remote sensing an economic way of meeting managerial needs.

Satellite sensing methods can provide considerable detail over wide areas. They can be expected to play an increasing part in aiding decisions concerning wildlife management on all scales ranging from global to regional and local.

16   *Crops and land use*

16.1 Introduction

As the world population continues to grow apace, one of the most pressing problems facing society is that of feeding and housing the people of the world. In order to tackle this problem land must be used to its full potential whilst seeking to avoid permanent environmental damage. In order to achieve such an aim it is necessary to have up to date information concerning present land use and crop yields, together with knowledge of the changes in land use and crop conditions.

At the present time agriculture dominates the land use of most countries, although forestry may play an important role in some. For example, in 1965 some 81.5% of the land in the United Kingdom was devoted to agriculture. However, there are now competing claims by recreation, forestry and urban growth for available land. Thus present trends indicate substantial changes in land use are likely to occur by the year 2000 (Table 16.1).

Although land use maps have been constructed for many parts of the world they have often shown inaccuracies by the time they were published. There is, therefore, a need for more up to date maps. Also one may note that for some regions of many of the developing countries there are no land use maps whatever at scales which are useful for planning purposes. There is, therefore, a great need for remote sensing techniques capable of identifying land use.

In the field of agriculture the main need is for early information to allow for marketing of produce and management of the farming industry. In an age when there is substantial government intervention in the form of support prices and subsidies there is a need for national and international data concerning agricultural production. Thus early acreage determination of main crops, early yield prediction, information on disease and pest attacks, quantification of the effects of moisture deficiency or surplus and continuous updating of inventories are some of the requirements of modern agriculture.

The Earth's surface consists partly of natural features ('land cover') such as natural vegetation, snow and ice, and partly of features due to the activities of man ('land use'). It is desirable that any inventory showing areas devoted to different purposes should be compared with the total land area concerned. Therefore care must be taken to account for land cover areas e.g. streams even though the main purpose may be to establish areas of crops. It is not necessary, of course, to categorize each of the land cover types in detail unless there is some justification for such additional information.

Table 16.1 Land use in the United Kingdom (estimated)

Land use	1965		2000	
	Area (10^3 hectare)	% Total	Area (10^3 hectare)	% Total
Agricultural land	19 624	81.5	18 122	75.25
Urban land	2 043	8.49	2 745	11.43
Forestry and woodland	1 817	7.54	2 617	10.86

Table 16.2 Land use and land cover classification system for use with remote sensor data. (Source: Anderson *et al.*, 1976)

Level I	Level II
1 Urban or built-up land	11 Residential.
	12 Commercial and services.
	13 Industrial.
	14 Transportation, communications, and utilities.
	15 Industrial and commercial complexes.
	16 Mixed urban or built-up land.
	17 Other urban or built-up land.
2 Agricultural land	21 Cropland and pasture.
	22 Orchards, groves, vineyards, nurseries, and ornamental horticultural areas.
	23 Confined feeding operations.
	24 Other agricultural land.
3 Rangeland	31 Herbaceous rangeland.
	32 Shrub and brush rangeland.
	33 Mixed rangeland.
4 Forest land	41 Deciduous forest land.
	42 Evergreen forest land.
	43 Mixed forest land.
5 Water	51 Streams and canals.
	52 Lakes.
	53 Reservoirs.
	54 Bays and estuaries.
6 Wetland	61 Forested wetland.
	62 Nonforested wetland.
7 Barren land	71 Dry salt flats.
	72 Beaches.
	73 Sandy areas other than beaches.
	74 Bare exposed rock.
	75 Strip mines. Quarries, and gravel pits.
	76 Transitional areas.
	77 Mixed barren land.
8 Tundra	81 Shrub and brush tundra.
	82 Herbaceous tundra.
	83 Bare ground tundra.
	84 Wet tundra.
	85 Mixed tundra.
9 Perennial snow or ice	91 Perennial snowfields.
	92 Glaciers.

A land use and land cover classification suitable for use with remote sensor data is shown in Table 16.2. Information relevant to many of the land cover categories can be found in sections of this book dealing with water in the environment (Chapter 12) ecological studies and resource management (Chapter 15) and the built environment (Chapter 17). Attention will be focussed here on those aspects of land use inventory of particular relevance to crop monitoring.

The general considerations of spectral response to vegetated areas have been discussed in Chapter 15. Since agricultural crops are simply cultivated versions of natural plants similar general considerations apply, although such factors as row direction, row tillage, irrigation practice and harvesting patterns have an important bearing on the response characteristics of cultivated crops.

16.2 Aerial photographic surveys

Early examples of the use of aerial photography in land use studies include experimentation with airphoto cover in Northern Rhodesia (Zimbabwe) in the mid-1930s. By the 1940s there were air-photo-interpretation studies of agriculture and forestry in many areas of the world. These early studies were limited to black and white photography which was becoming readily available in vertical line overlap form.

Photo-interpretation keys were developed for various crops for particular agricultural regions. A useful review of these early techniques in mapping agricultural land use from air-photos is given in the Manual of Photographic Interpretation published by the American Society for Photogrammetry, 1960. At this time features such as patterns of fields, tone, texture, shape and size of the cropped areas were used as aids to interpretation. Further evidence was sought where photos were available for different seasons. Changes in the photographic appearance of particular areas could often be related to conditions of tillage, crop growth, crop rotation or conservation measures. As in all manual interpretation work the best results were obtained by a combination of field knowledge and careful examination of the images. The same principles apply today in the visual interpretation of both aerial photography and satellite pictures.

At a very early stage American and British research workers recognized that aerial photography was useful for the detection and mapping of crop diseases and pest infestations. However, the development of colour, infrared colour and multispectral photography has added further impetus to this area of research in the last decade. Diseases such as potato blight, take-all disease in cereals, leaf spot in sugar beet and yellow dwarf disease in barley have been observed and their effect on crop conditions noted. Likewise the effects of nematode attacks (eelworm) and soil fungi may be detected and affect the spectral response of a particular crop. (Plates 16.1(a) and (b))

Where crops have been affected by disease or adverse environmental conditions dead portions of the crop can be readily identified using infrared colour photography. On this emulsion, healthy plants image in red whereas diseased and dead portions assume different colours. The colours of diseased plants often range from salmon pink to dark brown depending on the severity of the attack. Dead portions image in green or bluish grey.

Crop discrimination using infrared colour photography has been studied closely and it has been found that the percentage accuracy depends on time of year, location and environment. For example, in North Eastern Kansas in late July all crops were found to image red. In Britain, however, it is noted that the quality of grass pasture can often be detected by variations in the intensity of the red hue, and in some cases this is a reflection of the fertilizer treatment the field has received (Colour Plate 15.1).

16.3 Multispectral sensing of crops

In view of the different spectral responses observed for different plant types, a great deal of effort has been made to measure the spectral reflectance characteristics of a number of crops. A major source of data concerning crop signatures is that obtained under the Air Force Target Signatures Measurement Program in the USA. These were mostly measurements in the visible wavelengths, and much of the data was derived from laboratory investigations. Examples of the signatures obtained for certain crops are given in Fig 16.1. The spectral response of a field crop depends partly, however, on

the layering within the crop and so modelling of crop canopies is also important (see Section 15.2).

The principal objective now being pursued is that of automatic recognition of land use categories and crops by means of computer analyses of spectral reflectance. Such analyses can be applied to either airborne sensor data or satellite data.

There are broadly two approaches which may be adopted in mapping from spectral data. On the one hand a *training set* of data can be obtained from matching known crops to sets of data from particular sensor systems. This training set can then be entered into the computer so that the computer classifies all similar data as the crop in question. Such classifications are termed *supervised* classifications. Alternatively, it is possible to adopt an *unsupervised* classification approach. In this case the remote sensing data are grouped and mapped by clustering algorithms (see also Chapter 9).

The sequence of operation for crop surveys includes data preprocessing, training set selection, and classification by statistical pattern recognition. In the case of aircraft data a similar sequence is necessary as, for example, in the Laboratory for Agricultural Remote Sensing (LARS) method at Purdue University (Fig. 16.2).

Data preprocessing includes radiometric correction and registration of each picture element in geometric coincidence. The training sample selection phase aims to determine the separable classes and subclasses in a given data set. Pattern analysis systems have been developed which allow several methods of class selection to be employed as appropriate. Statistics can be computed for up to 30 wavelength bands and printed out in the form of histograms, correlation matrices, and coincident one sigma spectral plots. One of these types of output can then be used to group areas having similar spectral responses. Another method of class separation consists of using clustering techniques to group image points such that the overall variance of the resultant sets is minimized.

The fields for which ground truth data are available are then identified by line and column coordinates so that they can be entered together with multispectral data into the computer. Histograms and statistics are then computed and printed for each field. An adequate statistical sample (> 30) for each crop type is necessary at this stage. Samples of

CORN, Normal Stand (USAERDL Field Data)

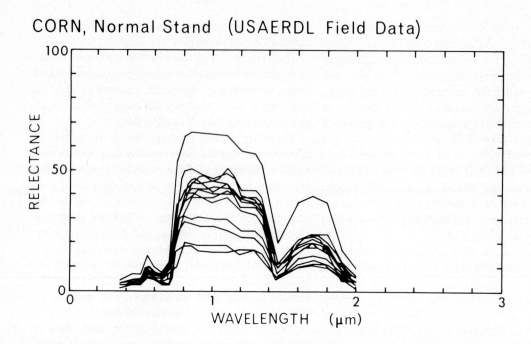

OATS, (USAERDL Field Data)

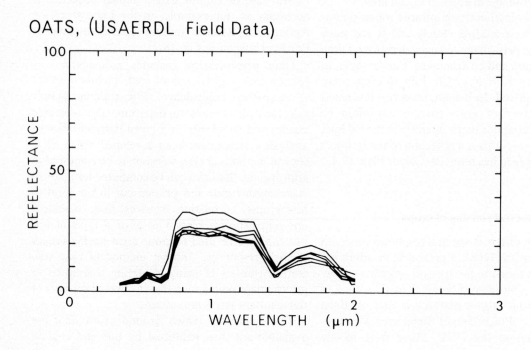

Fig. 16.1 Spectral signatures of corn, oats and alfafa. (Source: Leeman *et al.*, 1971.)

ALFALFA

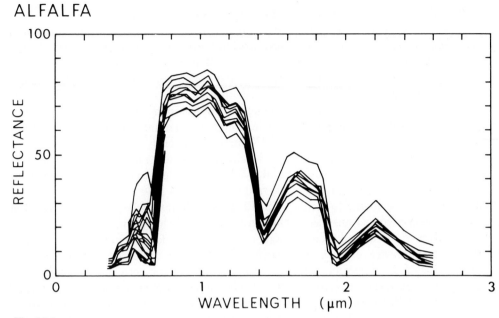

Fig. 16.1 cont.

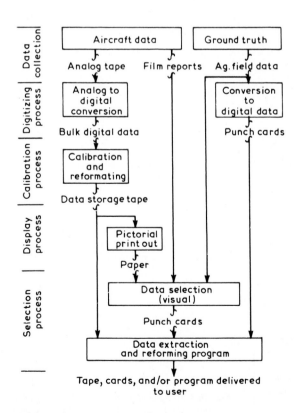

Fig. 16.2 LARS data flow system for crop studies.

data from each of the classes identified by the statistical process are then used for training the pattern classifier, e.g. the histograms for barley are used as training sets in respect of the automatic classification of barley.

In the pattern recognition stage two methods can be used to classify the multispectral imagery. One method classifies each image point into one of the defined classes. Alternatively an entire field can be classified as one decision. The latter technique, termed 'per field classification', has the advantage of speed. However it demands that the field co-ordinates must be known and fed into the classifier before any classification can be performed. Classification by the 'per point' method is time consuming since every resolution element in the image is classified separately. However it can classify any area without field boundaries being specified.

Following the pattern recognition phase, which produces automatic classification of the land use, it is necessary to evaluate the classification accuracy quantitatively. A large number of test fields are necessary for this purpose. These are located in the computer classification and the ground truth is compared with the computer result.

Supervised classification techniques of the type described above have certain drawbacks. There is the very considerable difficulty that spectral signatures show high variability and supervised techniques generally demand that reference signatures be collected directly from a training area lying within the survey area or nearby. The unsupervised classification technique avoids this difficulty by not requiring reference signatures in the data processing phase. Unsupervised techniques will group the multispectral data into a number of classes based on the same intrinsic similarity within each class. The meaning of each class in terms of land use category is then obtained after data processing by checking a small area belonging to each class. Preliminary tests of supervised and unsupervised classifications over areas in the United States suggest that the overall performance of each is comparable. If this conclusion is supported by future work it is likely that unsupervised clasifications will be preferred because they work from the remote sensing data to the group instead of vice versa.

One of the most extensive crop studies so far made by remote sensing techniques, using supervised classification based on ground truth, is that of the Corn Blight Watch Experiment, 1971. The corn (maize) leaf blight is caused by the fungus *Helminthosporium maydis* and is widespread in maize growing in tropical areas of the world. Until 1969 corn leaf blight was a minor problem in the USA but in 1970 there was extensive corn blight as a result of the development a new race of the fungus. As a result, yields are thought to have dropped by about 700 million bushels in 1970.

In 1971 two aircraft collected colour infrared photography and 12-channel multispectral scanner imagery. The ground data, flight logs and the image analysis results were combined and analysed to record the results of the corn blight survey (Fig. 16.3). The interpretation phase of this study required extensive correlations of field observation, photo-interpretation and automatic analysis of MSS data. Five categories of corn blight severity were

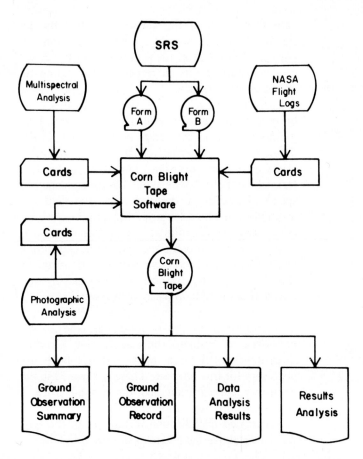

Fig. 16.3 Merging of ground data, flight logs and analysis results in 1971, Corn Blight Watch Experiment. (Source: MacDonald *et al.*, 1973.)

used in the field estimates as illustrated in Fig. 16.4(a), and the results (Fig. 16.4(b)) showed that the best correlations were achieved when the blight symptoms were well established (stages 2, 3 and 4). It also appears that the correlation between MSS analysis and field data became better at later dates in the study (Fig. 16.5(a) and (b)). This important study concluded that neither manual photo-interpretation of small scale photography nor machine made analysis of MSS data gave adequate detection

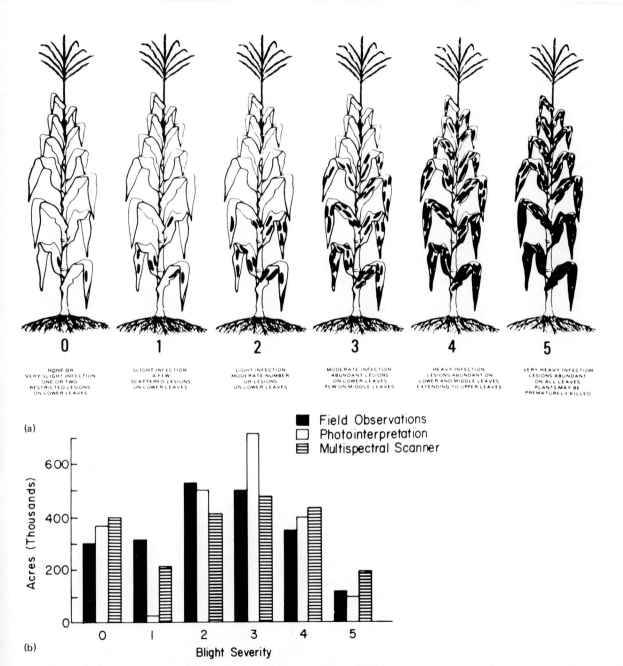

Fig. 16.4 (a) Scale for estimating southern corn leaf blight severity. (b) Comparison of field observation, photo-interpretation and machine analysis estimates of corn acreage in individual blight classes for an intensive study area, 23 August to 5 September. (Source: MacDonald *et al.*, 1973.)

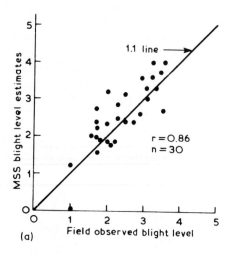

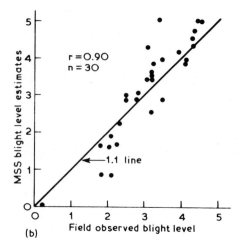

Fig. 16.5 Correlation of field observation and machine-assisted analysis of multispectral scanner data estimates of segment average blight severity levels, 23 August (left) and 6 September (right). (Source: MacDonald *et al.*, 1973.)

of Corn Leaf Blight during early stages of infection. Analysis of the data did, however, permit the detection of outbreaks of moderate to severe infection levels.

Laboratory investigations by other workers have shown that maize leaves infected with corn blight possess similar reflectivities to healthy leaves in the visible range (0.4–0.75 μm). In the near infrared range (0.8–2.60 μm) healthy leaves show significantly higher reflectivities than *Helminthosporium maydis* infected leaves. Thus further research studies should investigate the infrared wavelengths for corn blight surveys.

Mapping of the spread of those levels of infection over the region could be carried out with relatively high accuracy. There seems to be good evidence that land use, crop stress and crop disease can be mapped using satellite multispectral scanning data (Plate 16.2; colour plates). However, ways must be found for reducing the amount of ground truth data required for initial selection of training samples. This may, again, point to increasing use of unsupervised classification. The problem remains, however, to devise classificatory methods which can be applied over a wide range of environmental conditions and over large geographic distances. There is also a need for careful consideration of the effects of sun angle, observation angle and atmospheric effects on the spectral patterns of crops.

16.3.1 Advances in crop monitoring using multispectral data

It is interesting to note the steady advances made in the USA over two decades towards a crop monitoring system. The Agricultural Board of the National Research Council set up its Committee on Remote Sensing for Agricultural Purposes in 1961. In the 1960s the multispectral concept was developed and initial feasibility experiments were carried out. Improved airborne multispectral sensors were developed in the mid 1960s and the development of decision algorithms to analyse multispectral data soon followed. Thereafter there was a long period (1966–1972) when controlled field experiments were carried out together with overflights with an airborne scanner. Then in 1969 a Landsat 1 multispectral scanner was flown on Apollo 9. Subsequently the Cornblight watch and the launch of Landsat 1 in 1972 gave an added impetus to both analytical techniques and concepts. As a result in 1974 the Lacie (Large Area Crop Inventory Experiment) programme was initiated to test the technology and know-how developed by applying it to the assessment of crop production over several of the most important agricultural regions of the world (Fig. 16.6). The objectives of the Lacie experiment were to:

(a) Evaluate the applicability of space technology to the global monitoring of agricultural crops.

Plate 16.1 (a) Stem nematode infection in lucerne showing as dark patches. Background differential growth is due to soil variations.

(b) Integrate the necessary components into an experimental system to permit the acquisition and analysis of the required volume of data in a timely manner.

(c) Conduct the experiment in a quasi-operational manner to evaluate the technology under conditions representative of any future operational missions.

(d) Assess the suitability of the 'first-generation' technology and identify areas requiring improvement and suggested means of development.

(e) Meet an accuracy goal such that the estimates at harvest should, on average, be within ±10% of the true country production 90% of the time.

(f) A timeliness goal be set such that Landsat data should be reduced to acreage information within 14 days after acquisition.

The Lacie system data flow supporting the Lacie programme is shown in Fig. 16.7.

The Lacie study was focused on monitoring wheat production in the selected regions shown in Fig. 16.6 and the Lacie technology was evaluated over a nine-state 'yardstick', together with exploratory studies in India, China, Australia, Argentina and Brazil.

In order to estimate wheat production a country was subdivided into areas (strata) where yield and prevalence of wheat were fairly uniform. As shown in Fig. 16.8 the yield and the areal extent of wheat within each stratum were estimated by independent methods and then multiplied together to obtain production at the stratum level. The production estimates in each stratum were then added to obtain the production estimates at other geographical or political levels. In addition, area and yield were

Plate 16.1 cont. (b) Patches of stunted growth due to barley yellow dwarf virus infection differentially invaded by sooty mould fungus (*Cladosporium spp.*). (Courtesy, Ministry of Agriculture, Fisheries and Food, Cambridge.)

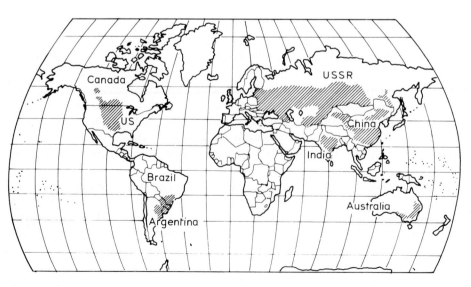

Fig. 16.6 Lacie study areas.

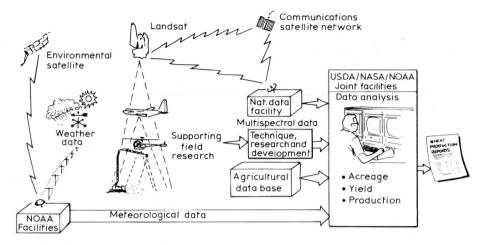

Fig. 16.7 The Lacie system data flow.

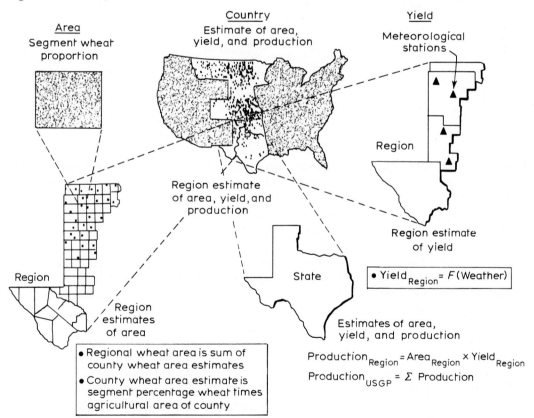

Fig. 16.8 Production estimation from sampling in the Lacie study.

aggregated to determine wheat area and yield at other hierarchical levels within the country.

The crop identification and mensuration of US 'yardstick' areas was carried out using a stratified random sampling strategy based on 5×6 nautical mile segments randomly allocated to strata according to the 1969 Census of wheat growing areas. The basic sample unit was termed a Lacie image and consisted of 196 picture elements (pixels) by 117 scan lines for each of the four MSS channels. These

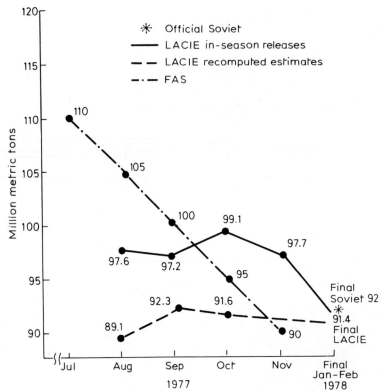

Fig. 16.9 Lacie estimates of 1977 Soviet wheat production.

segments were extracted from full frame Landsat images and four acquisitions of data representing four different stages of wheat growth were taken to classify wheat from non-wheat. These four channels (N) were merged into a multi-temporal (4N) image and then classified by sum-of-likelihoods classifier. The Lacie classifier was dependent on the concept of analogous areas and analyst/interpreter techniques.

In the period 1974–1977 the algorithms used were reviewed to allow for improved stratification of the inventory region and different acquisitions of data were programmed. Despite processing problems estimates were made of the USSR wheat production by the end of 1977. These were compared with the US Department of Agriculture estimates made by the Foreign Agricultural Service (FAS) (Fig. 16.9).

An investigation of the Landsat data and the yield model response at sub-regional levels showed that drought conditions observable by Landsat were accurately reflected by reduced yield estimates in the affected regions. The Lacie estimates met the accuracy goal at harvest and even achieved this in its forecasts made 1.5 to 2 months before harvest. It was noted, however, that in small field areas there

was greater confusion in the identification of crops and a resulting underestimation of spring small grains. It is considered that this problem will be largely overcome when the higher resolution sensors of Landsat D are operative.

The actual contact time required to analyse a Landsat segment, manually select training fields, compute training statistics, and computer process some 23 000 elements of a Landsat image was reduced progressively during the study. At first it was 10–12 h, then 6–8 h and in the final phases of the study 2–4 h. The investigators concluded, therefore, that the timeliness goal of 14 days from acquistion of the data to reporting the acreage could be realized in a future operational system.

As noted above, the Lacie experiment revealed that the accuracy of estimate of cropland acreage decreases where field sizes become smaller. This is a fundamental problem when crop identification is being attempted in Europe where field sizes may be less than one hectare in some cases. The crux of the matter is that from space altitudes many of the ground resolution elements are individually composed of a mixture of crop categories. Thus many of the picture elements generated by multispectral sensors are not characteristic of any one crop but relate to a mixture.

Morrison, in 1977, expressed a NASA view of the prospects for a Worldwide Crop Information System by writing 'I do not think we will be able to do this successfully until about the 1984 period. Present satellites do not have the resolution necessary for providing worldwide crop data in field situations. For this, we have to wait for the Thematic Mapper on Landsat-D.'

In Europe the problem of field size is reflected in the small size of the agricultural holdings (Table 16.3). It will be evident that about 46% of holdings are less than 20 hectares in size. In these small holdings several crops often divide the farm into parcels smaller than 1 hectare. It can be estimated that correct recognition of a 0.5 hectare parcel would require a pixel size of no more than 25×25 m. Furthermore, in Europe the cloud cover makes it difficult to obtain more than two repetitive scenes using visible sensors. Present experience indicates that to achieve an accuracy of 95% in acreage estimation, using only two multispectral scenes, would require a ground resolution of 20 m. It is for this reason that interest is being shown in the new generations of satellites which will provide this order of ground resolution. Also, one may note the interest in mixing MSS data with higher resolution data obtainable from RBV sensors on Landsat. An example of combining MSS and RBV data is shown in Plate 16.2 (see colour section). It is evident that greater definition of rural land use can be achieved by this means.

Crop identification is greatly assisted if remote sensing data are available for different times during the growing season. An example of such work is that of the phenological study made in the Agreste (1973–1977) project aimed at evaluating the potentiality of remote sensing techniques applied to agriculture and forestry problems under typical European conditions. The project was conducted in close collaboration between the Commission of the European Communities (in particular its Joint Research Centre at Ispra, Italy), and specialized national laboratories in Southern France and Northern Italy. Six test sites were chosen in Northern Italy and Southern France and one in Madagascar. The programme used various observation platforms – Landsat, MSS airborne sensors, spectrometers and radiometers installed on helicopters and Cherry-Pickers, classical air photography, and ground and laboratory support data.

Fig. 16.10 shows the application of remote sensing to the phenological cycle of rice. Landsat 2 coverage and traditional airborne MSS data for limited areas are indicated. Although the Landsat coverage was not optimum, the ground and helicopter observations partially filled this gap. The most important source of error in acreage determinations derived from poor sensor resolution, where the pixel element was of the same order of magnitude as the dimension of the fields. There was, therefore, confusion of signatures of competing crops and this was made more difficult by atmospheric attenuation of signals. Nevertheless, comparison of reference zones (Plate 16.3) with maximum likelihood classifications (Plate 16.4) showed an overall over-estimation of 9% for rice surfaces and 8% over-estimation of poplar areas.

16.4 Radar sensing of crops

Two aspects of crop quality lend themselves to radar monitoring: moisture status and the topology of the crop. A strong correlation between the radar backscattering coefficient and plant moisture content has been found by workers at the University of Kansas using a 9.4 GHz radar system mounted on a platform 26 m above the ground. Linear regression analyses of the radar backscattering coefficient ($\sigma°$) on plant moisture provided good correlations of about 0.9 and a slope of -0.275 dB per cent plant moisture with nadir observation. Also it was found that $\sigma°$ undergoes rapid variations shortly before

Table 16.3 Number and area of agricultural holdings with 1 hectare arable area and over by size groups in the EEC. (Source: Eurosat)

Size group	Quantity	Area
1–5 hectare	2.429×10^6	6.14×10^6 hectare
5–10 hectare	1.067×10^6	7.67×10^6 hectare
10–20 hectare	1.067×10^6	15.14×10^6 hectare
20–50 hectare	0.845×10^6	25.32×10^6 hectare
>50 hectare	0.2961×10^6	34.48×10^6 hectare

The % of the holdings with ≥ 20 hectare is 54%

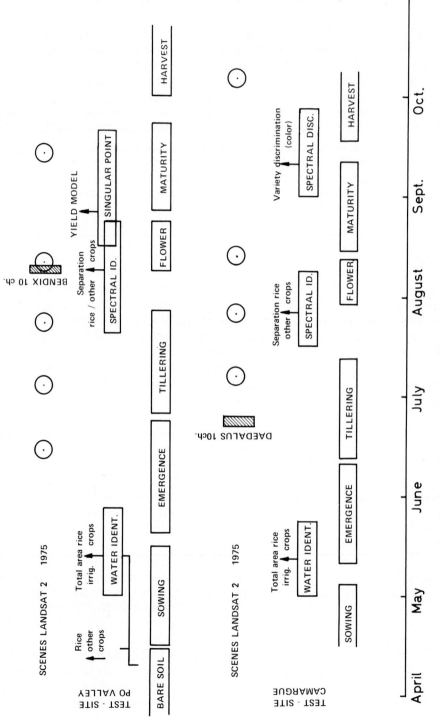

Fig. 16.10 Phenological cycle of rice (crop calendar). (Source: Fraysse, 1977.)

Plate 16.4 Maximum likelihood classification for area shown in air photograph Plate 16.3. Water streams in black, rice fields in dark grey, poplar fields in light grey. (Source: Fraysse, 1977.)

Plate 16.3 Aerial photography: reference area for an inventory of poplar and rice fields (3000 hectares). Adult poplar fields are represented in black, rice fields in dark grey. (Source: Fraysse, 1977.)

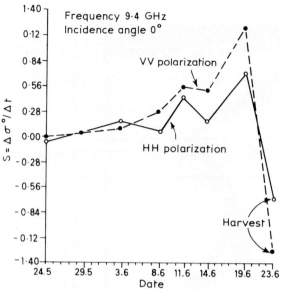

Fig. 16.11 Changes in radar response as a wheat crop matures in Kansas. The period of observation extends from 24 May to 23 June, 1974. Note the increase in response in the ripening stage before harvest and compare with Figure 16.13 for changes in visible reflectance in a cereal crop in Britain. (Source: Bush and Ulaby, 1975.)

and after wheat is harvested (Fig. 16.11). These variations suggest that radar could be used for estimating wheat maturity and for monitoring the progress of harvest.

In parts of Europe there may be only eight days in the year when conditions are ideal for Landsat MSS observations. In these circumstances the use of radar has many attractions. In cloudy regions throughout the world it may be the only sensor capable of providing data for time-discriminant analysis. Evaluation of radar has taken place in

Canada and Europe in the early 1980s and it is likely that the following decade will see heavy emphasis on radar studies. Perhaps the most promising avenues of research are concerned with ratioing polarized radar returns at different grazing angles and using the data in time-discriminant analysis (Fig. 16.12).

Much of the work of an experimental nature is being carried out under the auspices of the Joint Program for Agriculture and Resources Inventory Surveys Through Aerospace Remote Sensing (AgRISTARS) in the United States, the successor to the Lacie project.

16.5 Thermal infrared sensing and its role in crop yield studies

Thermal infrared sensors operating from aircraft can provide high resolution imagery capable of distinguishing hedgerows and individual trees (see Plate 4.5). Satellite infrared systems have a ground resolution capability of about 500 m, although this may be improved to about 100 m in the Landsat D sensor system. As a result of this, low resolution thermal data are of little use in crop *identification;* it is, however, important data when used in crop *yield* (productivity per unit area) models.

The two primary environmental determinants of crop yield are temperature and moisture. The technology required to remotely measure surface temperatures of soil and crops is well developed. Thus the crucial issue for the production of crop yields by remote sensing data remains the assessment of available soil moisture (see Chapter 13). Temperature measurements may, however, provide

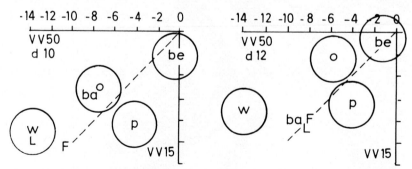

be beets, p potatoes, w wheat, o oats, ba barley,
L *Lolium*, F *Festuca*

Fig. 16.12 An example of the discrimination of different crops by a ground-based X-band radar observing at different angles of incidence (VV 50°: VV 15° on day 10 (d10) and day 12 (d12) of a series of observations. The circles are drawn at 2 decibels radius and VV indicates vertical polarization. (After Kasteren and Smit, 1977.)

valuable information and there have been efforts to model the relationships between air temperature, leaf temperature and crop yield. The underlying theory is that moisture deficiency in the crop leads to increase of leaf temperature above air temperature. Thus the term 'stress degree day' (SDD) has been coined to indicate those days when the crop is suffering from a shortage of water which will depress yield. The final yield of the crop (Y) can be deemed to be linearly related to the total SDDs accumulated over a given critical period. Thus a general least squares regression equation (intercept α and gradient β) can be expressed as:

$$Y = \alpha - \beta \left(\sum_{i=b}^{e} SDD_i \right)$$

where SDD_i = mid-afternoon (14.00 hrs) value of (leaf temperature) $-TA$ (air temperature) on day i.

   b = day on which summation begins.

   e = day on which summation ends.

It was first found that the best time over which to sum daily values of the SDD measurements is the period from the first appearance of the awns to the time when the head produces no more dry matter. This period can be determined by a discontinuity in the albedo data between first head emergence and maturity of the crop. For example, the changes in reflectance of a spring barley crop in England (Fig. 16.13) clearly reflect crop maturity. Albedo

measurements during the early stages of crop growth could also provide estimates of the green leaf area index at the time of head appearance. In this way the potential for head growth could be estimated from albedo measurements and then the growth during the crucial period of head growth be monitored by SDD values obtained by thermal sensing.

There is good reason to believe that temperatures of the canopy of the crop can be measured effectively. For example, a study of a wheat crop at Davis, California, showed that the correlation between temperatures measured on the ground with a PRT-5 Radiation Thermometer (10.5–12.5 μm bandpass) could be closely correlated with airborne measurements at 300 m altitude using a Texas Instruments R-25 infrared line scanner in the same bandwidth (Fig. 16.14).

Attempts have been made recently to combine the stress degree day concept(moisture factor) with the classical growing degree day (GDD) in order to improve crop yield modelling. The growing degree day can be defined as:

$$GDD = \frac{T_{max} + T_{min}}{2} - T_b$$

where T_{max} and T_{min} = daily maximum and minimum air temperatures

   T_b = base temperature (circa 5.5°C) below which physiological activity is inhibited.

Measurement of the climatic factors of GDD plus

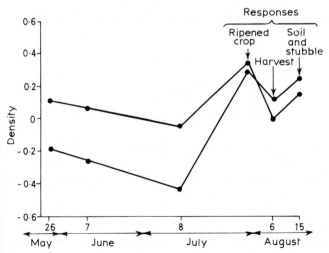

Fig. 16.13 Changes in reflectance of visible light during the growth of spring barley (variety Vada) on soils of the Sherborne Series, Badminton, Gloucestershire. (Adapted from Curtis, 1978.)

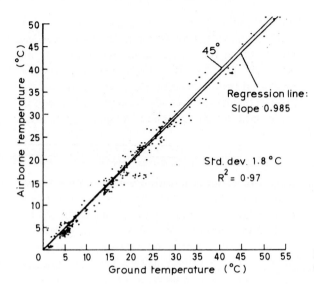

Fig. 16.14 Airborne versus ground-measured temperatures for an entire wheat growing season. (Source: Millard *et al.*, 1979.)

calculation of daylight minutes in the period between emergence of the crop and the appearance of heads and awns are now used as additional inputs for the estimation of crop yield.

The main sensor for crop studies remains the MSS in the range 0.4–2.5 μm. Improved interpretation and classification should be possible using bands as narrow as 0.04 μm. The choice of bands will be dependent on the growing experience of remote sensing experts. Present evidence suggests the following bands (or near variants) will be found useful: wavelength (μm) 0.53–0.59; 0.58–0.63; 0.62–0.68; 0.76–0.90; 2.00–2.60. As a supporting sensor the MSS will be supplemented by a thermal infrared channel in the range 10.5–12.5 μm. The desirable channel widths appear to be 100 nm for acreage determination, 40–50 nm for identification of the phenological stages and 2–4 μm for the thermal infrared.

16.6 Land use in the past

Aerial photographs have been widely used in studies of earlier civilizations and their settlement patterns; in fact air photography is now an accredited, almost veteran, aspect of the archaeological method. Nevertheless, its potential and limitations are still not widely appreciated.

In practice oblique photographs are widely used to illustrate the features of archaeological sites because they afford the advantages of a view comparable to that which could be obtained from a hill top in a hand picked position at close quarters. The vertical air photograph serves a different purpose in that it is the main tool for mapping features and detecting their planimetric relationships. When overlapping verticals are studied stereoscopically they offer many advantages in that the stereomodel enables the viewer to see the physiographic positions in which archaeological remains are found.

When air photographs are used to study former land use patterns there are two principal lines of evidence. First, there are vegetation markings which mainly occur within cropped land. Where buried features exist beneath the surface the soil may be shallow, e.g. over a buried wall. In this case the rooting depth of crops will be restricted and at times of water deficiency (usually late summer) the crops may show yellowing of the leaves or stunted growth. Conversely the buried feature may be in the nature of a ditch. Under these circumstances the soil over the ditch may be more moisture retentive and higher in nutrients. In consequence the crop may be more luxuriant and greener along the ditch line. Such features are sometimes referred to as positive (stronger growth) or negative (weaker growth) crop marks. They are usually best seen when the crop is mature and when the climatic conditions have produced moisture stress in the soil.

The direction of photography is of crucial importance in archaeological survey. Differences of

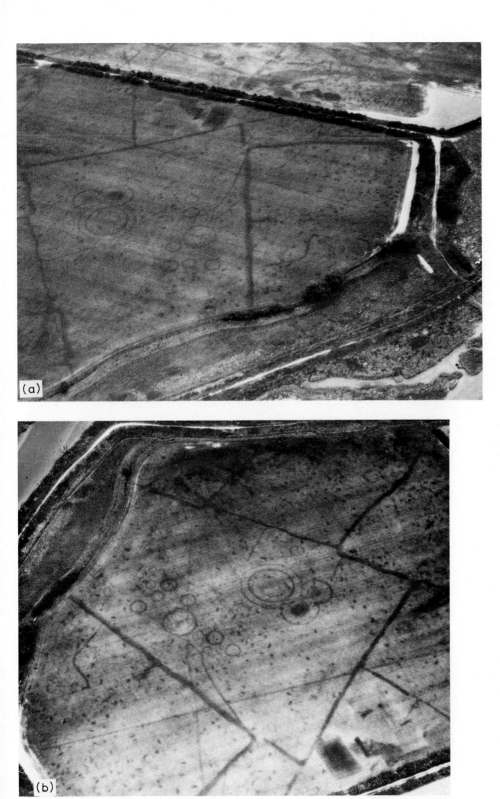

Plate 16.5 Crop marks of ring ditches at Lawford (Essex), June 1970: (a) looking into the sun; (b) with the sun. Tonal contrasts are entirely due to direction of view since photos were taken within a minute of each other. (Source: University of Cambridge; copyright reserved.)

contrast as a result of photography taken in different directions relative to sun azimuth can be striking (Plate 16.5).

The second type of evidence for prehistoric features consists of soil markings. These occur where the soil is bare or has only the thinnest covering of vegetation. The disturbed soil along field boundaries or foundations is different in structure, texture and colour from that of the undisturbed soil adjoining. These differences are often sufficient to induce diffferent reflectances from the surface which can be detected by remote sensors. In England some of the greatest contrasts have been seen on chalk and limestone soils. Usually

Plate 16.6 A pattern of medieval fields in ridge and furrow near Husbands Bosworth, Leicestershire. Note how the low sun angle casts shadows which serve to emphasize small changes in surface relief. (Source: University of Cambridge Collection; copyright reserved.)

soil markings are most clearly detected after a period of exposure to wind and rain rather than immediately after ploughing.

Some archaeological sites can best be identified as a result of transient surface climatic conditions. For example, dark bands of bare earth may reveal where snow or frost has melted first above the filling of buried ditches. A similar effect is sometimes seen where the lines of buried foundations (or perhaps more probably of trenches where the foundations have been dug out and robbed for later buildings) are picked out by early melting of hoar frost on grass.

The patterns of ancient field systems are often incorporated into present day field patterns. It is not always easy to distinguish old field boundaries from maps but aerial photography can often provide startling pictorial evidence for such features. For example the Roman field systems can often be observed in great detail. The Roman method of land partition was to allocate blocks of land which were usually in the form of a square (the centuria quadrata) of 710 m size. High altitude photographs show the grid pattern of centuriation well and in some cases lower altitude observations allow detection of earlier Greek systems to be identified. Similarly the patterns of ancient strip fields can be mapped from remote sensing data (Plate 16.6).

In this brief discussion of the use of remote sensing for the study of prehistoric land use attention has been focused on the photographic sensors. There seems little doubt, however, that the application of infrared scanning systems could provide additional material of interest in such studies. These sensors are sensitive to changes in surface temperature and moisture status and could provide evidence of buried features.

17 *The built environment*

17.1 General considerations

We remarked in the opening Chapter that man's environment is partly natural and partly of his own making. Although we have already considered some aspects of man's imprint on the surface of the Earth, this imprint is most complete in what we may call the 'Built Environment.' Whether our focus of attention is a single structure in a rural setting, or some great conurbation, it is here that we find the most unnatural features of the environment – but also some of the most important for man, because it is here that he lives, and from here that he organizes his use or manipulation of the rest of his surroundings.

This chapter is divided into four parts: the first three are concerned with built environments of increasing degrees of artificiality, in rural, urban, and industrial settings; the fourth is a different kind of overview concerned with population, population distribution and population change.

Since the scales of unitary features in the built environment are generally smaller than those in the natural environment satellite remote sensing is still, at present, less helpful in this connection than the more traditional remote sensing approaches of air photo-interpretation and photogrammetry. However, we shall see that significant use has been made already of Landsat imagery for analysis and assessment of the fabric of urban and industrial areas. Doubtless, 'Earth Resource' satellites will play an increasing part in monitoring of the Built Environment in the future.

17.2 Rural structures

It is convenient to differentiate between a number of contrasting geometrical components of the rural landscape. These are *point*, *line*, and *area* features.

Since we have discussed rural land use already in Chapter 15, and much of this involves open (i.e. unbuilt) countryside, our consideration of area features in this section will focus solely on settlements, and some specialized aggregates of point, line, and area features (e.g. airports). As in other contexts in the present chapter, remote sensing is a valuable tool for rural survey purposes because it permits relatively cheap, rapid and repetitive mapping even in areas to which access on the ground might be difficult or dangerous.

Key structures in the rural environment include:

(a) Point features, e.g. dwellings, storehouses and animal sheds.
(b) Linear features, e.g. field boundaries, pipes and power lines, roads, railways and canals.
(c) Area features, e.g. settlements (hamlets, villages and small towns), airports, military establishments, large schools, hospitals and penal institutes.

In the first instance, a remote sensing survey may be carried out as an aid to topographic mapping. If repeated later, air photography may be analysed to reveal rural change, perhaps for interpretation in economic terms. For example, is there evidence of changing rural wealth, perhaps through increasing dereliction of homes and outhouses, or through new or improved structures?

Field boundaries are also important barometers of rural culture and economic change. Different sizes and patterns of fields and field systems may be found in relation to different ethnic communities as well as in response to national boundaries (Plate 17.1), or important terrain and soil characteristics. Significant changes may result from the implementation of land reform policies, especially after a change of government. Air survey provides the

Plate 17.1 A clear example of an administrative boundary (the US–Canadian border), which is evidenced in Landsat MSS imagery through associated differences in land use and farming practice, except where topography over-rules. To the north of the border lie the grassy central Great Plains of Alberta and Saskatchewan; to the south, the agricultural lands of Montana, where small grains are grown by strip farming methods to reduce soil erosion. In the west the forested mountains of Sweetgrass Arch appear more 'Canadian' than 'American'. (Courtesy, EarthSat Corporation, Bethesda, Maryland.)

most convenient means whereby checks can be made on the progress and landscape effects of policies of agrarian reform.

The use of remote sensing in respect of existing communication lines (roads, railways and canals) is more restricted, for principal routes have always been mapped in detail. Furthermore, alterations to these networks are usually controlled by government or corporations which are answerable to the general public: changes can be incorporated into topographic and route system maps without recourse to additional data sources. However, remote sensing is fulfilling a vital role in respect of engineering surveys for new road and rail construction projects. Such surveys are widespread, and increasing in developing countries where there is a particularly high premium on a combination of survey accuracy and speed. Multispectral air

photographs and infrared linescan data are used by commercial remote sensing survey companies to appraise ground conditions for road and rail planning purposes. Using such data in conjunction with established terrain evaluation procedures it is possible to assess many factors affecting the cost of alternative alignments and types of construction methods. In conjunction with ground truth from selected sites, the evaluation of surface conditions in key localities can be extended along the proposed construction line(s), taking special account of features having practical engineering significance (e.g. slope stability, weathering, erosion, hydrology, etc.). Associated inventories of indigenous materials for road or railroad construction may also be vitally important if the final alignments are to represent the most advantageous and economic routes permitted by the terrain.

The use of aerial photographs in other civil engineering projects too is more than a future dream. For example, aerial photographs have been used successfully for many years in the development of water distribution services; such photographs provide quick indications of the numbers of premises to be served, the distances involved, and the types of land use in the service area. Sewage collection is also efficiently planned thereby. Remote sensing provides perhaps the most practical and economic method whereby gradients can be assessed, the layout of pipelines planned, and pumping stations and force mains appropriately sited. Photogrammetric surveys are always conducted in the USA for the laying of long-distance pipelines and high voltage transmission lines. Microwave transmission towers can be sited best by such means, and their heights selected to meet the necessary requirements for line-of-sight arrangement. In the United Kingdom the Central Electricity Generating Board regularly checks the condition of its major power-lines by helicopter-borne infrared line-scan.

Lastly, rural aggegations of point and linear features, in the form of small rural community settlements, may be mapped from air photographs in exploratory surveys of areas as yet unexplored on the ground (e.g. in tropical rain forest regions such as Amazonia and New Guinea), prior to initial attempts to contact primitive tribes. Similar surveys may be repeated to build up a realistic picture of the movement of rural populations still practising shifting cultivation in humid tropical environments. Radar imagery may be a useful supplement to visible and infrared photography for such purposes, for there is a high incidence of cloud cover in the tropics. It has been shown that radar images of forested areas possess internal textural variations which stem from the local history of 'slash and burn' economy. Indeed, radar is being seen as an increasingly valuable system for delimiting and tracing many man-made features in the rural environment, especially when the principles of *polarization* and *cross-polarization* are exploited, (see also Chapter 4).

All electromagnetic waves are polarized, that is to say, once propagated, they continue to move at a given angle measured against a standard plane of reference – unless some outside force or object

changes that angle. When radar waves are transmitted horizontally or vertically and are received at the same angle or polarization, they are termed 'like-polarized'. If they are received at a different angle, they are said to be 'cross-polarized' (see p. 233). In rural areas, frequently-recurrent characteristics of elements of the built environment (e.g. flatness, sharpness of corners and unusual material composition) combine to yield stronger radar signal returns from man-made structures than those received from more natural features of the landscape.

In general, cross-polarized signals produce grainier images than the like-polarized, but under certain circumstances some objects in the rural environment are revealed more clearly through them. For example, detecting and tracing communication nets is performed most easily, completely and accurately using cross-polarized imagery when the net traverses the flight path. On like-polarized imagery the best results are obtained when its components are parallel to the direction of flight. Therefore, the most efficient system seems to be one in which both like- and cross-polarized components can be assessed. The most useful and obvious applications of such a system are in topographic mapping in developing countries. It is also possible that, with suitable improvements and refinements, urban road systems may be investigated by such means, with special reference to their surface materials. It is to urban areas that we may now turn our attention in greater detail.

17.3 Urban areas

Whilst urban centres afford the greatest concentrations of facilities for living and working, they also present man with a wide range of attendant problems. These most notably involve *internal* ('intra-urban') problems concerned with the provision of acceptable residences, transportation networks, and places of employment, in conditions of suitable freshness and cleanliness, and *external* ('inter-urban') problems concerned with the maintenance of acceptable relationships between one city and another, and between the cities as a group and the intervening countryside. It has been suggested that many of the contemporary problems of cities may be traced to:

(a) The reasons for city foundation and growth, and the history of their evolution.
(b) The on-going, competitive and conflicting processes of arrangement and rearrangement of land use and functional areas within cities.
(c) The scales, complexities, and patterns of concentrations of intra-urban activities.

Today there is widespread interest in the development of 'urban information systems' for the collation, inter-relation, and practical utilization of urban data from a wide variety of different sources. Urban geographers and planners already recognize remote sensing data as a vital part of the total information pool. Remote sensing data contribute to two classes of information in particular. These involve *static phenomena* and *dynamic phenomena* as follows:

(a) Static phenomena. These include such things as city size; the number, pattern and capacity of roads; building sizes and types; and the characteristics of types of neighbourhoods (e.g. industrial, residential, commercial).
(b) Dynamic phenomena. These include variables which cannot be observed directly, either because they change so rapidly, or because they are not physically visible, e.g. population statistics, traffic flow data, and socioeconomic conditions.

We may illustrate and exemplify the value of remote sensing as a source of valuable data for urban information systems by reference to a number of different aspects of the urban environment.

17.3.1 Urban spatial structure and setting

Here the focus of interest is the way in which cities are laid out, and how they are related to each other, and the smaller intervening communities. Cities can be analysed in terms of *hierarchies*, which take account of the rank orders of cities, and the distances between them on the ground. Small-scale satellite imagery can be used to establish the general hierarchical relationships between cities and towns. For example, night-time visible imagery from a low-light intensifier camera system on some of the early Defense Meteorological Satellite Program (DMSP) weather satellites of the early 1970s have been interpreted in this way over areas of sub-continental scale. Major features of contrast between cities and

their surroundings have been investigated through analyses of satellite infrared images, for example the 'heat island' patterns which develop with the growth of cities as well-known features of urban climates.

17.3.2 Urban subregions and the internal structures of cities

Here the purpose is to identify, classify, and evaluate important component features and areas within each city, in order to understand better their roles, functions and associated activities. The steps in analysing the subregions of a city may be summarized as follows:

(a) Locate the major features, structures and areas with basically similar characteristics and appearances. These may include the Central Business District, interior and suburban residential communities, inner and suburban commercial centres, industrial complexes, transportation hubs, etc. Air photo-interpretation has long proved useful in these respects. Today, increasing use is being made of airborne radar, and Landsat MSS and RBV data for such purposes.
(b) Classify major features, structures and areas into a suitable number of categories for systematic mapping. Plate 17.2 exemplifies urban mapping from aerial photographs. Supervised computer classifications based on Landsat MSS data are being developed and applied widely at the present time.
(c) Count and/or measure important static and physically visible attributes of the cityscape (e.g. size and area of subregions and subareas, numbers and dimensions of buildings and transportation systems elements).
(d) Generate and collate data on dynamic and other urban phenomena for which the image data evidences are inferential, not direct. Such information 'surrogates' are used, for example, in large-scale air-photo and radar image data analyses for the differentiation of residential areas in terms of socioeconomic class, population densities, and associated quality of life.
(e) Derive information and calculate indices to represent dynamic urban features for which clear visible evidence exists, e.g. car parking distributions and habits, availability of land for development, etc.

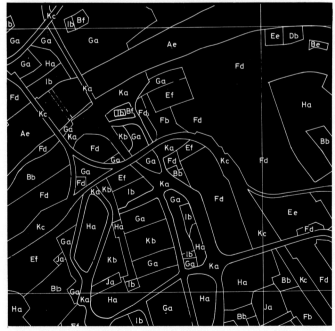

Plate 17.2 An example of aerial photo-interpretation of an urban area: the centre of Newcastle astride the River Tyne. (Courtesy, Fairey Surveys.)

17.3.3 Transportation systems

Aerial photography has been used successfully for many years in the field of transport studies. In the USA, for example, a survey of major highway organizations has revealed that nearly three-quarters of them have made use of aerial surveys in highway planning, although only one-quarter have used them 'extensively'. Four types of studies have been undertaken, namely:

1. Road planning. Photogrammetric analyses of air photographs have provided engineers with both qualitative and quantitative data, especially in the early stages of highway routing and design. In some countries sources of road-building materials have been located from the air.
2. Traffic studies. Aerial photography has been used to pinpoint areas and causes of traffic congestion, and to provide information on traffic flow, both for future road design purposes and immediate ameliorative action. Closed circuit television is used on some urban motorways, for example on the Chiswick Flyover in West London, to assist speedy breakdown and recovery operations and help maintain free movement of traffic.
3. Parking assessments. Aerial photography can reveal where the heaviest concentrations of automobiles tend to build up, and where additional parking spaces might be most beneficial.
4. Highway inspection. Air surveys are a quick and convenient means of assessing the states of road surfaces which may be in need of repair. In winter the successfulness of snow removal may be adjusted in a similar way.

Although the data from Landsat and Skylab are generally rather coarse for transportation studies, the applicability of spacecraft observing systems to transportation geography and linkage analysis has been carefully assessed. In 1965 a broad-based conference held at the NASA Manned Spacecraft Centre at Houston, Texas, called for data relating to three broad groups of transportation facts:

1. Network information. Ideally this necessitates infrared and colour photographic systems, chemical devices, and multispectral sensors, capable of providing data with resolutions of between one and ten metres. From such data, maps of the physical linkages and terminal facilities in a network could be compiled. Pattern analyses by computer processes could be carried out once the information had been transformed into matrix and vector form.
2. Flow phenomena and associated problems. Here infrared and panchromatic photography and radar could be used, giving an optimal resolution of one-half metre. Matters commensurate to investigation would include origin-destination patterns and daily 'tidal' flows of traffic as a whole. These would have useful applications in road design and traffic control.
3. Relationships between transportation and land use. These necessitate infrared, colour and panchromatic photography, and chemical sensing devices particularly sensitive to phosphorus and nitrogen. The optimal resolution would be one metre. Matters such as the relationships between land-use intensity and distance between road and rail links, land-use capability, and the positions of areas within their larger economic regions would be amenable to study.

Clearly the ranges and resolutions of data generally available to the geographic community are not yet adequate for the achievement of such aims. Nevertheless the list indicates something of the potential of the satellite in transportation studies – as well as some of the hopes and aspirations of would-be data users.

17.4 Industrial complexes

In view of the great economic, social, and political significance of manufacturing and extractive industries, and the dominant role they play in many urban areas, we may consider these in more detail both for their own sakes, and in order to exemplify in greater detail methods of application of remote sensing data to the built environment. Three aspects of the industrial components of urban areas may be elucidated most usefully by aircraft or satellite imagery. These are:

1. The present location of industry, and different types of industrial land.
2. Heat loss and the spread of atmospheric and water pollutants away from an industrial complex.

3. The opportunities that exist for the establishment of new industries for the redevelopment of existing land under industrial use.

Historically, the analysis of industrial activity from remote sensing imagery can be traced to World War 2, when air photography was used for the identification of tactical military targets and the selection of military objectives for aerial attack. Today, air photo-interpretation of industrial activity has many peaceful uses, not only in mapping industrial areas and their changes through time, but also updating inventories of stock-piled materials.

Aerial photography is still the most commonly used remote sensing imagery for industrial analysis: high-resolution data are usually essential for the adequate interpretation and assessment of features of the industrial landscape. Aerial photography of industrial areas is frequently flown at scales larger than 1:10 000, even in some cases larger than 1:5 000. Colour film is sometimes used in preference to panchromatic film on account of the higher definition it provides. Often photo-interpretation keys for use in industrial areas are very detailed and technical; their compilation is a highly-skilled business, requiring intimate knowledge of industrial structures and processes.

Stock-pile inventories may be made more accurately and expeditiously from air photographs than from the ground where raw materials or finished goods such as coal, ores and other minerals, pulpwood and lumber are accumulated in the open air. Neatly-stacked materials can be assessed very accurately, albeit from air photographs at scales of 1:1000 or better. Where stockpiles are of less regular shapes more variance is to be expected in the periodic estimates. However, some experts claim to be accurate within 2%, and assert that the lowest-cost ground estimates have an average accuracy of only 15%. The same experts underline the fact that air survey costs can be substantially below those of surveys on the ground. Since both methods follow essentially the same set of procedures, they can be used together or interchangeably. It is commonplace to:

(a) Map the main stockpile with closely-spaced contours (often 1 or 2 apart).
(b) Map the tops of the piles (whose areal dimensions are smaller) in greater detail.

(c) Determine stockpile volumes by planimetering areas between successive contours and multiplying by the depth of material.
(d) Convert the cubic unit volumes into weights, using tables for specific minerals.
(e) Adjust the initial estimates of weights using compaction factors to allow for settling of material in the pile(s).

Similar kinds of measurements can be made in support of extractive industries, e.g. of the amounts of material removed from an open-cast mine through selected periods of time. The resulting data can be of great value to mining companies, and, potentially to governments seeking an accurate, low-cost means of monitoring mineral industries for taxation on a weight basis. Since the first major rise in oil prices in 1974, increasing interest has been shown in the use of infrared line-scan imagery for the mapping and assessment of heat loss (see Plate 17.3, colour section). Many commercial remote sensing firms today provide airborne heat-loss survey services. Indeed, these are widely used not only by industrialists, but also bodies such as central and local government, housing authorities, hospital management boards etc. Poorly-insulated buildings are often quite clearly evident even in the raw infrared image products. By thermographic mapping, values can be placed upon the rates of heat-loss through every structure covered by a survey.

When we broaden the discussion to embrace the whole field of environment pollution, it emerges that a wide range of remote sensing techniques have now been tested in the search for efficient methods to measure the concentration and spread of pollutants. Examples include:

(a) Visible waveband photography. This has been used for such studies as the behaviour of chimney plumes and the appearance and intensification of urban, grassland and forest fires.
(b) Infrared imagery. This has been used as the basis for studies as diverse as the monitoring of volcanic eruptions, fire detection, and the delineation of warm effluents in rivers, lakes or the sea.
(c) Microwave techniques. These are being used to map oil spills in coastal and deeper waters.
(d) Multispectral processing techniques. These can be used to assess water quality through the

different levels of light penetration in selected regions of the thermal emission spectrum.

In general it has been suggested that remote sensing of such aspects of environmental quality in and around urban centres and elsewhere may have two significant contributions to make; namely through the provision of a more complete spatial inventory of areas of atmospheric and water pollution to a single standard throughout a nation, or indeed, the world, and through the determination of regional, national and global burdens of pollution.

Obviously such broad developments can be contemplated only with satellites, not aircraft, as with the sensor platforms. Early experience with Landsat produced encouraging results:

(a) Air pollution. It has been demonstrated that satellite remote sensing can detect particles emanating from both point (e.g. industrial) and mobile (e.g. aircraft) sources.

(b) Water pollution. It has been possible to map large-scale patterns of turbidity in rivers and oceans using the types and resolutions of data supplied by the Landsat satellites. Water pollution from domestic, municipal and industrial sources has been observed and differentiated.

(c) Land pollution. Large scale landscape problems can be identified, including strip mines, tailing piles and dereliction. However, it is in this area that limitations imposed by the spatial resolution of the Landsat system are most apparent; for example, areas of solid waste disposal are usually quite small.

Turning lastly to the social needs of redeveloping old industrial land it is clear that we must deal with a scale of imagery much better than that achieved by any environmental satellite yet flown, and finer even than much obtained from past aircraft missions. Much old industrial land is rather poor in quality, often as a direct result of its exploitation for industrial purposes. In many cases it is not sufficient just to know the extent of such areas. The precise nature and quality of the land may be equally vital in influencing the new use to which it might be put.

In the United Kingdom, for example, studies of derelict industrial land have been made in the West Riding of Yorkshire (see Table 17.1), using aerial photography at a scale of 1:10500. Field checks revealed that a very high degree of accuracy was achieved in the identification of different types of derelict areas within the study area, which measured 200 km². The detailed stereoscopic examination of the prints took approximately 15 man hours. In contrast 15 man weeks would probably have been required for a field survey to collect and map the same information. There is much scope for such studies in the industrialized world, especially where land reclamation is proposed.

17.5 Demography and social change

17.5.1 Population studies

An important use of remote sensing data today is in *population estimation*. In developing countries such as India and Nigeria where full censuses may be difficult to organize, use has been made of air photographs and Landsat imagery for the classification of both rural and urban settlements in terms of types and densities of dwellings so that more complete estimates of population can be made. In the developed world, it has been shown that remote sensing data can be used to provide valuable information on population changes in intercensal periods. A pioneering study of this kind was undertaken in the 1960s in the Tennessee River Valley region for which aerial photographs were available for 1953 and 1963. Four hypotheses were treated:

(a) The population of an urban area is positively related to the number of links it has with other urban areas.

(b) The population of an urban area is positively related to the population of the nearest larger urban area.

(c) The population of an urban area is inversely related to the distance to the nearest urban area.

(d) The population of an urban area is proportional to the observable area of occupied space of such a population.

These hypotheses were tested using stepwise linear regression. It emerged that, except for urban area, the order of the independent variables differed from time to time, the values of the various coefficients differed significantly, and for one variable in particular (the distance to the nearest larger urban area) reversals of the direction or sign of the co-

Table 17.1 An air photo-interpretation scheme for the identification and mapping of derelict land in the UK using panchromatic photographs. (Source: Bush and Collins, 1974)

No. in air-photo key	Code in derelict land key	Brief description of derelict item
1r	A1ia	Ridge tip
1f	A1ia	Low flat tip
1c	A1ia	Conical coal tip
2	A1ib	Coal dump
3	A1if	Degraded land above ground level associated with coal mining
4	A1ig	Degraded land above ground level peripheral to a coal mine
5	A1ij	Coal sludge above ground level
6	A2ia	A tip of domestic refuse
7	B1ic	Open-cast coal workings
8	B1ih	Open-cast coal workings not yet 'excavations'
9	B1iic	A dry brick clay quarry
10	B1iid	A wet brick clay quarry
11	B1viic	A dry sand and gravel excavation
12	B1viid	A wet sand and gravel excavation
13	B1viie	Degraded land below ground level, resulting from sand and gravel workings, but partially restored
14	B1viih	Sand and gravel workings not yet 'pits'
15	C1ib	Coal dump site at ground level
16	C1ie	Degraded land at ground level, resulting from coal workings
17	C1if	Degraded land at ground level, associated with coal mining
18	C1ig	Degraded land at ground level, peripheral to a coal mine
19	C1ij	Coal sludge at ground level
20	C1viif	Degraded land at ground level, associated with sand and gravel workings
21	C1viij	Sand and gravel sludge at ground level
22	C3if	Degraded land at ground level, associated with a brickworks
23	C3vj	Power station waste at ground level
23a	B1i/3v	Power station waste used to fill a coal excavation
23b	B1vii/3v	Power station waste used to fill a sand and gravel excavation
24	C3vij	Sewage sludge at ground level
25	C4iiif	Railway dereliction at ground level
26	D3i	A disused brickworks
27	l1i	A coal mine
28	l1vii	A sand and gravel works
29	l3i	A brickworks
30	l3v	A power station
31	l3vi	A sewage works

efficient occurred. However, notwithstanding these complications, the multiple correlation coefficients between actual and estimated population for 40 selected central places were 0.95 for 1953 and 0.88 for 1963. These indicated that over 91%, and 77% respectively of the variations in population of the urban areas is explained on the bases of the selected independent variables. Because of the high correlation between urban area and population (see Fig. 17.1) the inclusion of the remaining independent variables adds little to the explanation of the variation that remained.

Other studies in the USA have attempted to estimate population densities and totals for small areas

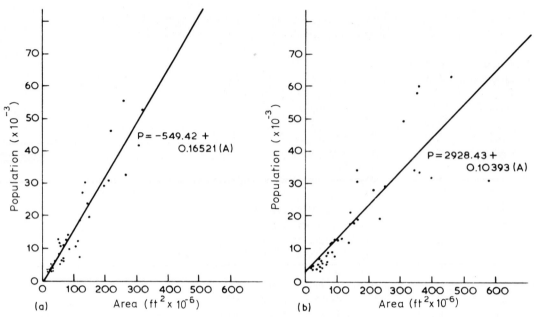

Fig. 17.1 Observed relationships between population and area of towns in the Tennessee River Valley, in (a) 1953 and (b) 1963. (Source: Holz *et al.*, 1969.)

within larger urban centres, using multispectral air photographs and ground truth from sample areas. One such model was applied to Washington, DC in 1970. It was formulated as:

$$Y = f(x_1, x_2, \ldots, x_n) \qquad (17.1)$$

where Y represents housing unit counts or population (1970 tract statistics) and $x_1 \ldots x_n$ are imagery-derived variables such as the number of single-family structures, the number of multiple-family structures, distance from the central business district etc. Relationships between x and Y were developed by multiple regression analysis.

In practice, residential land use was identified on 1:50 000 scale aerial photographs, and the residential land area was computed. A block-by-block count was then conducted to establish the number of dwelling units per structure according to a four-category classification (single-family housing units, 2–5, 6–14, and more than 15 housing units). Multiple regression coefficients of 0.65 and 0.54 were obtained for central city and suburban tracts respectively. These are encouraging, but point to the need for a number of improvements before such a scheme becomes operational. These include the provision of remote sensing imagery of a type in which tree cover poses less of a problem for the location and classifi-

cation of housing units, plus a greater range of ground truth data for the formulation of the model and the checking of its output.

We may conclude that remote sensing, even at a coarse scale, seems destined to provide valuable information concerning population levels and changes, in most areas of the world. The application of aerial photography to dwelling-unit and population estimation would appear to have a particular potential in developing countries where rates of urban and population growth are especially great. Demographic data in such countries are frequently less than adequate for planning purposes, since national censuses are difficult to undertake and often yield inaccurate results.

A partly-related question of considerable significance to government and municipal authorities is that of *housing quality*. For example, concerted attacks on the problems of urban poverty neighbourhoods cannot be planned – still less carried out – until such areas have been defined, and located on the ground. Conventional methods of amassing data on housing quality distributions are extremely time-consuming. Once again remote sensing techniques may provide new and better data than surveys on the ground, while providing such data more quickly and frequently.

Table 17.2 A comparison of housing areas in Austin, Texas, based on selected environmental criteria amenable to identification on remote sensing imagery. (Source: Davis *et al.*, 1973)

Criteria	Low-income areas	Middle-income areas
House size, (ft^2)	380–1220 (average, 731)	110–1560 (average, 1305)
Placement of house and lot (distance from street in feet)	12–42 (average, 27)	34–45 (average, 40)
Potential landholding per housing unit (ft^2)	5670	7500
Building density (%)	14.2	17.7
Average lot size and frontage (ft^2)	4337 Narrow frontage	7376 Wide frontage
Image, pattern, and texture	Lack of uniformity Irregular	Uniform Regular
Houses with driveway (%)	8.3	97.0
Houses with garage (%)	3.0	97.0
Number of visible autos per house	0.20	0.76
Unpaved street (%)	65	0
Street width (ft)	12–24 (average, 18)	24–31 (average, 28)
Quality of curbing	Generally lacking	Intact
Quality of vegetation	Not as vigorous	Cultivated, vigorous
Housekeeping	Presence of debris	Lack of debris
House orientation to street	Short side	Long side
City block pattern	Irregular, dead end twisting streets	Regular
Vacant lots per city block	0.6	0
Proximity to manufacturing and retail activity	Close to manufacturing Remote from shopping outlets	Remote from manufacturing Close to retail outlets

Table 17.2 provides a comparison of housing areas in Austin, Texas, based on selected environmental criteria. Most of these features may be assessed from remote sensing imagery under normal viewing conditions. We may appraise housing quality, study certain socioeconomic aspects of a neighbourhood, and estimate the level and distribution of family income on the basis of such a range of urban characteristics.

17.5.2 Changes in the built environment

We have seen how remote sensing systems may be exploited for both rural and urban land use detection; without doubt, remote sensing data are of greatest value when analysed for landscape change. It should be noted, too, that because such data are unselective (given a particular image type with its attendant spectral and spatial resolutions), the remote sensing view of a built environment reveals it as a single system, organically related to the open land within and around it. Since cities are the most powerful built environments we may conclude this chapter with reference to a range of methods and projects designed to monitor their growth and changing impacts on their own environs. Although remote sensing data can be analysed manually to provide rough estimates of urban growth and associated landscape change there is a trend towards the development of increasingly sophisticated automatic interpretation procedures for such purposes. One early and ambitious project to evaluate the usefulness of photography from high-altitude aircraft and Earth-orbiting satellites for urban land use change detection is the Census Cities Project. This has been organized by NASA in conjunction with the Geographic Applications Program (GAP) of the US Geological Service (USGS). Originally 26 cities were named as test cases, and the US Air Force Weather Service and NASA's Manned Spacecraft Center acquired multispectral, high-altitude photography from twenty of them. The year chosen for this survey was 1970, the time of the ten-year US Census. The basic remote sensing data were colour infrared photographs. This film type is especially valuable for use over towns on account of its high haze-penetration capability. A further useful characteristic of this type of film is that it distinguishes clearly between vegetation (reddish in colour) and cultural features (which have a blue appearance on the film). This makes it especially useful where cities have a strong urban structure/woodland mix. The ultimate goal of the Census Cities Project was the production of an Atlas of Urban and Regional Change, accommodating many types of data presentation, including photomosaics, conventional maps, computer-printed maps, tabulated data and text. In a pilot study for the city of Boston, a 24-category land-use classification was employed. The minimum cell size was about 10 acres. The land use data were computer-processed to make them compatible with the 1970

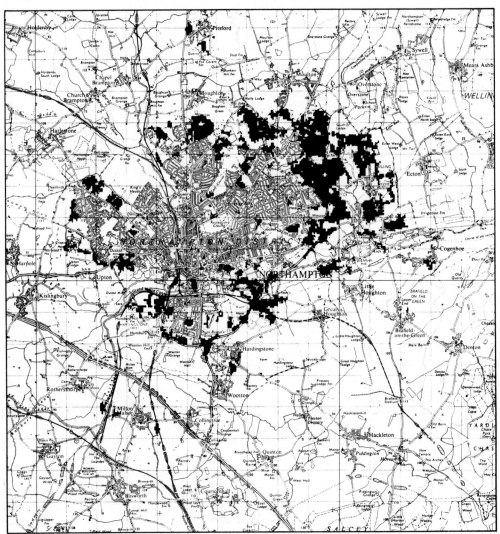

Fig. 17.2 Urban growth (shaded black) around the English Midlands town of Northampton, beyond the 1969 developed area boundary, as determined from Landsat (1975) imagery. (Courtesy UK Atomic Energy Authority and the Department of the Environment.)

census data. Consequently they can be retrieved, either singly, or in selected combinations, by the census tracts.

On a smaller scale, Landsat imagery has been processed to give land-use maps of urban change at scales from 1:250 000 upwards. One such programme has brought together the Planning Intelligence Department of the UK Department of the Environment (DoE), and the Image Analysis Group at the UK Atomic Energy Authority (UKAEA) at Harwell, Berks. For planning purposes, land-use information is needed especially around urban areas, but, at present, there is no single source of comprehensive data in the UK which can provide up-to-date information of this kind on a national scale. Consequently studies have been undertaken to investigate the potential of Landsat as a suitable information source. Using imagery obtained from the European receiving station at Fucino, Italy, a

NASA software package has been further developed for quick and efficient geometrical rectification. Each Landsat scene is computer processed using a multi-category supervised classification in multi-dimensional measurement space. To verify the suitability of Landsat data for monitoring urban growth, four test areas (each about 35 km²) covering a wide range of conditions have been examined in detail. The accuracy of the computer classification of the Landsat data has been checked against ground truth obtained from aerial photography and Ordnance Survey maps. Taking Northampton, an important English Midlands town (population 130 000) as an example, over 90% of the new development that has taken place from 1969–1975 has been accurately identified (see Fig. 17.2). It is clear that Landsat can be a very valuable tool for the assessment and mapping of urban change, even in the more developed countries of the world. Since

Table 17.3 Some user requirements for land-use data: user type versus area size and land use classification. (Source: Horton, 1974)

User type	Functional type	Areal unit	L.U. classification*
National	Community development	Cities over 2500 pop. Counties	one-digit
	Economic development	Cities over 2500 pop. Counties	one-digit
	Human resources	Cities over 2500 pop. Counties	one-digit
	Natural resources	Counties	one-digit urban two-digit nonurban
State	Community development	Townships	two-digit urban one-digit nonurban
	Economic development	Cities over 100 pop.	two-digit
	Human resources	Census tracts for cities over 25 000 pop.	two-digit urban one-digit nonurban
	Natural resources	40-acre parcels (1/16 section)	one-digit urban two-digit nonurban
Local	Community development	City block	four-digit urban two-digit rural
	Economic development	40-acre parcel (1/16 section)	four-digit
	Human resources	City block	four-digit residential two-digit other
	Natural resources	40-acre parcel (1/16 section)	two-digit urban four-digit non-residential

*One-digit classification divides land use into general categories, such as residential, industrial, commercial, agricultural, etc.; two-digit classification of land use breaks each general category into subdivisions such as single-family residence, two-family structures, multifamily structures, etc.; four-digit classification provides an extremely detailed breakdown such as distinguishing among retail uses.

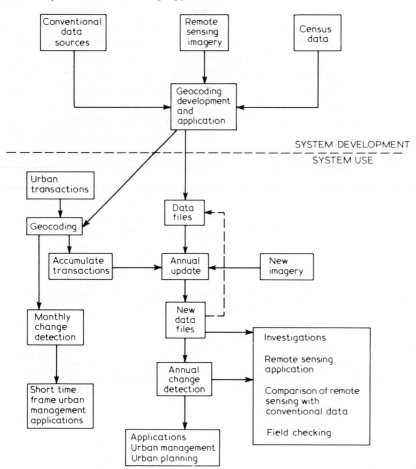

Fig. 17.3 A flow diagram for an operational urban change detection system. (Source: Horton, 1974.)

information obtained in this way is in digital form it can be compared and integrated easily with digital data of other kinds (e.g. from population censuses) for many central and local governmental planning applications.

Looking even more broadly at the change-detection potentialities of remote sensing data sources, Table 17.3 summarizes general user requirements for land-use data in the urban environment. Of particular significance are the implications that, for different types of user, different scales of analyses are necessary, and, therefore, different types of remote sensing data must be used. For example, if urban land-use information is required only on a grid square or block basis, then a relatively low-resolution system will provide adequate information and be advantageous in eliminating the finer details of urban morphology, which may be regarded as a form of picture 'noise'. Aerial photography at a scale as small as 1:100 000 may then suffice. On the other

hand, if details of land use at the plot or parcel level are sought, then much larger scale photography is necessary, supplemented by additional types of information.

In time we may expect governments to set up formal procedures for the operational inclusion of remote sensing data in urban and regional land-use change detection schemes. The form such a scheme may be expected to take is suggested by Fig. 17.3. Remote sensing is vital to it, since this provides the 'status report' information involved in the 'update' section of the scheme in operational use. Clearly, frequent snapshot data from high-altitude platforms should increase the dynamic element in studies of urban land use and morphology, with resulting benefits to the whole population through more economic and rational control, and for development of the complex urban/rural systems of which the modern nation is comprised.

18   *Hazards and disasters*

18.1 Definitions

We live in an age of growing sensitivity to environmental traumas. As population densities continue to rise, living standards and their expectations improve, and the value of human property chronically inflates, so the impact of abnormal and extreme events in the environment becomes ever more serious and costly. Not infrequently environmental traumas are man-induced, for example, acts of war and terrorism. Some are caused unconsciously or indirectly by man, for example landslides resulting from slope destabilization by the excavation of road or rail cuttings, and undesirable soil changes resulting from vegetation clearance and subsequent agricultural activities, However, natural traumas still constitute the biggest threat to man and his economy. It is these that we will concentrate on here.

First we must define our terms.

(a) A *hazard* is a condition (natural or anthropogenic) of the environment which can exert an adverse influence on human life, property or activity. Since man has to a large extent geared himself to the normal state of his environment, the most significant hazards usually accompany environmental states or phenomena which are liable to high degrees of variance about the mean.

(b) A *disaster* is a serious, damaging effect on human life, property or activity which results from the impact of a hazard which has exceeded its critical level(s).

It was estimated in 1978 that the cost to the global economy of natural disasters was at least $US 40 billion – and on a rising trend. Some $15 billion from that astonishing total was expenditure on disaster prevention, including the monitoring of potentially serious environmental hazards. It is clear that, whilst hazard monitoring and disaster assessment procedures will play an increasingly vital role on the world stage in the late twentieth century every effort must be made to ensure that they are carried out not only as efficiently, but also as economically, as possible. Remote sensing is promising in both respects.

18.2 Disaster assessment

Today most large nations, and many international agencies and organizations, are specially equipped to respond to disaster situations. The United Nations itself has a practical interest in such matters, spearheaded by UNDRO, the United Nations Disaster Relief Organisation, with headquarters in Geneva, Switzerland. Once a disaster has occurred, the priorities for action include *assessment* of the damage caused, *relief* for the affected population, and *planning* of measures necessary to return life to normal in the disaster area. The use of remote sensing in such connections is already well-established. Aircraft surveillance of stricken areas is often the most rapid and convenient way of building up a realistic picture of the scale and the intensity of a disaster situation. Air survey and photogrammetry is often useful in the formulation of plans to facilitate safe reoccupation and reuse of the affected region. Fig. 18.1 illustrates the value of aircraft remote sensing in disaster situations. Often this is the only way of obtaining a quick synoptic view of the troubled zone, for surface communications may be severely affected.

Although there is increasing interest in the potential of satellite remote sensing systems for

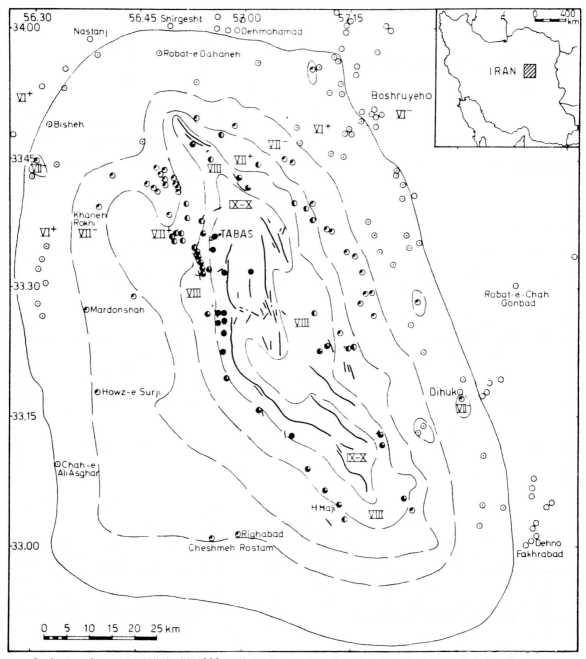

Surface rupture associated with 1968 earthquake ;   Isoseismal on Modified Mercally Intensity scale:
● 80–100 % destruction   ◕ 50–80 %   ◑ 35–50 %   ◔ 20–35 %   ◎ 5–20 %   ○ 5 % destruction

Fig. 18.1 Isoseismal map and aircraft-assessed damage distribution in the epicentral region of Tabas-e-Golshan (Iran) earthquake of 16 September 1978. (From Berberian, 1978.)

Table 18.1 Hazards in the north-eastern USA which are routinely assessed by US Weather Bureau offices from the evidence of geostationary satellite images. (After Wasserman, 1977; from Barrett and Hamilton, 1980)

Type of Hazard	Applicability of satellite data analysis	Key satellite evidence or indications
1. Heavy rain	Direct	Significant, slow-moving convective activity within 100 km of area of interest, or upslope flow of moisture of marine origin.
2. Flash flooding	Indirect	Heavy rain, as in (1).
3. River flooding	Indirect	Heavy rain as in (1).
4. Heavy snow	Direct	Significant precipitation within 100 km of area of interest, and/or upslope flow of moisture of marine origin. Areas of heaviest snow indicated by imagery.
5. Freezing precipitation	Direct	Liquid precipitation in or approaching the area of interest if the surface temperature is near or below freezing, or expected to fall below freezing.
6. Poor visibility (highway, aviation, marine)	Direct	Fog and stratus over the area of responsibility, and possible advection of thick fog into a critical area.
7. High winds (other than those associated with severe convection)	Direct	Movement and change in intensity of a storm that may produce high winds.
8. Severe thunderstorms and tornadoes (high winds, hail, lightning, heavy showers)	Direct	Significant convective activity within 100 km of area of interest or a triggering mechanism approaching an unstable area.
9. Low-cloud ceilings at aviation terminals	Direct	Development and change in areal coverage of low clouds.
10. Strong low-level wind-shear	Direct	The presence of mountain wave clouds, or boundaries of sharp discontinuities in wind flow.
11. Aircraft turbulence (clear air or mountain wave)	Indirect	Cloud features as in (10).
12. Aircraft icing	Direct	Movement of clouds associated with aircraft icing reports.
13. Coastal or lake flooding	Direct	Movement and change in intensity of storms with which flooding may be associated.
14. High coastal or lake waves	Direct	The presence of intense convection cells over marine areas, and movement and change in intensity of storms that may cause high waves.
15. Frost, freeze and cold wave	Direct	Night-time cloud-free areas favourable for radiational cooling. Clues to falling temperatures in cloud-free areas in enhanced i.r. images.
16. Heat wave	—	—
17. Air pollution potential	—	—
18. Fine weather potential	—	—
19. Beach erosion potential	Direct	Movement and change in intensity of a storm that could cause beach erosion.

disaster assessment, that potential is still largely unrealized. Present weather satellites lack both the high spatial resolution capability which is often required for such purposes, and the all-weather capability which would enable them (if equipped with microwave sensing systems) to penetrate cloud cover and reveal the situation on the ground irrespective of the state of the sky. Further, present Earth Resource satellite types lack the high temporal frequency of imaging which is necessary for the evaluation of disasters, which typically occur with great suddenness. Perhaps the best hope for the serious use of satellites in disaster assessment in the forseeable future involves the notion of steerable high-resolution sensor systems on geostationary satellites; these could be pointed towards any area of key significance at the behest of a Command and Data Acquisition station. The polar-orbiting French satellite 'SPOT' (Satellite Probation Observation Terrain) may be equipped with a steerable sensor of this kind (see Chapter 19).

It would appear, however, that satellites have a much more immediate and extensive part to play in monitoring hazards – and the international community appreciates the overwhelming logic of improving hazard monitoring in order to reduce the impact of hazards when they 'go critical' and a disaster threatens. At the same time, of course, it is clear that aircraft are not attractive propositions as remote sensing platforms for general monitoring of environmental hazards. Aircraft operations are notoriously expensive, ranging at present from about $US 500–5000 per hour in the air. Although a few specially-equipped aircraft are maintained for certain key applications, for example hurricane and severe weather monitoring in some tropical and sub-tropical regions, and ice monitoring in high-latitudes, the cost factor is a very effective limit on the use of aircraft in hazard monitoring as a whole.

We may conclude that, if aircraft are still the best remote sensing platforms for the assessment of disasters, satellites are better for the monitoring of many environmental hazards – though it would also be true to say that relatively little of the potential of satellites in such respects has yet been exploited in operational programmes.

18.3 Hazard monitoring

Two different types of approach to the use of

Table 18.2 Examples of agricultural hazards which are amenable to monitoring from satellites, classified in terms of the time periods required for them to become effective. (After Howard *et al.*, 1979)

Class of hazard	Types of hazard
Short-term impact (hours to weeks)	Severe storms and hurricanes Flash floods Forest fires Wind Frost Heatwaves Hailstorms
Short-term cumulative impact (weeks to years)	Flooding from prolonged rainfall Drought Ice and snow cover Adverse synoptic weather Persistent cloud cover Insect infestations Disease
Long-term impact (years to centuries)	Climatic change Soil degradation Desertification Pollution

satellites for hazard monitoring are appearing in the literature:

1. A regional approach, exemplified by Table 18.1.
2. A thematic approach, as illustrated by Table 18.2.

Amongst the more common hazards, earthquakes and volcanic eruptions are, at least potentially, as dramatic as any in their surface impacts. However, the majority of natural hazards are not of geological, but of atmospheric, origin: these lend themselves more readily to satellite remote sensing. Consequently, we will focus our attention in the remainder of this chapter on the use of satellites in monitoring hazards directly or indirectly related to the state and behaviour of the atmosphere.

Authorities in the hazard monitoring field have established that the most frequent natural disasters are associated with flooding, whether slow or sudden in developing. Some 60% of all natural disasters are said to be of this type. On the other hand, the most frequent stimuli of disastrous loss of life are severe tropical storms and hurricanes (alternatively known as tropical cyclones or typhoons). In view of the significance of such events, and the detailed attention which they have been accorded in remote sensing circles, we may profitably take severe convective storms (the instigators of most flash floods) and hurricanes as the two case studies, to illustrate the value of satellites in hazard monitoring. Our third case study will involve insect infestations; interesting and varied research has been undertaken and is in progress in respect of several insect pests. Already our ability to monitor and control some large insect populations has been significantly improved using satellite data.

18.4 Severe convective storms

Certain parts of the world are susceptible to severe convective storms, which may cause considerable loss of life and damage to property through associated high wind speeds, short duration, high-intensity rains and even tornadoes – the most destructive atmospheric phenomena on Earth, judged by the fury of their impact in relation to their size. The USA, open in summer to very warm, moist, convectively-unstable air from the Gulf of Mexico, has a very significant severe-storm

problem. This is magnified by the high property values of such an advanced nation. Conventional activity over the coterminous United States is of special concern to the US Weather Bureau, who make intensive use of geostationary imagery from Goes satellites in support of conventional data for monitoring severe convective situations. Of special interest is the work of the Synoptic Analysis Branch of NESS, who began supporting the Quantitative Precipitation Branch (QPB) of the National Weather Service in 1977 with estimates of severe storm precipitation derived from satellite data by so-called Scofield–Oliver technique (see p. 183). There are three particular reasons why such an approach is deemed essential in a country well-equipped with surface weather-observing stations, including a good weather-radar network:

1. During storms normal communications may be interrupted, and/or conventional weather stations damaged, so that the normal *in situ* data flow is impaired.
2. The satellite provides a more complete areal view than the conventional observing network can give. This is often of special significance in short-term forecasting of severe weather activity.
3. Much of the USA is susceptible to flash-flooding if high-intensity rains occur. Since the exact location and path of movement of a severe convective storm may be very significant in terms of the river basins it may affect, the conventional station network is often not close enough to give the spatial detail desired in such cases.

An appreciation of the scale of the flash-flood threat can be gained from the following examples, all of which occurred during one period of a little over 12 months in 1976–1977:

(a) The Big Thompson Canyon flood in Colorado, 31 July, 1976: 305 mm (12 inches) of rain falling, in parts, in only 4 hours caused 139 deaths and an estimated $35 million damage.
(b) The Johnstown (Pennsylvania) flood, 19–20 July, 1977: 305 mm (12 inches) of rain in 10 hours led to 77 deaths and some $200 million damage. The same community had previously suffered from one of the worst flash floods in US history: 220 known deaths in 1889.
(c) The Kansas City flood, 12 September, 1977: 203

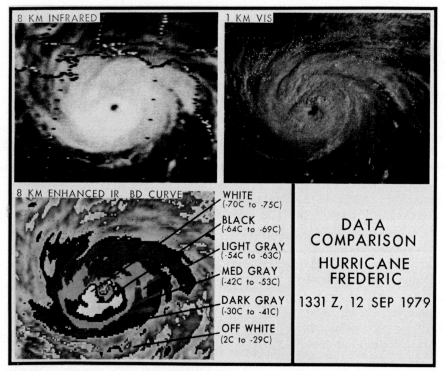

Plate 18.1 Comparison of raw infrared and visible images of Hurricane Frederic from an American geostationary satellite at 1331 GMT, 12 September 1979, with an enhanced infrared product in which radiation brightnesses have been translated into temperatures of radiating surfaces. These can be interpreted in terms of heights above sea-level (colder cloud tops have higher radiating tops). Enhanced products are often physically more informative than unenhanced images. (Courtesy, NOAA.)

mm (8 inches) of rain in 6 hours resulted in 25 deaths, and an estimated damage to property of $90 million.

The Scofield–Oliver technique employs visible, infrared, and enhanced infrared images (cf. Plate 18.1), analysed through a rather complex decision-tree interpretation scheme, to establish suitable rainfall estimates for sensitive areas, assessed on a small grid-square basis (down to a few km²). Of special significance is the enhancement process which contours cold convective cloud tops in terms of temperature (or height). Warmer portions of the imaged areas are displayed by a nearly-linear relationship between radiation temperature and picture brightness (dark for warm, white for cold), up to the edges of cumulonimbus anvils (at about −30°C). Within the anvils lower temperatures are represented by a further set of grey shades from black to white. Essentially the Scofield–Oliver method is comprised of three stages. In these the analyst seeks to:

(a) Identify the active portions of each powerful convective system.

(b) Form an initial rainfall estimate through the indications of the enhanced infrared imagery, and knowledge of the typical performance of such convective systems locally.

(c) Examine successive pairs of visible and infrared images for evidence of heavier rainfall. This most commonly involves:

   (i) Quasi-stationary cumulonimbus systems, and or such systems regenerating *in situ*;

  (ii) Cold, rapidly-expanding anvils;

 (iii) Cumulonimbus mergers;

 (iv) Merging lines of cumulonimbus;

  (v) Overshooting tops (rapid upward growth of cumulonimbus towers through cumulonimbus anvils).

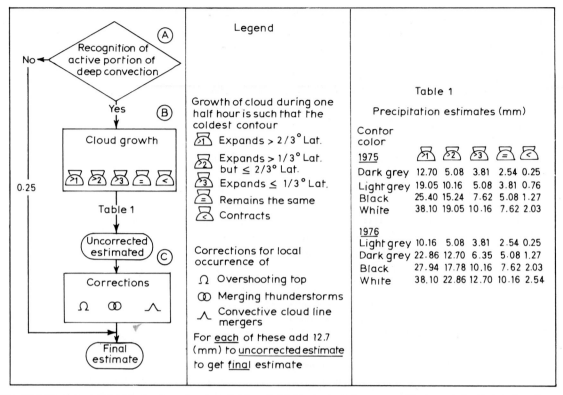

Fig. 18.2 Flow chart and table for the estimation of point rainfall from geosynchronous satellite imagery for S. America. (Source: Ingraham and Amarocho, 1977.)

Although this type of approach is complex, and depends heavily on the skill of the analyst, it is in operational use in the USA, and has yielded much helpful information in potential summer flash-flood weather situations. It is also in use now in a modified form to give precipitation estimates for significant convective outbreaks in winter. Further development of both forms is under way in the Applications Division of the National Earth Satellite Service, where higher levels of automation are being sought, and in the hydrological offices of other countries of Central and South America (see Fig. 18.2).

Already, numerical methods have been developed to extract cloud dimensions and locations from the geostationary imagery; these may be taken to represent an instantaneous observation of convection. An indicator of convective growth or decay is then derived by tracking cloud area locations over several picture intervals, and recording the changes in cloud dimensions. Cloud tops within cumulo-nimbus anvils are compared visually with isohyetal analyses of rainfall based on surface data, and

adjustments are made by the cursor, on the basis of the satellite evidence.

Finally, the short-period (30 or 60 min) improved rainfall totals are integrated areally and temporally by the computer to provide final rainfall maps covering the periods required. Such an 'interactive' approach to the severe-storm monitoring problem seems the best that can be envisaged for the foreseeable future, and is of great potential value for the assessment of related hazard levels and the mitigation or aversion of associated disasters in any part of the world where intense convection is a threat.

18.5 Hurricanes

In the previous section we considered the use of weather satellite data in the monitoring of severe weather which is organized characteristically at the meso-scale (10–100 km in diameter). Here we consider much larger, more strongly-organized severe weather systems capable of inflicting very severe damage to human life and property, not only *directly* through high intensities of wind and rain, but also *indirectly* through associated surface effects, especially in low-lying coastal zones. Synoptic-scale revolving or rotating tropical vortices (see Plate 18.1) – variously known as tropical cyclones, hurricanes, or typhoons in different regions of the world – are legendary for their destructive potential. Adjudged by the scale and breadth of the impact they often inflict on certain tropical and sub-tropical lands, these are the most destructive weather phenomena on Earth.

The hurricane is so significant a threat to man, and so clearly-defined in terms of its vortex of dense, organized convective cloud, that the very first weather satellites were conceived largely as means whereby these storms might be more effectively monitored: most of their lives are spent over remote oceans and seas whence conventional weather data are very sparse. The monitoring of hurricanes from satellites has been so overwhelmingly successful that no significant tropical storm or hurricane has gone unseen since the first operational polar-orbiting weather satellite system was inaugurated by two Essa satellites in February 1966. Concerted efforts have been made, especially in the USA, to interpret weather satellite imagery as fully as possible to improve both the monitoring of existing hurricanes and the forecasting of their likely movements and changes in the future. Currently, satellite imagery are used to:

(a) Monitor hurricane tracks.
(b) Assess hurricane intensities, and estimate maximum wind speeds.
(c) Identify areas of influence of each hurricane.
(d) Arrive at a reasonable estimate of associated volumes of rain, although schemes for this are the most tentative of the four.

It is instructive to consider each of these in more detail.

18.5.1 Hurricane tracks

In some cases it is easy to identify the centre of a hurricane because a well-defined, relatively cloud-free 'eye' is plainly evident in or near the middle of the dense clouds of the vortex (see Plate 18.1). However, on many occasions, especially in sub- or post-mature situations, centre location is difficult because no eye is visible and/or the vortical cloud is fragmentary in appearance. Here it is necessary to distinguish between the *central features* (CF) of the hurricane cloud system, and peripheral *banding features* (BF) which are arcs of cloud tending to point or spiral towards the centre of the storm circulation in the lower troposphere.

Through careful tracing of cloud arcs, a convergence centre can be established with reasonable confidence in many cases. Results compiled from many hurricane seasons show that tracking is usually within ¼–½° (28–55 km) for the location of the centre of each storm. Only occasionally – in respect of particularly ill-organized storms – are centre locations 1–2°, or more, astray. For forecasting purposes, recognition of significant changes of vortical cloudiness is often profitable (see Fig. 18.3).

18.5.2 Hurricane intensities

Early classifications of tropical storms and hurricanes revealed that, through considerations of their sizes and stages of development, good estimates could be made of sustained maximum wind speeds (MWS). Such work has led to the development of a highly-detailed classification scheme which can be used both diagnostically and predictively. This is illustrated by Fig. 18.4, which categorizes common tropical cyclone patterns in terms of 'T-numbers' which are largely (but not entirely) determined by the sum of values allotted to CF and BF characteristics. Comparisons with surface and aircraft observations have shown that Current Intensity (CI) numbers (T-numbers modified to take account of influential factors not directly related to cloud features) can be interpreted diagnostically in terms of MWS as shown by Fig. 18.5. The T-numbers alone can then be used predictively to give forecasts of hurricane wind speeds for up to a few days ahead, once it has become clear whether the storm in question is following a rapid, normal or slow development curve. Verification tests have indicated that assessments of hurricane intensities expressed in terms of MWS are 90% correct within two T-number categories (i.e. ±1.0 T) which, translated into wind speeds, is ±15 knots. This is reasonable, remembering that hurricane winds often substantially exceed 100 knots.

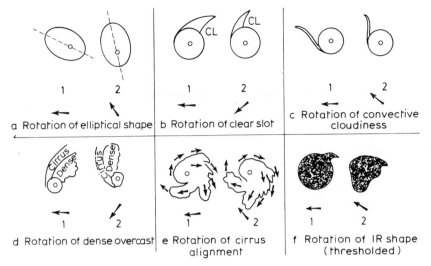

Fig. 18.3 Characteristic changes in cloud patterns associated with directional changes of motion of tropical cyclones. (Source: WMO, 1979.)

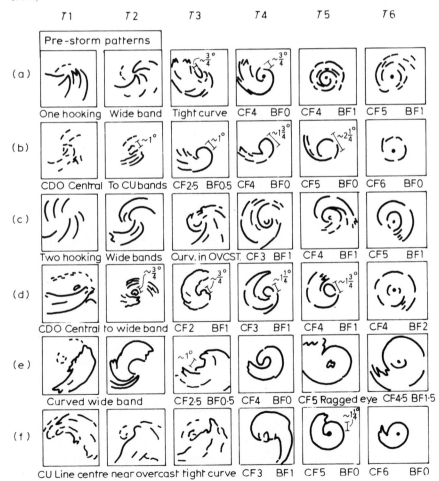

Fig. 18.4 Common tropical cyclone patterns and their corresponding T-numbers. The T-number shown must be adjusted for cloud systems displaying unusual size. The patterns may be rotated to fit particular cyclone pictures. (Source: WMO, 1977.)

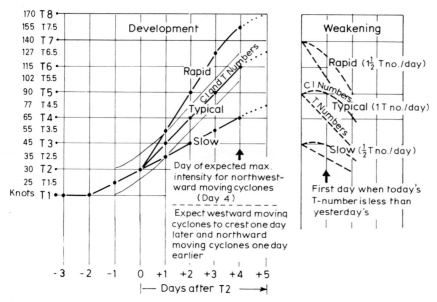

Fig. 18.5 Current intensity change curves of the tropical cyclone model. The shaded area surrounding the typical curve represents 'intensity' as a zone, one T-number in width. (Source: WMO, 1977.)

18.5.3 Areas of hurricane influence

Here the aim is to establish the corridor likely to be affected by strong winds and heavy rain, especially as a hurricane makes a landfall and tracks inland. The central dense overcast (CDO) is of greatest significance for associated rain. Banding features may also have to be taken into account in order to outline the total region of possible wind damage.

18.5.4 Hurricane rainfall

This is much more difficult to estimate accurately, due both to the extreme rainfall intensities which hurricanes may bring, and the propensity for hurricane rainfall to vary spatially to a high degree. Experience with hurricanes in the American sectors of the North Atlantic and North Pacific Oceans indicates that the following rainfall rates may be assumed as representative of a transect marked by the storm centre path:

(a) Wall Cloud area (around the eye, within 20 nautical miles of the storm centre): 51 mm h^{-1} (± 25 mm).

(b) Inner Central Dense Overcast (CDO) area (within 50 n.mi. radius of the storm centre,

adjustable on the evidence of enhanced infrared imagery): 25 mm h^{-1} (± 12 mm).

(c) Outer CDO area (radius defined by the outer edge of the CDO): 1–2 mm h^{-1}. (If there are embedded convective bands the rates there will be higher, between those of (b) and (c)).

Since the total volume of rain falling from a hurricane is often even more significant, e.g. for basin hydrology monitoring and flood potential assessment, efforts are being made to develop schemes more suited to these needs. The following relationship has been proposed for hurricanes in the Caribbean Sea (Gulf of Mexico region):

$$\hat{R} = (\delta/v)<R>$$

Where $\hat{R}$ represents total rain potential, δ is the width of a storm along its path relative to a calibration station or stations, v is the speed of movement of the storm, and $<R>$ its average rain-rate. Schemes of this kind are necessary, but further testing and development of them is required.

We may conclude that the satellite is a vital tool for use in hurricane monitoring and forecasting generally, and that, as Chapter 1 suggested, this use of satellite data alone justifies the operation of satellite systems for many countries in the tropical and sub-tropical world.

18.6 Insect infestations

Researchers in remote sensing have long pursued the possibility that conditions conducive to up-surges of several major insect pests might be monitored from aircraft. Perhaps the best example is the mosquito family, whose breeding habitats have been extensively defined using multiband aerial photography. More recently, attention has turned rather to the potential of satellite remote sensing for improved pest monitoring and control. This has involved a variety of insect pests, which are influenced by a range of different environmental parameters and characteristics. Here we will con-sider three pests which seem especially amenable to monitoring from satellite altitudes.

18.6.1 The Screwworm (Cochliomyia hominovorax)

This is a devastating cattle pest infecting, for example, Central and Southern North America, being endemic in Mexico, and expanding into the USA in summer. It is temperature controlled in its adult stage, the fly being especially sensitive to both cold and very hot weather (for this reason there are no over-wintering populations in the USA). The distinctive features of the species are that:

(a) The larvae feed only on living tissue.
(b) The population density is relatively small.
(c) The female fly mates only once in her lifetime.

Point (c) renders the fly vulnerable to the intro-duction of sterile males, which are used as an effec-tive control measure. Without this, annual losses in the USA could exceed $200 million.

In the Screwworm Eradication Program of NASA high-resolution infrared data from Noaa satellites have been processed in conjunction with correction factors for atmospheric attenuation to give daily air-temperature maps registered to a common geo-graphic grid. Longer-term temperature maps and maps of degree days have also been produced. Tests in Northern Mexico have been adjudged successful, sterile flies having been released by aircraft over significant areas of Screwworm activity. Unfor-tunately, as project staff reported, 'The logistics of

handling images on computer tapes turned out to be a major problem, and the extensive processing required to digitize, register, calibrate, and produce the various products was expensive.' Further model and programme development seems necessary, es-pecially in developing countries where the need is greatest.

18.6.2 The Desert Locust (Schistocerca gregaria)

The Desert Locust is both a traditional scourge of, and a continuing threat to, the agricultural economies of many countries stretching across the desert and semi-desert belt of the Old World, from western Africa through the Middle East to south-central Asia. To counter and contain this threat a multinational organization, the Desert Locust Commission, has been set up. This meets annually in the headquarters of FAO in Rome, which helps coordinate the work of the area Desert Locust Control Commissions, and organizes research into new methods of survey and control. In particular, detailed research has been undertaken since 1975, focussed initially on the area of the North-West African DLCC (covering Algeria, Libya, Morocco, and Tunisia), to develop an operational method for satellite-improved Locust monitoring and control (see Fig. 18.6).

The Desert Locust is ideally suited to the exploi-tation of favourable conditions for breeding as and when they arise. In periods of adversity there is a basal population of locusts spread thinly throughout the affected zone. Some of these 'solitary locusts' become concentrated in rain-fed areas, due largely to the vacuuming effect of converging winds. Here, the solitary locusts change their shape, colour and pattern of behaviour: their feared 'gregarious phase' begins.

The Desert Locust is capable of considerable migratory movement in search of food and breeding grounds. In suitable areas its reproductive rate achieves geometrical proportions. Suitable condi-tions for breeding include:

(a) A sandy-silty soil for easy egg-laying.
(b) Sufficient soil moisture in the top layer of the soil to encourage eggs to hatch.
(c) Green vegetation for the immature locusts to feed on after hatching.

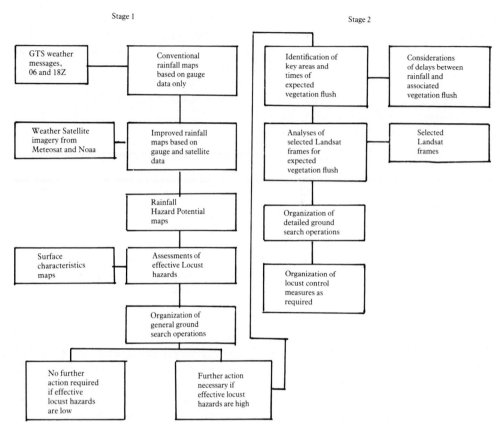

Fig. 18.6 Flow-diagram for Desert Locust monitoring involving satellite (Meteosat, Noaa and Landsat) imagery. This scheme is being adopted for use by the Desert Locust Commission and FAO, beginning with North-West Africa.

Since (b) and (c) are determined principally by the distribution of rainfall events, improved monitoring of rainfall is a primary objective of a satellite-assisted Desert Locust control operation. However, (a) is determined by desert morphology and pedology, and both (b) and (c) are also influenced by surface characteristics. Consequently Landsat imagery has been used for the compilation of surface characteristic maps of North-West Africa which are much more detailed than existing topographic maps. Weather satellite imagery are used regularly in support of surface weather station data to give the best possible maps of rainfall for periods of 12 h and upwards. These maps are interpretable in terms of related 'locust hazard levels', for two important practical applications:

1. The identification of areas where significant rain is thought to have fallen in areas whose surface characteristics may be conducive to vegetation

flush and locust breeding; post-rainfall Landsat scenes are ordered for such areas. False colour composites of MSS Bands 4, 5 and 7 reveal the extent and density of any significant growths of vegetation. These should then be inspected in more detail from the air or on the ground.

2. The planning of routes to be followed by ground inspection teams, so that these can be directed more efficiently and economically than before to areas of greatest locust breeding hazard.

Fig. 18.7 illustrates the types of results which have been obtained from the operational research programme in North-West Africa. Since vegetation flush post-dates rainfall by periods from a few hours in spring to weeks or months in autumn and winter, and the locust life-cycle from egg-laying to the first flight of a new generation spreads over about 21 days, there is ample time for the collection, analysis and interpretation of the satellite data and the

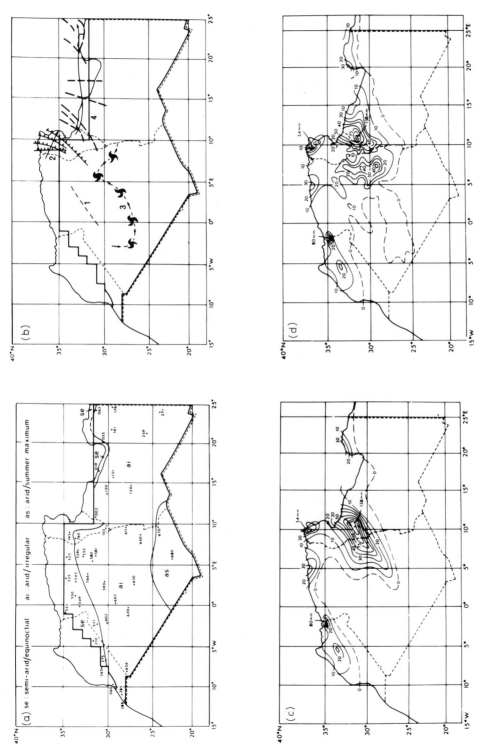

Fig. 18.7 Examples of the use of satellite-improved rainfall monitoring in North-West Africa in support of Desert Locust Control Commission operations: (a) the scatter of GTS rainfall stations, and morphoclimation regionalization for satellite rainfall estimation, (b) plots of significant raincloud systems from satellite evidence, (c) rainfall analysis from gauge data only, and (d) rainfall analysis from gauge plus satellite data, all for 29 March–4 April, 1979.

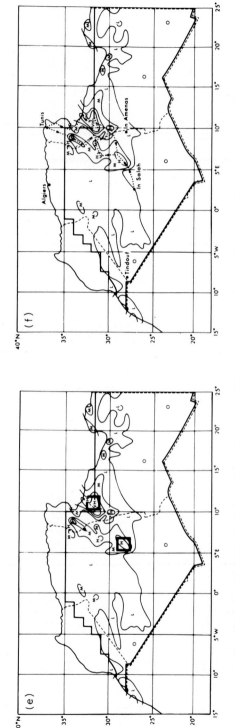

Fig. 18.7 cont. (e) selection of Landsat frames for post-rainfall vegetation assessment, and (f) selection of preferred routes for Ground Inspection Teams, both on the basis of 'Hazard Maps' (zero, low, medium high and very high locust hazard) drawn from satellite-assisted rainfall analysis for Spring 1977. (After Barrett, 1981.)

planning of detailed 'search and destroy' operations before local population dynamics become critical for the development of plagues of the Desert Locust. Indeed, it seems that most important infestations of recent years have developed not so much because of lack of knowledge of them, but more because key areas have been closed to control teams for political reasons and, in several cases, associated hostilities. The satellite approach to the monitoring of Desert Locust habitats and changing environmental conditions, as developed within FAO, merits a much wider application than in North-West Africa alone. There is cause to hope that satellite monitoring techniques may soon be applied to most, if not all, the susceptible areas from the coasts of West Africa to Southern Asia.

18.6.3 The Armyworm (Spodoptera exempta)

This is the larval stage of a migratory moth which inflicts considerable damage each year to agricultural crops, especially in East Africa anywhere from the Red Sea to Mozambique. Whilst the seasonal movements of the adult moths are generally well-known and moderately predictable, many uncertainties remain; it is not without reason that this pest is colloquially known as the 'Mystery Worm.' In particular there is uncertainty as to: (a) the exact length of the breeding cycle, (b) the distribution and cause of the quite strongly-localized breeding areas which lead to the destructive infestations of armyworm larvae, and (c) the ability of individual moths to contribute significantly to the long-distance seasonal migrations which the populations are known to make. It is thought possible that plagues of larvae may develop following massive cumulonimbus convection in areas of large moth populations; strong wind convergences may concentrate the moths, which are beaten down by heavy rain and so lay their eggs in areas favourable for vigorous vegetation growth.

Research investigating evidences of important mesoscale convection within the ITCZ as imaged by Meteosat is being undertaken jointly by the University of Bristol and the Centre for Overseas Pest Research in London to test these hypotheses. Ground truth on moth populations and larval infestations is being supplied by government organizations in countries of East Africa. The aim is to provide a satellite-improved forecasting scheme for the Armyworm; if achieved, this could be almost as valuable as the satellite-improved scheme for Desert Locust monitoring and control, as outlined in Section 18.6.2. Almost certainly the use of satellites in pest monitoring and control is still in its infancy, and further pests will be added to the list of satellite scalps in the near- to mid-term future, probably including the tsetse fly of central Africa, and sandflies of the semi-arid zones.

18.7 Conclusions

Weather satellite data have been an integral part of the data pool for operational meteorology since the mid-1960s. The addition of geostationary satellites, especially since the mid-1970s, when the first SMS/Goes satellites were put into orbit, has greatly enhanced the contribution made by data from space. Stimulated to a large degree by research undertaken by NOAA, especially in its 'Applications Group', American meteorological centres at both national and regional levels undoubtedly make more thorough use of satellite data in weather forecasting and associated environmental assessment than any other organizations throughout the world.

It is the considered opinion of a staff member of one US National Weather Service regional office – the Eastern Region Headquarters, Garden City, NY – that '*Satellite pictures, especially when used together with other types of observations, such as radar, are now one of the most important aides we have in detecting hazardous weather in its earliest stages of development.*' As we saw in Table 18.1, set procedures have been drawn up for the duty forecaster to consider the value of the available satellite data in the preparation of special advisories or 'nowcasts' to warn of current or expected weather and weather-related hazards.

More specialized but, perhaps, even more important nowcasts and warnings based largely on satellite data analyses are issued by the Hurricane Forecasting Center in Miami, Florida, and by many other national weather offices in hurricane-susceptible areas.

Increasingly, satellite data are being used in a widening range of other environmental contexts, especially related to agricultural activity. Crop hazards can all too easily lead to shortages of food, increased food prices, and even famine. Longer-term results may even involve changes in the agricultural policies or internal and external politics of nations or groups of nations.

The Food and Agriculture Organisation of the United Nations, with its headquarters in Rome, Italy, is vitally concerned with the global availability of food crops, and crop shortages. Unfortunately the intelligence required for international technical assistance and relief activities is often not available from conventional (ground) sources as quickly as it needs to be if major human disasters are to be averted. However, the satellite, equipped to observe changing conditions and events frequently and very widely over the surface of the Earth could help greatly in the detection and monitoring of acute hazards to threatening agricultural practice and production.

We may conclude that there is tremendous potential in satellite remote sensing for the more efficient and economic monitoring of environmental hazards, and, therefore, for the increased avoidance or mitigation of natural disasters. What is needed now is a *concerted research effort* to establish the optimum techniques of data analysis with such ends in mind, and a *strengthening of those organizational structures which would foster such approaches* in addition to, or in place of, the existing procedures where these are inadequate to meet the present needs. In particular, it would seem desirable to *centralize* more fully within many countries or groups of countries, the acquisition, processing, analysis and interpretation of the new data types so that their costs might be shared and a higher degree of quality control exercised over the whole operation – and a higher degree of urgency evidenced through the timely issuing of appropriate advice to the general public.

19   *Problems and prospects*

19.1 The limitation of costs on satellite remote sensing studies

The Space Age can be said to have begun on 4 October 1957 when Sputnik 1, the Russian automatic satellite, circled the Earth sending back locational bleep signals. Since then there has been a tremendous advance in space technology, mainly as a result of work in the United States and the Soviet Union. Thus, the Russian programme provided the first man in space (Gagarin) and the first space walker (Leonov) whilst the American programme culminated in the exciting Apollo moon programme.

All the major advances in space technology have been either Russian or American, though it should be noted that several satellites have been sent up by other nations, chiefly on American and Russian launchers. There has also been the development of major programmes of space research by the European Space Research Organisation (now ESA)*. Nevertheless, to the uncommitted onlooker it has seemed to be something of a space race with political undertones, with prestige as a reward. It has also become clear that such programmes are very expensive indeed. Thus the question is raised as to whether such massive expenditure on space research is beneficial to mankind when investment in other forms of industry is thereby limited. Furthermore, the world needs for programmes of education, social and medical welfare cry out for investment of available funds. Competition in space may, therefore, be seen by some as wasteful. It could be argued that the needs of the world community would be better served if the space nations collaborated and pursued

*The European Space Research Organisation (ESRO) and the European Launcher Development Organisation (ELDO) were disbanded in 1975 and a new organization named the European Space Agency (ESA) was formed.

complementary research activities. Even when the desirability of space research is accepted the expenditure is seen by some as being unacceptably high; therefore, the most urgent need seems to be to reduce the cost of space transportation (cost per kg to low Earth orbit).

The next decade will be an important one in space endeavour because it is during this period that attempts will be made to transform the space frontier of the 1970s into familiar territory, easily accessible to man in the 1980s. In order to achieve this it will be necessary to reduce the cost of transportation by as much as a whole order of magnitude. Important savings can be achieved (perhaps cutting costs by two thirds) if space vehicles can be re-used. Further savings could be made if running maintenance were carried out in orbit. With such objectives and possible means of achieving them in mind, present trends in space programmes can be reviewed.

First, one may note that the American Post-Apollo programme incorporates the Space Shuttle as a key element for the 1980s. This is a re-usable transportation system which consists of a vehicle that has features in common with an aircraft, a rocket and a satellite. Its fuselage and wing dimensions are comparable with those of a medium-size twin engined commercial jet aircraft, e.g. the DC9. It can be brought back from space and land on a runway. The Space Shuttle is designed for rapid re-conditioning and repeated use, with a life of 100 sorties or more. The gross lift-off weight is more than 2000 tons and after a certain time the solid-propellant boosters are jettisoned. The fall of the boosters is checked by parachutes and retro-rockets so that they can be recovered from the sea, then refurbished and re-fuelled for re-use (see Chapter 5).

In addition to the Space Shuttle an unmanned propulsive vehicle termed the 'Space Tug' has been designed. The Space Tug would be carried into space by the Space Shuttle. It would then be used to transfer a payload from one orbit to another, rendezvous and dock with satellites and with the Shuttle, and be brought back to Earth by the Shuttle for refurbishment and re-use. At the cost of several flights the shuttle could build 'trains' of tugs. Space payloads could then be given the very high velocities necessary for lunar and planetary exploration e.g. around Mercury, Jupiter and Saturn. It has been ascertained that such missions could be achieved more economically by such means than is at present attainable by expensive Saturn rockets.

The Post-Apollo programme is important because the re-usable Space Shuttle can be used to place a whole series of unmanned satellites in orbit. It also enables man to work in space under technical and economic conditions more favourable than hitherto. It is possible to envisage a modular-type (i.e. sectionalized) space station constructed in space from separate elements carried up by the Shuttle and then forming a set of working laboratories. In addition there may be a family of 'Research and Application Modules' (RAM) complementing the activities of the station and working with it in close liaison. At present the programme is scheduled to provide the modular space station with associated RAMs by the mid 1980s. This programme seeks, therefore, to provide for some reduction in the costs of space activity, whilst increasing the flexibility of operations in space.

The development of the Post-Apollo programme has been accompanied by interesting examples of international cooperation in space research. First, one may note the European participation in the Space Shuttle programme, which was agreed between ESRO and NASA. The Space Shuttle programme is also a collaborative European–American venture. Manufacture of the Spacelab is being undertaken by VFW-Fokker ERNO as the Spacelab contractors. The current position is that the USA is responsible for developing, building and operating the Shuttle, and will make it available for European use on either a cooperative (non-cost) or cost-reimbursable basis. Europe, on the other hand is responsible for the design, development and manufacture of Spacelab and associated equipment. The general scheme of management for the co-operative programme is outlined in Fig. 19.1.

The motivation for participation in the Post-Apollo programme by Europe can be seen to be twofold – the *experience* and the *spin-off*. The experience promised by this venture emanates from actual European involvement in an advanced manned space programme with all its attendant status, prestige and challenge. In fact, the decision by the Americans to internationalize the Space Shuttle programme has given Europe an opportunity to play a part in an advanced technological development which would not otherwise have been possible because of the high overall cost. The spin-off, which might be very large, is of both a technological and scientific nature.

Alongside this example of international cooperation between Europe and the USA there have been parallel develoments of proposals for Soviet–American joint activity in space; in particular, the first tests have been made of docking procedures and rendezvous methods in space. Such collaboration, if brought to full fruition, could lead to collaborative manning of space stations and also mutual support on occasions of space emergencies.

The above summary of the emerging programmes for remote sensing from space illustrates the great potential in future work. However, there remain some severe problems to be overcome before the full benefits of remote sensing technology can be enjoyed by the ordinary man. These problems lie especially in the broad fields of administration and data processing, and at the interface between user and technologist. Of these it may well be that the problems of administration will place the greatest constraints on the applications of remote sensing techniques in the foreseable future.

19.2 Security restrictions

When one examines the general administration of remote sensing applications at the present time there are several apparent sources of difficulty. Amongst these there is the continuous restraint, placed by security rulings of various kinds, on civilian use of new tools in remote sensing. Many sensors and sensing interpretation systems are classified as secret and are, therefore, available only to military users. In

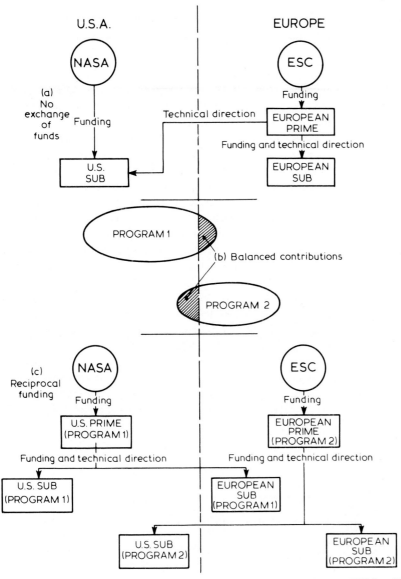

Fig. 19.1 General schemes of management for a cooperative programme between ESRO and NASA. (Source: ESRO, 1972.)

some cases current security classification rulings seriously inhibit the exploitation of new observation methods for peaceful purposes. The inhibiting effects of security ratings may be summarized as follows:

(a) They restrict the availability of existing data and equipment for present programmes.
(b) They limit the education of scientific and engineering personnel who could use new tools in the future for civilian applications.

(c) They hinder the formulation and development of new research projects in natural science and fruitful interdisciplinary efforts.
(d) They interpose a barrier between large numbers of scientists knowledgeable in fields where interpretation of sensor output can be useful to both advances in natural science as well as to military reconnaissance applications.
(e) They retard the free flow of technical information within the scientific community, encouraging duplication of effort and the reporting of uncriticized findings.

(f) They restrict the development of civilian markets so that system and component costs remain high.

Clearly, this problem is one of balancing potential gains to civilian programmes against potential losses in national defence. The related judgements are always difficult to make but it seems clear that the classification of remote sensing devices as secret should be brought under constant review. Likewise periodic reviews of the rules of classification should be made, as has been done by the Directorate for Classification Management of the Department of Defense, USA.

Interwoven with this problem of the security wraps placed on remote sensing systems, there is the other administrative problem of whether 'open sky' observations can be made or not. Some countries are unwilling to allow observations to be made from overflying aircraft and satellite platforms. It is, indeed, praiseworthy that NASA has made weather satellite and Landsat imagery available to Earth scientists with a minimum of constraints. One could wish that more of the imagery obtained by the Soviet Union were as easily available to environmental scientists.

19.3 Handling large quantities of data

Additional problems arise in the field of data processing as a result of the vast quantities of data that can be gathered by modern remote sensing systems (Fig. 19.2). The Landsat multispectral scanner, for example, produces 15 megabits s^{-1} but new systems (e.g. the Thematic Mapper) will generate at least 10^4 megabits s^{-1} and such high data rates cannot be utilized to the full even by the total global user community. In fact some authorities consider that an operational Earth-observation system, similar to a weather forecast system, should aim at average data streams far below 5 megabits s^{-1}. As a result there is a need for studies of *data compression* and the extent to which proposed data might be redundant. This requires study of the information content of images, and the establishment of methods of user-oriented redundancy reduction. It will be especially difficult to define the low-redundancy thresholds in terms of physical quantities like spatial limit, frequencies, dynamic range etc.

Another approach to the problem of the quantity of data is to adopt some means of generalizing data subsequent to its collection. For example, some workers have adopted unit test areas (UTAs) of 50

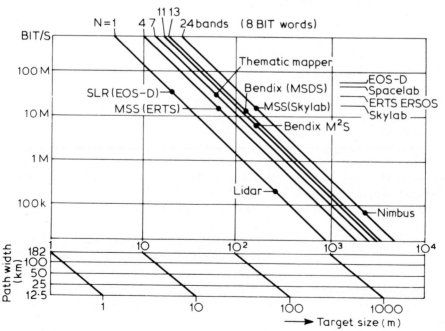

Fig. 19.2 Data rate versus target size compared with high-rate sensors and Earth-observation systems. (Source: Davidts and Loffler, 1974.)

km² for generalization purposes. The data for each UTA is reduced to a single number for each spectral band. Preliminary results suggest that for most environmental applications this would be too drastic a reduction in the information content of the data.

19.4 The interface between technologist and user

Another major problem area is that of the interface between the remote sensing technologists and the users of remote sensing data. Defining the term 'user' is somewhat difficult in practice because there are many users with different objectives. Broadly speaking, however, we may say that there are two main groups of users, the academic and the operational. In the academic category one can include Earth scientists in universities and other research institutes. On the other hand, operational users are typified by National Parks, agricultural departments, meteorological services, geological surveys and mineral companies. Each of these groups of users finds difficulty to a greater or lesser extent in:

(a) Comprehending the nature and significance of remote sensing data obtained by non-photographic sensors.
(b) Defining its own requirements in terms of the engineering and physical properties of sensors used in remote sensing.

It is also apparent that the space technologist does not always fully comprehend the dynamics or properties of the environmental surface being sensed. Furthermore, as increasing use is made of non-visible waveband imagery rather than conventional photography, the problems of communication between the remote sensing technologist and the general user become more severe. To the uninstructed person an image obtained by infrared or microwave systems looks like a photograph. Its record is, however, an entirely different information array from that obtained by photography. For example in microwave studies the measured response (image) obtained from a land surface is affected by complex factors which can be grouped under two headings:

(a) Characteristics of the microwave system: polarization direction, observation angle and frequency of bands used.

(b) Characteristics of the surface sensed: electrical and thermal properties, surface roughness and size and temperature and its distribution.

In these circumstances it is clearly desirable that there should be a flow of information from the physicist/technologist to the Earth science user. However, it is equally true that many physicist/technologist personnel have very limited conceptual and mathematical models of the Earth phenomena being sensed. This has sometimes resulted in exaggerated claims being made for the usefulness of remote sensing systems. For instance, some early writers on air photography and soil studies led potential users to think that soil mapping could be achieved easily by air-photo interpretation. Such statements were based on a lack of knowledge of soil classification and how classificatory techniques affected the objectives of air-photo interpretation. Lack of environmental training has also resulted in some researchers collecting second rate data concerning the land surface conditions (e.g. in respect of soil moisture – see Chapter 13) for comparison with highly accurate remote sensing data.

It is highly desirable that the teams working in remote sensing should be of a multi-disciplinary character from the outset. If the eventual users were involved at the beginning they could help to solve many of the technical problems, such as the type of output required. Such participation would also allow the user to make an objective evaluation of how to make best use of the information. The need for such involvement by the users is underlined by this comment made in respect of crop inventory studies during a symposium on results from Landsat 1: *'From the standpoint of applications, almost every investigation lacked complete definitions of techniques and procedures which would allow a quasi-operational project to be undertaken'*.

So, it seems that remote sensing systems have grown largely without regard to the needs of the community which may use them. Perhaps this is understandable at the early stages of a massive research programme with so many ramifications. It becomes less acceptable when a system which could bring great benefits to man has failed to achieve its full potential. There surely comes a point at which potential users should be given an opportunity to mould the system towards their own ends and also have an opportunity to adapt their administrative and social programmes to allow use to be made of the remote sensing data collected.

Some potential user agencies have been criticized in the past because they do not appear to recognize and understand the uses of remote sensing data. This criticism will doubtless continue to be voiced so long as the planning and execution of remote sensing studies does not involve the users at an early stage. It should be recognized that fruitful use of remote sensing data depends not only on successful scientific advance but also on budgetary and administrative considerations within the user organizations. This often shows itself when studies of cost effectiveness are made. At a late stage in an investigation, rather than at its beginning, one frequently hears the question: 'would it not be cheaper and easier to send out a man in an automobile to collect information?' Quite often the answer to the question is 'no', but sometimes the economics of the case are less clear and then the accusation of 'technology-chasing users' seems apt. It seems fair to add, however, to those who make such comments, that innovation can scarcely ever be proved to be cost effective. Thus, the first move towards progressive technology is almost certainly more expensive than status quo.

Clearly these issues are matters of judgement and the decisions are often difficult to make. In these circumstances it is important that facilities should be available to educate the decision makers and inform the general public about the possibilities and limitations of remote sensing techniques. The educational facilities are rather unevenly spread at the present time. University students are more likely to have been instructed in some aspects of remote sensing studies than school leavers. Geography departments are likely to include some formal studies of remote sensing for environmental monitoring purposes, but teaching of remote sensing is likely to be fragmentary in other disciplines.

19.5 Future developments

It is fitting that this book should conclude with a brief glimpse further into the future – and that it should identify matters which will demand the most careful attention if environmental remote sensing is to play an ever more important part in operational monitoring, as well as in scientific research of this planet and other objects and phenomena in the Universe. Four present trends are likely to continue to be of special significance:

1. National and international commitments to remote sensing will increase and spread and ground receiving stations for satellite data will become more numerous, often linked to centralized data-processing facilities.
2. More new satellite series and concepts will be designed and tested, e.g. the American ice Experiment (ICEX) and Stormsat, and the ESA Earth Resources Satellite (ERS). These will expand the range of aspects of man's environment receiving special attention from dedicated spacecraft systems.
3. More nations will join the once-select club operating environmental satellites of their own. Already the USA and USSR have been joined by Western European nations (through ESA activities), Japan (through GMS 1 and 2 and the proposed Maritime Observation Satellite, MOS), and India (through 'Insat', the Indian National Satellite and Bhaskara); soon France (through 'SPOT', Fig. 19.3) and others will operate environmental satellites too.
4. Satellite data analyses will increasingly combine data from different satellite systems and/or satellites and conventional sources for maximum benefits to the user community. The 'FRONTIERS' project (combining Meteosat and radar imagery) is a current example (see Colour Plate 19.1).

However, many current aspirations of remote sensing may not be fulfilled unless present problems and difficulties can be satisfactorily overcome. It should concern the whole remote sensing community – from systems engineers through hardware designers and manufacturers, through software and data-handling specialists to analysts, interpreters, and ultimate users of the results of remote sensing programmes – that, outside meteorology, progress towards fully-operational satellite systems has been, at best, very erratic and fraught with doubts and uncertainties. Even meteorology itself has not had all its most basic needs for operational systems fully met. An

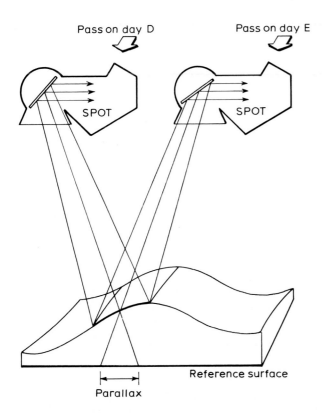

Pass on day D

Pass on day E

SPOT

SPOT

Reference surface

Parallax

Fig. 19.3 The French satellite SPOT will carry two High Resolution Visible (HRV) imaging systems designed to permit stereoscopic viewing of a given scene by making two observations on successive days such that the two images correspond to pointing angles on either side of the vertical. Stereoscopic imagery from SPOT, or the proposed American satellite Stereosat, will be very useful in photo-interpretation and photogrammetry for a wide range of practical applications.

intending operational user of satellite remote sensing systems needs, above all, promises of specified data and service continuity through sufficiently long periods ahead to give him the chance to justify heavy capital expenditure on equipment and, even more importantly, reliance on a satellite source of information, before he can decide to organize his activities accordingly.

The end of the 1970s and the beginning of the 1980s has been a particularly inauspicious period for environmental satellites, with relatively few dependable systems in operation. Goodwill for remote sensing, built up slowly through the 1960s and 1970s in the face of considerable scepticism, is a precious commodity that must be conserved and encouraged if the undoubted ability of satellites in environmental monitoring is to be turned into practical programmes for the good of all mankind. It is fitting, therefore, that we should conclude with some questions with which you, our reader, may have to grapple as you go further in remote sensing of the environment:

(a) Given suitable remote sensing systems for operational applications, will the data be available at prices most users can afford? Budgetary constraints are prompting reviews of the pricing policies of many satellite systems, e.g. Landsat and Meteosat; it must be feared that many researchers will not be able to afford the data basic to their work. Thus, the advance of remote sensing may be slowed down.

(b) Given the increasing volume of information from satellites, will 'local' users (Universities, individual agencies, even small nations) be able to afford data analysis systems suitable for the extraction of the information they require? If not, and the products in which they are interested cannot be obtained from elsewhere, such users will cease to invoke satellite remote sensing for their purposes.

(c) If data from different satellites are to be combined, can an agreement be reached soon on a suitable geographic referencing system to which such data will be automatically conformed

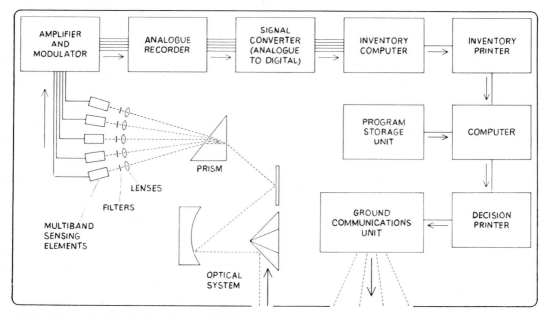

Fig. 19.4 Computerized satellite system for the future. It would sense resources in several wave bands, automatically identify them, weigh them against previously programmed data on the cost effectiveness of various management possibilities, and send to the ground a decision on what should be done. It also could be used to monitor developing situations, such as a forest fire, suggesting how ground crews might fight it, and to perform automatically such tasks as turning irrigation valves on and off as required. (Source: Holz, 1973.)

before they are supplied to the user? So far discussions at the highest levels have been unproductive, resulting in massive inefficiencies in many monitoring programmes using satellite imagery.

Although it is probably inevitable that environmental remote sensing will increasingly adopt the mantle of a 'big science' operation, it is essential that it listens to, and takes account of, the needs of the 'small' user. In recent years, the environmentalist has been involved more in satellite design than in earlier years. For example, it was claimed that Seasat was the first satellite designed primarily by the user for the user. However, it is all too easy for the leaders in the field (the 'big users') to make assumptions of the needs of others which do not match reality amongst most members of the world community of nations. The celebrated microchip should help bring the benefits of space technology within the purchasing power of many potential users less interested in multi-purpose analysis systems than hard-wired ('dedicated') processors designed to give specific answers to narrowly-framed questions. Beyond that, the ultimate goal for environmental remote sensing on the global scale may be a fully-automatic, general purpose, computerized satellite system as shown in Fig. 19.4., but this will not easily be achieved. If it is, one may wonder whether man is fully-equipped himself to use the knowledge (and leisure) such an automated environmental control system might offer him.

This book has sought to provide a base from which the reader can proceed towards greater understanding of the part that remote sensing techniques can play in our lives. If it has opened some windows on the opportunities and pitfalls that lie ahead in the use of remote sensing for Earth resource development, the authors will be well pleased.

Bibliography

Chapter 1

Abiodun, A.A. (1978), The economic implications of remote sensing from space for the developing countries, in *Earth observation from space and management of planetary resources*, **ESA SP-134,** European Space Agency, Paris, 575–584.

American Society of Photogrammetry (1975), *Manual of Remote Sensing* (Vols. I and II), Falls Church, Virginia.

Barrett, E.C. (1971), *Geography from Space*, Pergamon, Oxford.

Barrett, E.C. and Curtis, L.F. (eds.), (1974), *Environmental Remote Sensing; Applications and Achievements*, Edward Arnold, London.

Barrett, E.C. and Curtis, L.F. (eds.), (1977), *Environmental Remote Sensing: Practices and Problems*, Edward Arnold, London.

Estes, J.E. and Senger, L.W. (1974), *Remote Sensing: Techniques for Environmental Analysis*, Hamilton Publishing Company, Santa Barbara.

Haggett, P. (1975), *Geography: a modern synthesis*, 2nd edn., Harper and Row, New York.

Lagarde, J.B. (1980), Space and meteorology – an intricate cost/benefit ratio, in *Proceedings of the International Colloquium on Economic Effects of Space and Other Advanced Technologies*, Strasbourg, 28–30 April, **ESA SP-151,** European Space Agency, Paris, 175–183.

MacQuillan, A.K. and Clough, D.J. (1978), *Benefit of spaceborne remote sensing for ocean surveillance, in Earth observation from space and management of planetary resources*, **ESA SP-134,** European Space Agency, Paris, 585–596.

Peel, R.F., Curtis, L.F. and Barrett, E.C. (eds.), (1977), *Remote Sensing of the Terrestrial Environment*, Butterworths, London.

Richardson, B.F. (ed.) (1978), *Introduction to Remote Sensing of the Environment*, Kendall/Hunt, Dubuque, Iowa.

Stone, K. (1974), Developing geographical remote sensing, in *Remote Sensing: Techniques for Environmental Analysis*, Estes, J.E. and Senger, L.W. (eds.), Hamilton Publishing Company, Santa Barbara, pp. 1–14.

Chapter 2

Feinberg, G. (1968), Light, *Sci. Am.*, **219,** 50–58.

Holz, R.K. (ed.), (1973), *The Surveillant Science: Remote Sensing of the Environment*, Part 1, The Electromagnetic Spectrum – energy for information transfer, Houghton Mifflin, Boston, pp. 1–27.

McAllister, L.G. and Pollard, J.R. (1969), Acoustic sounding of the lower atmosphere, *Proceedings of the Sixth International Symposium on Remote Sensing of Environment*, Ann Arbor, Michigan pp. 436–450.

Slater, P.N. (1980), *Remote Sensing, Optics and Optical Systems*, Addison–Wesley, Reading, Massachusetts.

Chapter 3

Aracon (1971), *The Best of Nimbus*, Contract No. **NAS-5-10343,** Aracon, Concord, Massachusetts.

Barnett, J.J. and Walshaw, C.D. (1974), Temperature measurement from a satellite, in *Environmental Remote Sensing; Applications and Achievements*, Barrett, E.C. and Curtis, L.F. (eds.), Edward Arnold, London, pp. 185–214.

Barrett, E.C. (1974), *Climatology from Satellites*, Methuen, London, pp. 19–74.

Fleagle, R.G. and Businger, J.A. (1963), *An Introduction to Atmospheric Physics*, Academic Press, New York.

Laing, W. (1971), Earth resources satellites, in *A Guide to Earth Satellites*, Fishlock, D. (ed), Macdonald, London and Elsevier, New York, pp. 69–91.

Lockwood, J.G. (1974), *World Climatology: an Environmental Approach*, Edward Arnold, London.

Peel, R.F., Curtis, L.F. and Barrett, E.C. (eds.), (1977), *Remote Sensing of the Terrestrial Environment*, Butterworths, London.

Polcyn, F.C., Spansail, N.A. and Malida, W.A. (1969), How multispectral sensing can help the ecologist, in *Remote Sensing in Ecology*, Johnson, P.L. (ed.), University of Georgia Press, Athens, Georgia, pp. 194–218.

Schanda, E. (ed.), (1976), *Remote Sensing for Environmental Sciences*, Springer–Verlag, Berlin.

Schmugge, T. *et al.* (1973), *Microwave Signatures of Snow and Fresh Water Ice*, Publication No. X-652-73-335, NASA, Greenbelt, Maryland.

Sellers, W.D. (1965), *Physical Climatology*, University of Chicago Press, Chicago.

Chapter 4

Brock, G.C. (1970), *Image evaluation for Aerial Photography*, Focal Press, London and New York.

Corless, G.C. (1977), Remote sensing by radar, in *Remote Sensing of the Terrestrial Environment*, Peel, R.F., Curtis L.F., and Barrett, E.C. (eds.), Butterworths, London.

Curtis, L.F. (1973), The application of photography to soil mapping from the air, in *Photographic Techniques in Scientific Research*, **Vol. 1**, Cruise, J. and Newman, A.A. (eds.), pp. 57–110.

EMI Electronics Ltd. (1973), *Handbook of Remote Sensing Techniques*, Department of Trade and Industry, London.

Gjessing, D.T. (1978), *Remote Surveillance by Electromagnetic Waves*, Ann Arbor Science Publishers Inc., Ann Arbor, Michigan.

Grant, K. (1974), Side looking radar systems and their potential application to Earth resources surveys, *ELDO/ESRO Scientific and Technical Review*, **6**, 117–136.

Laird, A.G. (1977), Passive infrared sensing of the environment, in *Remote Sensing of the Terrestrial Environment*, Peel, R.F., Curtis, L.F. and Barrett, E.C. (eds.), Butterworths, London.

Lintz, J. and Simonett, D.S. (1976), *Remote Sensing of Environment*, Addison–Wesley, Reading, Massachusetts.

Natural Environment Research Council (1974), *Remote Sensing Evaluation Flights, 1971*, Curtis, L.F. and Mayer, A.E.S. (eds.) Publication Series C, no. 12.

Ohlsson, E. (1972) *Summary Report on a Study of Passive Microwave Radiometry and its Potential Applications to Earth Resources Surveys*, European Space Research Organisation Publication, CR-116, Paris.

Plevin, J. and Honvault, C. (1980), The ESA remote sensing programme, *International Journal of Remote Sensing*, **1(1)**, 53–67.

Schanda, E. (ed.), (1976), *Remote Sensing for Environmental Sciences*, Springer–Verlag, Berlin.

Smith, J.T. (ed.), (1968), *Manual of Color Aerial Photography*, American Society of Photogrammetry, Falls Church, Virginia.

Welch, R. (1972), Quality and applications of Aerospace Imagery, *Photogrammetric Engineering*, **38**, 379–398.

Chapter 5

European Space Agency (1979), *Spacelab Users Manual*, ESA Scientific and Technical Publications Office, European Space Agency, Paris.

NASA (1976), *Landsat data users handbook*, Document No. **76SDS-4258,** Goddard Space Flight Center, Greenbelt, Maryland, USA.

Plevin, J. and Honvault, C. (1980), The ESA remote sensing programme, *International Journal of Remote Sensing*, **1(1)**, 53–67.

Sabins, F.F. (1978), *Remote Sensing: Principles and Interpretation*, W.H. Freeman, San Francisco.

Smolders, P. (1973), *Soviets in Space*, Lutterworth Press, London.

Velten, E. (1976) MEOS, *Study on geosynchronous multidisciplinary Earth observation satellite*, Final Report, ESA Contract No. **SC/127/76/HQ**, Dornier/BAC/Sodetag.

Chapter 6

Beckett, P.H.T. (1974), The statistical assessment of resource surveys by remote sensors, in *Environmental Remote Sensing; Applications and Achievements*, Barrett, E.C. and Curtis, L.F. (eds.), Edward Arnold, London, pp. 11–27.

Curtis, L.F. (1971), *Soils of Exmoor Forest*, Soil Survey Gt. Britain, Rothamsted Experimental Station, Harpenden.

Curtis, L.F. (1973), The application of photography to soil mapping from the air, in *Photographic Techniques in Scientific Research*, Vol. **1**, Cruise, J. and Newman, A.A. (eds.), pp. 57–110.

Curtis, L.F. (1974), Remote sensing for environmental planning surveys, in *Environmental Remote Sensing*, Barrett, E.C. and Curtis, L.F. (eds.), Edward Arnold, London, pp. 88–109.

Curtis, L.F. and Hooper, A.J. (1974), Ground truth measurement in relation to aircraft and satellite studies of agricultural land use and land classification in Britain, *Proceedings Frascati Symposium on European Earth Resources Satellite Experiments*, European Space Research Organisation, 405–415.

Sorensen, B.M. (1979), *The North Sea Ocean Color Scanner Experiment, 1977*, Joint Research Centre, Ispra, Italy.

Townshend, J.R.G. (ed.) (1981), *Terrain Analysis and Remote Sensing*, George Allen and Unwin, London.

Chapter 7

Bryant, M. (1974), *Digital Image Processing*, Optronics International Inc., Publication No. **146,** Chelmsford, Massachusetts.

Curtis, L.F. (1973), The application of photography to soil mapping from the air, in *Photographic Techniques in Scientific Research*, Vol. **1**, Cruise, J. and Newman, A.A. (eds.), pp. 57–110.

Estes, J.E. and Senger, L.W. (eds.), (1974), *Remote Sensing: Techniques for Environmental Analysis*, Hamilton Publishing Company, Santa Barbara.

Reeves, R.G. (ed.), (1975), *Manual of Remote Sensing*, American Society of Photogrammetry, Falls Church, Virginia.

Ross, D.S. (1976), Image modification for aiding interpretation and additive color display, in *CENTO Workshop on Applications of Remote Sensing Data and Methods, Proceedings, Istanbul, Turkey*, US Geological Survey, 170–212.

Slama, C.C. (ed.) (1980), *Manual of Photogrammetry*, American Society of Photogrammetry, Falls Church, Virginia.

Smith, J.T. (ed.), (1968), *Manual of Color Aerial Photography*, American Society of Photogrammetry, Falls Church, Virginia.

Chapter 8

Alford, M., Tuley, P., Hailstone, E. and Hailstone, J. (1974), The measurement and mapping of land resource data by point sampling on aerial photographs, in *Environmental Remote Sensing: Applications and Achievements*, Barrett, E.C., and Curtis, L.F. (eds.), Edward Arnold, London.

Bressanin, G. and Erickson, J. (1973), *Data Processing Systems for Earth Resource Surveys*, European Space Research Organisation, Noordwijk, Netherlands.

Lillesand, T.M. and Kiefer, R.W. (1979), *Remote Sensing and Image Interpretation*, John Wiley and Sons, New York.

Lo, C.P. (1976), *Geographical Applications of Aerial Photography*, Crane Russak, New York; David and Charles, Newton Abbot, Devon.

Reeves, R.G. (ed.), (1975), *Manual of Remote Sensing*, American Society of Photogrammetry, Falls Church, Virginia.

Slama, C.C. (ed.) (1980), *Manual of Photogrammetry*, American Society of Photogrammetry, Falls Church, Virginia.

Swain, P.H. and Davis, S.M. (eds.) (1978), *Remote Sensing: The Quantitative Approach*, McGraw Hill, New York.

Chapter 9

Bristor, C.L. (1968), *Computer processing of satellite cloud pictures*, ESSA Technical Memorandum, **NESCTM-3**, US Dept. of Commerce, Washington, DC.

Browning, K. (1979), The FRONTIERS Plan, *Meteorological Magazine*, **108,** 161–174.

Conlan, E.F. (1973), *Operational products from ITOS scanning radiometer data*, NOAA Technical Memorandum, **NESS 52,** Washington, DC.

EMI (1973), *Handbook of Remote Sensing Techniques*, DTI Contract No. **K46A/59,** Technical Reports Centre, Orpington, Kent.

Estes, J.E. and Senger, L.W. (eds.), (1974), *Remote Sensing: Techniques for Environmental Analysis*, Hamilton Publishing Co., Santa Barbara, California.

Gonzalez, R.C. and Wintz, P. (1977), *Digital Image Processing*, Addison–Wesley, Reading, Massachusetts.

Lillesand, T.M. and Kiefer, R.W. (1979), *Remote Sensing and Image Interpretation*, John Wiley, New York.

Lintz, J. and Simonett, D.S. (eds.), (1976), *Remote Sensing of Environment*, Addison–Wesley, Reading, Massachusetts.

Miller, D.B. (1971), Automated production of global cloud climatology based on satellite data, *Air Weather Services (MAC)*, Technical Report 242, USAF, pp. 291–306.

Sabatini, R.R., Rabchevsky, G.A. and Sissala, J.E. (1971), *Nimbus Earth Resources Observations*, Technical Report No. **2,** Contract No. **NAS 5-21617,** Aracon, Concord, Massachusetts.

Swain, P.H. and Davis, S.M. (eds.) (1978), *Remote Sensing: the Quantitative Approach*, McGraw-Hill, New York.

Taylor, J.I. and Stingelin, R.W. (1969), Infrared imaging for water resources studies, *J. Hydraulics Division, Proceedings of the American Society of Civil Engineers*, **95,** 175–187.

Chapter 10

Allison, L.J., Wexler, R., Laughlin, C.R. and Bandeen, W.R. (1978), *Remote Sensing of the Atmosphere from Environmental Satellites*, Authorized Reprint from Special Technical Publication **653,** American Society for Testing and Materials, Philadelphia.

Anderson R.K. and Veltischev, N.F. (1973), *The use of satellite pictures in weather analysis and forecasting*, World Meteorological Organization, Technical Note No. **124,** WMO, Geneva.

Barrett, E.C. (1970), Rethinking climatology: an introduction to the uses of weather satellite photographic data in climatological studies, *Progress in Geography*, **2,** 153–205.

Barrett, E.C. (1974), *Climatology from Satellites*, Methuen, London, pp. 61–146.

Barrett, E.C. and Hamilton, M.G. (1980), *The use of Meteosat data for remote sensing applications*, Final

Report, Contract **E.343/1979,** Dept. of Industry, London.

Barrett, E.C. and Watson, I.D. (1977), Problems in analysing and interpreting data from meteorological satellites, in *Remote Sensing: Practices and Problems,* Barrett, E.C. and Curtis, L.F. (eds.), Edward Arnold, London, pp. 276–303.

Battan, L. (1973), *Radar Observation of the Atmosphere,* Chicago University Press, Chicago.

Fusco, L., Lunnon, R., Mason, B. and Tomassini, C. (1980), *Operational production of sea-surface temperature from Meteosat image data,* ESA bulletin, **21,** pp. 38–43.

Harris, R. and Barrett, E.C. (1975), An improved satellite nephanalysis, *Meteorological Magazine,* **104,** 9.

Hoppe, E.R. and Ruiz, A.L. (1974), *Catalog of operational satellite products,* NOAA technical Memorandum, NESS **53,** US Department of Commerce, Washington.

Schwalb, A. (1978), *The Tiros-N/Noaa A-G Satellite series,* NOAA Technical Memorandum, NESS **95,** US Department of Commerce, Washington, DC.

Shenk, W.E. and Holub, R.J. (1973), *A multispectral cloud type identification method using Nimbus-3 MRIR measurements,* presented to the Conference on Atmospheric Radiation, Fort Collins, Colorado.

Slater, P.N. (1980), *Remote sensing: optics and optical systems,* Addison–Wesley, Reading, Massachusetts, p. 575.

Spalding, T.R. (1974), Satellite data for tropical weather forecasting, in *Environmental Remote Sensing: Applications and Achievements,* Barrett, E.C. and Curtis, L.F. (eds.), Edward Arnold, London, pp. 215–240.

Chapter 11

Allison, L.J. (1972), *Air-sea interaction in the tropical Pacific Ocean,* NASA Technical Note, **TN D-6684.**

Barnett, J.J. *et al.* (1973), The first year of the selective chopper radiometer on Nimbus-4, *Q.J. Roy. Met. Soc.,* **98,** 32.

Barrett, E.C. (1974), *Climatology from Satellites,* Methuen, London, pp. 147–313.

Barrett, E.C. (1975), Analyses of image data from meteorological satellites, in *Processes in Physical and Human Geography,* Peel, R.F., Chisholm, M. and Haggett, P. (eds.), Heinemann, London, pp. 169–196.

Barrett, E.C. and Martin, D.W. (1981), *The use of satellites in rainfall monitoring,* Academic Press, London.

Harwood, R.S. (1977), Some recent investigations of the upper atmosphere by remote sounding satellites, in *Remote Sensing of the Terrestrial Environment,* Peel, R.F., Curtis, L.F. and Barrett, E.C. (eds.), Butterworths, London, pp. 111–124.

Miller, D.B. (1971), *Global atlas of relative cloud cover, 1967–70, based on data from meteorological satellites,* U.S.

Air Force/US Department of Commerce, Washington, DC.

Streten, N.A. (1973), Some characteristics of satellite observed bands of persistent cloudiness over the Southern Ocean, *Monthly Weather Review,* **101,** 486–498.

Streten, N.A. and Troup, A.J. (1973), A synoptic climatology of satellite-observed cloud vortices over the southern hemisphere, *Q.J. Roy. Met. Soc.,* **99,** 56–72.

Vonder Haar, T.H. and Suomi, V.E. (1971), Measurements of the Earth's radiation budget from satellites during a five year period, Part I: extended time and space means, *J. Atmos. Sci.,* **28,** 305–314.

Chapter 12

Allison, L.J. and Schmugge, T.J. (1979), A hydrological analysis of East Australian floods using Nimbus-5 Electrically Scanning Radiometer data, *Bull. Amer. Met. Soc.,* **60,** 1414–1420.

Barrett, E.C. and Martin, D.W. (1981), *The Use of Satellites in Rainfall Monitoring,* Academic Press, London.

Ferguson, H.L., Deutsch, M. and Kruus, J. (1980). Applications to floods of remote sensing from satellites, in *The contribution of space observations to water resources management,* Salomonson, V.V. and Bhavsar, P.D. (eds.), Pergamon Press, Oxford, p. 195.

Fraysse, G. (ed.), (1980), *Remote Sensing in Agriculture and Hydrology,* Balkema, Rotterdam.

Grinstead, J. (1974), The measurement of areal rainfall by the use of radar, in *Environmental Remote Sensing: Applications and Achievements,* Barrett, E.C. and Curtis, L.F. (eds.), Edward Arnold, London, pp. 267–284.

Nagle, R.E. and Serebreny, S.M. (1962), Radar precipitation echo and satellite cloud observations of a maritime cyclone, *J. Appl. Met.,* **1,** 279–286.

Painter, R.B. (1974), Some present uses of remote sensing in monitoring hydrological variables, in *Environmental Remote Sensing: Applications and Achievements,* Barrett, E.C. and Curtis, L.F. (eds.), Edward Arnold, London, pp. 285–298.

Plessey Radar (1971), *The Dee Weather Radar Project,* Publication No. **6776,** The Plessey Company Limited, Weybridge.

Rango, A. (1980), Remote sensing applications in hydrometeorology, in *The contribution of space observations to water resources management,* Salomonson, V.V. and Bhavsar, P.D. (eds.), Pergamon Press, Oxford, pp. 59–66.

Sabins, F.J. (1978), *Remote Sensing: Principles and Interpretation,* Freeman, San Francisco.

Salomonson, V.V. and Bhavsar, P.D. (eds.) (1980), *The Contribution of Space Observations to Water Resources Management,* Pergamon Press, Oxford and New York.

Szekielda, K.H., Allison, L.F. and Salomonson, V. (1970), *Seasonal sea surface temperature variations in the Persian Gulf*, Goddard Space Flight Center, Report No. **X-651-70-416,** NASA, Greenbelt, Maryland.

Viksne, A., Liston, T.C. and Sapp, C.D. (1969), SLR reconnaissance of Panama, *Geophysics*, **34,** 54–69.

Warnecke, G.L., McMillin, M. and Allison, L.J. (1969), *Ocean current and sea surface temperature observations from meteorological satellites*, NASA Technical Note, **D-5142,** NASA, Washington, DC.

Williamson, E.J. and Wilkinson, G.G. (1980), Sea-surface temperature measurements from Meteosat, in *Report of Second Meteosat Scientific User Meeting*, March 1980, ESA, Paris, 51–511.

Chapter 13

Barringer Research Ltd. (1975), *Final Report on Remote Sensing of Soil Moisture*. Centre for Remote Sensing, Ottawa, Canada, 89 pp.

Basiinski, J.J. (1959), The Russian approach to soil classification and its recent development, *J. Soil Sci.*, **10,** 14–26.

Beckett, P.H.T. (1974), The statistical assessment of resource surveys by remote sensors, in *Environmental Remote Sensing: Applications and Achievements*, Barrett, E.C. and Curtis, L.F. (eds.). Edward Arnold, London pp. 11–29.

Belyakova, G.M. *et al.* (1971), 'Study of microwave radiation from the satellite Cosmos 243 over agricultural landscape', *Coklady Akademic Nauk*, USSR, **201,** 837–842.

Bradley, G.A. and Ulaby, F.T. (1980), *Aircraft Radar Response to Soil Moisture*, Technical Report SM-KO-04005, NAG 5-30, Remote Sensing Laboratory, University of Kansas Center for Research, 35 pp.

Condit, H.R. (1970), The spectral reflectance of American soils, *Photogrammetric Engineering*, **36,** 955–966.

Curran, P.J. (1978), A photographic method for the recording of polarized visible light for soil surface moisture indication, *Remote Sensing of Environment*, **7,** 305–322.

Curran, P.J. (1981), Remote sensing: the use of polarized visible light to estimate surface soil moisture, *Applied Geography*, **1,** 41–53.

Curtis, L.F. (1974), Remote Sensing for Environmental Planning Surveys, in *Environmental Remote Sensing; Applications and Achievements*, Barrett, E.C. and Curtis, L.F. (eds.), Edward Arnold, London, pp. 87–109.

Curtis, L.F. (1976), The Mapping of Soil Associations by Remote Sensing Techniques, in *CENTO Workshop on Applications of Remote Sensing Data and Methods, Proceedings, Istanbul, Turkey*, US Geological Survey, 40–47.

Curtis, L.F. (1977), Remote sensing of soil moisture: user requirements and present prospects, in *Remote Sensing of the Terrestrial Environment*, Peel, R.F., Curtis, L.F. and Barrett, E.C. (eds.), Butterworths, London, pp. 143–158.

Gerbermann, A.H., Gausman, H.W. and Wiegand, C.L. (1971), Color and color IR films for soil identification, *Photogrammetric Engineering*, **37,** 359–364.

Gurney, R.J. (1979), The estimation of soil moisture content and actual evapotranspiration using thermal infrared remote sensing, in *Remote Sensing and National Mapping*, Allan, J.A. and Harris, R. (eds.), Remote Sensing Society, pp. 101–109.

Jenkins, D.S., Belcher, D.J., Gregg, L.E. and Woods, K.B. (1964), Technical Development Report 52, US Dept. Commerce, Civil Aeronautics Administration, Washington, DC.

Kroll, C. (1973), *Remote Monitoring of Soil Moisture using Airborne Microwave Radiometers*, Texas University Remote Sensing Center.

Leeman, V., Earing, D., Vincent, R.K. and Ladd, S. (1971), *NASA Earth Resources Spectral Information System: a data compilation*, NASA, CR-31650-24-T.

Meyer, M.P. and Maklin, H.A. (1969), Photointerpretation techniques for Ektachrome IR transparencies', *Photogrammetric Engineering*, **35,** 1111–1114.

Mintzer, O.W. (1968), Soils, in *Manual of Color Aerial Photography*, Smith, J. (ed.), American Society of Photogrammetry.

Myers, V.I. and Heilman, M.D. (1969), Thermal infrared for soil temperature studies, *Photogrammetric Engineering*, **10,** 1024–1032.

NASA, (1973), *Symposium on significant results from the ERTS-1*, Vols. **I** and **II,** NASA.

Parks, W.L. and Bodenheimer, R.E. (1973), Delineation of major soil associations using ERTS-1 imagery, *Symposium on significant results from ERTS-1*, pp. 121–125.

Parry, D.E. (1978), Some examples of the use of satellite imagery (Landsat) for natural resource mapping in Western Sudan, in *Remote Sensing Applications in Developing Countries*, Collins, W.G. and van Genderen, J.L. (eds.), Remote Sensing Society, pp. 3–12.

Parry, J.T., Cowan, W.R. and Heginbottom, J.A. (1969), Soil studies using color photos, *Photogrammetric Engineering*, **35,** 44–56.

Pomerening, J.A. and Cline, M.G. (1953), The accuracy of soil maps prepared by various methods that use aerial photographic interpretation, *Photogrammetric Engineering*, **19,** 809–817.

Romanova, M.A. (1968), Spectral luminance of sand deposits as a tool in land evaluation, in *Land Evaluation*, Stewart, G.A. (ed.), Macmillan, Australia, pp. 342–348.

Seevers, P.M. and Drew, J.V. (1973), Evaluation of ERTS-1 imagery in mapping and managing soil and range resources in the sand hills region of Nebraska, *Symposium on significant results from ERTS-1*, NASA, pp. 87–95.

Simakova, M.S. (1964), *Soil Mapping by Colour Aerial Photography*, Israel Programme for Scientific Translations, Jerusalem.

Steg, L. and Frost, R.T. (1971), *Visible polarization signature for remote sensing of soil surface moisture*, COSPAR Plenary Meeting, Leningrad, USSR (1970), p. 15.

Stockhoff, E.H. and Frost, R.T. (1972), Polarization of Light Reflected by Moist Soils, *Proceedings 7th Symposium on Remote Sensing*, ERIM Ann Arbor, Michigan, 345–364.

Thomson, F.J. and Roller, E.G. (1973), Terrain classification maps of Yellowstone National Park, *Symposium on significant results from ERTS-1*, pp. 1091–1095.

Townshend, J.R.G. (ed.) (1981), *Terrain analysis and remote sensing*, George Allen and Unwin, London.

Ulaby, F.T. (1980), Active microwave sensing of soil moisture: synopsis and prognosis, *Proceedings Workshop on Microwave Remote Sensing on Bare Soil*, European Association of Remote Sensing Laboratories, Toulouse, pp. 2–28.

Way, D.S. (1973), *Terrain analysis*, Dowden, Hutchinson and Ross, Stroudsburg, Pennsylvania, USA.

Webster, R. and Wong, I.F.T. (1969), A numerical procedure for testing soil boundaries interpreted from air photographs, *Photogrammetria*, **24**, 59–72.

Welch, R. (1966), A comparison of aerial films in the study of the Breidamerkur glacier area, Iceland, *Photogrammetric Record*, **5**, 289–306.

Werner, H.D., Schmer, F.A., Horton, M.L. and Waltz, F.A. (1973), *Application of Remote Sensing Techniques to Monitoring Soil Moisture*, Environmental Research Institute of Michigan, pp. 1245–1258.

Weston, F.C. and Myers, V.I. (1973), Identification of soil associations in Western South Dakota on ERTS-1 imagery, *Symposium on significant results from ERTS-1*, NASA, pp. 965–972.

Chapter 14

Billingsley, F.C. and Goetz, A.F.H. (1973), Computer techniques used for some enhancements of ERTS images, *Symposium on significant results from ERTS-1*, NASA, pp. 1159–1167.

Blodget, H.W. and Anderson, A.T. (1973), A comparison of Gemini and ERTS imagery obtained over Southern Morocco, *Symposium on significant results from ERTS-1*, NASA, pp. 265–272.

Bodechtel, J. and Haydn, R. (1977), Analog and digital processing of multispectral data for geologic application in *Remote Sensing of the Terrestrial Environment*, Peel, R.F., Curtis, L.F. and Barrett, E.C. (eds.), Butterworths, London and Boston, 159–168.

Bodechtel, J., Nithack, J. and Haydn, R. (1974), Geologic evaluation of Central Italy from ERTS-1 and Skylab data, *Proceedings Frascati Symposium on European Earth Resources Satellite Experiments*, ESRO, Paris, 205–215.

Boriani, A., Marino, C.M. and Sacchi, R. (1974), Geological features on ERTS-1 images of a test area in West-Central Alps, in *Proc. Symp. on European Earth Resources Satellite Experiments*, ESRO SP-100.

Carter, W.D. and Meyer, R.F. (1969), Geological analysis of a multispectrally processed Apollo space photograph, *New Horizons in Colour Areal Photography*, American Society of Photogrammetry, pp. 59–64.

Cassinis, R., Lechi, G.M. and Tonelli, A.M. (1974), Contributions of space platforms to a ground and airborne remote sensing programme over active Italian volcanoes, *Proc. Symp. on European Earth Resources Satellite Experiments*, ESRO SP 100, Paris, pp. 185–197.

Cole, M.M. (1973), Geobotanical and biogeochemical investigations in the sclerophyllous woodland and scrub associations of the Eastern Goldfields are of Western Australia, *J. Appl. Ecol.*, **10**, 269–284.

Cole, M.M., Owen-Jones, E.S. and Custance, N.D.E. (1974), Remote Sensing in mineral exploration, in *Environmental Remote Sensing; Applications and Achievements*, Barrett, E.C. and Curtis, L.F. (eds.), Edward Arnold, London, pp. 49–66.

Froelich, A.J. and Kleinkampl, F.J. (1960), Botanical prospecting for uranium in the Deer Flat area, White Canyon District, San Juan County, Utah, *US Geological Survey Bulletin*, 1085-B.

Gregory, A.F., (1973) Preliminary assessment of geological applications of ERTS-1 imagery from selected areas of the Canadian Arctic, *Symposium on significant results from ERTS-1*, NASA, pp. 329–344.

Laird, A.G. (1977), Passive infrared sensing of the environment, in *Remote Sensing of the Terrestrial Environment*, Peel, R.F., Curtis, L.F., and Barrett, E.C. (eds.), Butterworths, London and Boston, pp. 26–37.

Lathram, E.H., Taillevr, I.L. and Patton, W.W. (1973), Preliminary geologic application of ERTS-1 imagery in Alaska, *Symposium on significant results from ERTS-1*, NASA, pp. 257–264.

Martin-Kaye, P. (1974), Application of side looking radar in earth-resource surveys, in *Environmental Remote Sensing: Applications and Achievements*, Barrett, E.C. and Curtis, L.F. (eds.), Edward Arnold, London, pp. 29–48.

Sabins, Floyd, F. (1978), *Remote Sensing: Principles and Interpretation*, W.H. Freeman, San Francisco.

Schmidt, R., Clark, B.B. and Bernstein, R. (1975), A

search for sulfide-bearing areas using Landsat-1 data and digital image-processing techniques, *Nasa Earth Resources Survey Symposium*, NASA TM X-58168, v. 1-B, 1013–1027.

Short, N.M. (1973), Mineral resources, geological structure and landform surveys, *Symposium on significant results obtained from ERTS-1*, Vol. **111**, NASA, pp. 30–46.

Siegal, B.S. and Gillespie, A.R. (eds.) (1980), *Remote sensing for geologists*, John Wiley and Sons, New York.

Vincent, R.K. (1973), Ratio maps of iron ore deposits, Atlantic City District, Wyoming', *Symposium on significant results from ERTS-1*, NASA, pp. 379–386.

Vinogradov, B.V., Grigoryev, A.A., Lipatov, V.B. and Chernenko, A.P. (1972), Thermal structure of the sand desert from the data of IR aerophotography, *Proceedings of the 8th International Symposium Remote Sensing of Environment*, Ann Arbor, Michigan, pp. 729–737.

Wobber, F.J. and Martin, K.R. (1973), Exploitation of ERTS-1 imagery utilising snow enhancement techniques, *Symposium on significant results from ERTS-1*, NASA, pp. 345–351.

Chapter 15

Carneggie, D.M. and De Gloria, S.D. (1973), Monitoring California's forage resource using ERTS-1 and supporting aircraft data, *Symposium on significant results from ERTS-1*, NASA, pp. 91–95.

Colwell, R.N. *et al.* (1963), Basic matter and energy relationships involved in Remote Reconnaissance, *Photogrammetric Engineering*, **29**, 761.

Colwell, R.N. (1969), *Analysis of Remote Sensing Data for Evaluating Forest and Range Resources*, School of Forestry and Conservation, University of California, p. 207.

Curtis, L.F. (1978), Remote Sensing systems for Monitoring Crops and Vegetation, *Progress in Physical Geography*, **2**(1), 55–79.

Curtis, L.F. and Walker, A.J. (1980), Exmoor: A problem of Landscape Planning and Management, *Landscape Design*, **130**, 7–13.

Dethier, B.E., Ashley, M.D., Blair, B. and Hopp. R.J. (1973), Phenology satellite experiment, *Symposium on significant results from ERTS-1*, NASA, pp. 157–165.

Fraysse, G. (ed.) (1980), *Remote Sensing Applications in Agriculture and Forestry*, Balkema Rotterdam, The Netherlands.

Heller, R.C. (1968), Large scale color photography samples forest insect damage, in *Manual of Color Aerial Photography*, Smith, J. (ed.), p. 394.

Hilborn, W.H. (1978), Application of Remote Sensing in Forestry, in *Introduction to Remote Sensing of the Environment*, Richardson, B.F. (ed.), Kendall/Hunt, Dubuque, Iowa.

Howard, J.A. (1970), *Aerial Photo-Ecology*, Faber, London.

Jensen, C.E. (1948) *Dot-type scale for measuring tree crown diameters on aerial photographs*, U.S. Forest Service, Central States Forest Experiment Station, Note No. 48.

Johnson, P.L. (ed.) (1969), *Remote Sensing in Ecology*, University of Georgia Press.

Kalmbach, E.R. (1949), A scanning device useful in wildlife work, *J. Wildlife Management*, **13**, 226–236.

Leedy, D.L. (1968), The inventorying of wildlife, in *Manual of Colour Aerial Photography*, Smith, J. (ed.), p. 422.

Leeman, V., Earing, D., Vincent, R.K. and Ladd, S. (1971), *The NASA Spectral Information System: a data compilation*, NASA, Houston, Texas.

Lyons, T.R. and Avery, T.E. (1977), *Remote Sensing: A handbook for Archaeologists and Cultural Resource Managers*, National Park Service, US Department of Interior.

Parry, D.E. and Trevett, J.W. (1979), mapping Nigeria's Vegetation from Radar, *Geographical Journal*, **145**(2), 265–281.

Pedgley, D.E. and Symmons, P.M. (1968), Weather and the locust upsurge, *Weather*, **23**, 484.

Pedgley, D.E. (1974), Use of satellites and radar in locust control; in *Environmental Remote Sensing: Applications and Achievements*. Barrett, E.C. and Curtis, L.F. (eds.), Edward Arnold, London, pp. 143–152.

Roberts, H. and Colwell, R.N. (1968), *The Application of Remote Sensing to the Inventory of Livestock and Identification of crops*, School of Forestry and Conservation, University of California, p. 20.

Roffey, J. (1969), *Radar studies on the Desert Locust*, Anti-Locust Research Centre Occasional Report, 17/69.

Rogers, E.J. (1949), Estimating tree heights from shadows on vertical aerial photographs, *J. Forestry*, **47**, 182–190.

Safir, G.R. and Myers, W.L. (1973), Application of ERTS-1 Data to analysis of agricultural crops and forests in Michigan, *Symposium significant results from ERTS-1*, NASA, pp. 173–180.

Schaefer, G.W. (1972), Radar detection of individual locusts and swarms, *Proceedings of the International Study conference on the Current and Future Problems of Aridology*, London, 1970, pp. 379–380.

Seeley, H.E. (1948), *The Pole Scale*, Dominion Forest Service (Canada) Forest Air Survey Leaflet No. 1.

Shantz, H.L. (1954), The place of grasslands in the Earth's cover of vegetation, *Ecology*, **35**, 143–155.

Spurr, S.H. (1960), *Photogrammetry and Photointerpretation*, Ronald Press, New York.

Thorley, G.A. (1968) Some uses of color aerial photography in Forestry in *Manual of Color Aerial Photography*, Smith, J. (ed.), p. 393.

Watson, R.M. (1977), Air photography in East African Game management, in *The Uses of Air Photography*, St. Joseph, J.K.K. (ed.), John Baker, London.

Worley, D.P. and Meyer, H.A. (1955), Measurement of Crown Diameter and Crown Cover and their accuracy on 1: 12,000 scale photographs, *Photogrammetric Engineering*, **21**, 372–386.

You-Ching, F. (1980), Aerial Photo and Landsat Image Use in Forest Inventory in China, *Photogrammetric Engineering and Remote Sensing*, **46**, 1421–1430.

Chapter 16

Anderson, J.R., Hardy, E.E., Roach, J.T. and Witmer, R.E. (1976), *A land use and land cover classification system for use with remote sensor data*, US Geological Survey Professional Paper 964.

Bell, T.S. (1974), Remote sensing for the identification of crops and diseases, in *Environmental Remote Sensing: Applications and Achievements*, Barrett, E.C. and Curtis, L.F. (ed.), Edward Arnold, London, pp. 154–166.

Brooner, W.G. and Simonett, D.S. (1971), Crop discrimination with color infrared photography: a study in Douglas County, Kansas, *Remote Sensing of Environment*, **2**, 21–35.

Bush, T.F. and Ulaby, F.T. (1975), *Remotely sensing Wheat Maturation with Radar*, Technical Report 177–55, University of Kansas Center for Research, 120 pp.

Curtis, L.F. and Hooper, A.J. (1974), Ground truth measurements in relation to aircraft and satellite studies of agricultural land use classification in Britain, *Proceedings Frascati Symposium on European Earth-Resources Satellite Experiments*, European Space Research Organisation, pp. 405–415.

Curtis, L.F. (1978), Remote sensing systems for monitoring crops and vegetation, *Progress in Physical Geography*, **2**(1), 55–79.

Fraysse, G. (1977) Perspectives offered by remote sensing in agricultural resources management in *Remote Sensing of the Terrestrial Environment*, Peel, R.F., Curtis, L.F. and Barrett, E.C. (ed.), Butterworths.

Haralick, R.M. and Shanmugam, K.S. (1973), Combined Spectral and Spatial Processing of ERTS Imagery Data, *Remote Sensing of Environment*, **3**, 3–13.

Idso, S.B., Hatfield, J.L., Jackson, R.D. and Reginato, R.J. (1979), Grain Yield Prediction: Extending the Stress-Degree-Day Approach to Accomodate Climatic Variability, *Remote Sensing of the Environment*, **8**, 267–272.

Kasteren, H.W.J. van and Smit, M.K. (1977), Measurements on the back scatter of X-band radiation

of seven crops, *NIWARS publication 47*, Netherlands Interdepartmental Working Community for Applications of Remote Sensing Techniques, 115 pp.

LARS (1968), *Remote multispectral sensing in agriculture*, Laboratory for Agricultural Remote Sensing, Purdue University, Research bulletin, 844.

Macdonald, R.B., Bauer, M.E., Allen, R.D., Clifton, J.W., Erickson, J.A. and Landgrebe, D.A. (1973), Results of the 1971 Corn Blight Watch Experiment, *Proceedings of the 8th International Symposium on Remote Sensing of Environment*, Ann Arbor, Michigan, **1**, 157–189.

Macdonald, R.B. and Hall, F.G. (1978), The Lacie Experience: A Summary, *Proceedings International Symposium on Remote Sensing for Observation and Inventory of Earth Resources and Endangered Environment*, 26 pp.

Meyer, M.P. and Calpouzos, L. (1968), Detection of crop diseases, *Photogrammetric Engineering*, **34**, 554–556.

Millard, J.P., Hatfield, J.L. and Goettelman, R.C. (1979), Equivalence of Airborne and Ground Acquired Wheat Canopy Temperatures, *Remote Sensing of the Environment*, **8**, 273–275.

Morrison, J. (1977), NASA Earth Resources Survey Program: Problems and Prospects, *European Symposium on Remote sensing of the Earth from Space*, Strasbourg.

Nagy, G. and Shelton, G. and Tolaba, J. (1971), Procedural questions in signature analysis, *Proceedings of the 7th International Symposium on Remote Sensing of Environment*, Ann Arbor, Michigan.

Sisam, J.W.B. (1947), *The Use of Aerial Survey in Forestry and Agriculture*, Imperial Forestry Bureau, Oxford.

Su, M.Y. and Cummings, R.E. (1972), An unsupervised classification technique for multispectral remote sensing data, *Proceedings of the 8th International Symposium Remote Sensing of Environment*, Ann Arbor, Michigan, pp. 861–879.

Thompson, D.R. and Weymanen, O.A. (1980), Using Landsat digital data to detect moisture stress in Corn-Soybean growing regions, *Photogrammetric Engineering and remote Sensing*, **46**(8).

Wilson, D.R. (ed.) (1975), in *Aerial reconnaissance for archaeology*, Council for British Archaeology, London, p. 158.

Chapter 17

Bush, P.W. and Collins, W.G. (1974), The application of aerial photography to surveys of derelict land in the United Kingdom, in *Environmental Remote Sensing; Applications and Achievements*, Barrett, E.C. and Curtis, L.F. (eds.), Edward Arnold, London, pp. 167–183.

Davies, S., Tuyahov, A. and Holz, R.K. (1973), Use of remote sensing to determine urban poverty neighbourhoods, *Landscape*, pp. 72–81.

Holz, R.K., Huff, D.L. and Mayfield, R.C. (1969), Urban spatial structure based on remote sensing imagery, *Proceedings of the Sixth International Symposium on remote Sensing of the Environment*, Univeristy of Wisconsin, Ann Arbor, Michigan, pp. 819–830.

Horton, F. (1974), Remote sensing techniques and urban data acquisition, in *Remote Sensing: Techniques for Environmental Analysis*, Estes, J.E. and Senger, L.W. (eds.), Hamilton Publishing Co., Santa Barbara, pp. 243–276.

Richardson, B.F. (ed.) (1978), *Introduction to Remote Sensing of the Environment*, Kendall/Hunt, Dubuque, Iowa.

Chapter 18

Barrett, E.C. (1981), Satellite-improved rainfall monitoring by cloud indexing methods: operational experience in support of desert locust survey and control, In *Proceedings of the AWRA Symposium on Satellite Hydrology*, Sioux Falls, SD, American Water Resources Association, Minneapolis, Minnesota.

Barrett, E.C. and Hamilton, M.G. (1980), *The use of Meteosat data for remote sensing applications*, Final Report, Contract E. 343/1979, Dept. of Industry, London.

Barrett, E.C. and Martin, D.W. (1981), *The Use of Satellites in Rainfall Monitoring*, Academic Press, London.

Berberian, M. (1978) Tabas-e-Golshan (Iran) catastrophic earthquake of 16 Sept. 1978: a preliminary report, *Disasters*, **4**, 207–219.

Burton, I., Kates, R.W. and White, G.F. (1978), *The Environment as Hazard*, Oxford University Press, Oxford.

Giddings, L. (1976), *Extension of weather data by use of meteorological satellites*, Technical Memorandum, LEC-8377, Lockheed Corporation, Houston, Texas.

Hielkema, J.U. (1980), *Remote sensing techniques and methodologies for monitoring ecological conditions for desert locust population development*. Final Technical Report, FAO/USAID, CGP/INT/349/USA, FAO, Rome.

Howard, J.A., Barrett, E.C. and Hielkema, J.U. (1978), The application of satellite remote sensing to monitoring of agricultural disasters, *Disasters*, **4**, 231–240.

Ingraham, D. and Amorocho, J. (1977), Preliminary rainfall estimates in Venezuela and Colombia from Goes satellite image, In *Preprints 2nd Conference on Hydrometeorology*, Toronto, 25–27 August, American meteorological Society, 316–323.

UN Disaster Relief Organisation (1978), *Disaster Prevention and Mitigation: A Compendium of Current Knowledge* (in several parts), United Nations, New York.

Wasserman, S.E. (1977), The availability and use of satellite pictures in recognizing hazardous weather, In *Earth observation systems for resource management and environmental control*, D.J. Clough and L.W. Morley (eds.), Plenum Press, New York and London, pp. 419–436.

World Meteorological Organization (1977), *The use of satellite imagery in tropical cyclone analysis*, Technical Note No. 153, WMO, Geneva.

World Meteorological Organization (1979), *Operational techniques for forecasting tropical cyclone intensity and movement*. WMO No. 528, Geneva.

Chapter 19

Atlas, D., Bandeen, W.R., Shenk, W., Gatlin, J.A. and Maxwell, M. (1978), Visions of the future operational meteorological satellite system, *Proceedings of the Electronics and Aerospace Systems Conference*, Arlington, Virginia, USA.

Browning, K. (1979), The FRONTIERS plan, *Met. Mag.* **108**, 161–174.

Clough, D.J. and Morley, L.W. (1977), *Earth Observation Systems for Resource Management and Environmental Control*, Plenum Press, New York and London.

CNES (1981), *SPOT: satellite-based remote sensing system*, Centre National d'Etudes Spatiales, Toulouse.

Davidts, D. and Loffler, A. (1974), Reduction of information redundancy in ERTS-1 and EREP Data, *Proceedings Frascati Symposium on European Earth Resources Satellite Experiments*, European Space Research Organisation, pp. 81–91.

ESA (1978), *Earth observation from space, and management of planetary resources*. ESA SP-134, Paris.

ESA (1978), *Applications of Earth Resources satellite data to development aid programmes*, ESA SP-1010, Paris.

Holz, R.K. (ed.), (1973), *The Surveillant Science*, Houghton Mifflin, Boston.

Plevin, J. (1977), A European Earth resources space programmes, In *Remote Sensing of the Terrestrial environment*, R.F. Peel, L.F. Curtis and E.C. Barret (eds.), pp. 263–275.

Suits, G.H. (1966), Declassification of infrared devices, *Photogrammetric Engineering*, **32**, 988–992.

Index

Note: the arrangement of entries is word-by-word alphabetization. **Main entries** are indicated by **bold** numbers, *illustrations* or *photographs* by *italic* numbers. 'Remote Sensing' is abbreviated to 'RS'.